P9-CRO-967

Defense of the Scientific Hypothesis

Defense of the Scientific Hypothesis

From Reproducibility Crisis to Big Data

BRADLEY E. ALGER, PHD

Professor Emeritus
Department of Physiology
Program in Neuroscience
University of Maryland School of Medicine, Baltimore

UNIVERSITY PRESS

Oxford University Press is a department of the University of Oxford. It furthers the University's objective of excellence in research, scholarship, and education by publishing worldwide. Oxford is a registered trade mark of Oxford University Press in the UK and certain other countries.

Published in the United States of America by Oxford University Press
198 Madison Avenue, New York, NY 10016, United States of America.

Library of Congress Cataloging-in-Publication Data
Names: Alger, Bradley E., author.
Title: Defense of the scientific hypothesis : from reproducibility crisis to big data / Bradley E. Alger, Ph.D., Professor Emeritus, Department of Physiology, Program in Neuroscience, University of Maryland School of Medicine.
Description: New York, NY : Oxford University Press, [2020] |
Includes bibliographical references.
Identifiers: LCCN 2019017418 | ISBN 9780190881481 (hardback) |
ISBN 9780190881498 (updf) | ISBN 9780190881504 (epub)
Subjects: LCSH: Science—Methodology.
Classification: LCC Q175.A455 2019 | DDC 501—dc23
LC record available at https://lccn.loc.gov/2019017418

1 3 5 7 9 8 6 4 2

Printed by Sheridan Books, Inc., United States of America

To my wife Lindsay for her love and support.

Contents

PART III. PRESENT POLICIES AND THE FUTURE

Figures

Boxes

Acknowledgments

Throughout my career as a research scientist and teacher of graduate students, I was unaware of most of the material in this book. I could have done a better job in the lab and the classroom if I'd known more. Bringing together information about the scientific hypothesis, the most versatile thinking tool that a scientist has, is my attempt to make things easier for others.

I thank my friends and colleagues whose many thoughtful comments on drafts of these chapters markedly improved the book. That it is not better is entirely my responsibility. In particular, I thank Iris Lindberg, whose detailed and carefully considered criticisms helped make many chapters more readable and informative than they had been. Discussions with Tom Abrams provided much of the inspiration to undertake the project in the first place, and I am very grateful to him, Asaf Keller, Bruce Krueger, and Paul Shepard for critical comments on several chapters. I thank graduate students Jon Van Ryzin and Sarah Ransom Metzbower for their unique perspectives. Mordy Blaustein, Justine Forrester, Joe Kao, Peg McCarthy, Frank Margolis, Brian Polster, and Scott Thompson read and provided feedback on some of the dreadful early drafts; I'm glad we're still friends. I benefited greatly from interesting discussions with Tom Abrams, Mordy Blaustein, Soren Bentzen, Tom Blanpied, Joe Kao, Asaf Keller, Bruce Krueger, Brian Polster, Paul Shepard, and Scott Thompson. I thank Bill Hilgartner, Justine Forrester, and Bill Forrester for allowing me access to pre-college textbooks. The folks at the Writing Center at the University of Maryland, Baltimore, offered supportive tips, and Jocelyn Broadwick provided excellent initial editorial input. David Linden gave encouragement and advice on practical matters related to book publishing.

I thank my editor at Oxford, Joan Bossert, for supporting my early efforts as well as the anonymous reviewers whose insightful reading and suggestions were enormously helpful. Special thanks are due to Gerd Gigerenzer for his valuable recommendations and suggestions throughout the project.

Introduction

I.A Motivation for the Book

Why write a book about the scientific hypothesis? In particular, why a book about defending the hypothesis?

First, most of us, ordinary citizens and scientists alike, do not really know what a hypothesis is and yet it touches nearly every aspect of science. If you want to understand what science is truly "about," you can't do better than to understand the hypothesis.

Second, despite its traditional place in science, the hypothesis today is suffering simultaneously from neglect and attacks from critics who will tell you that it is no longer relevant; that, in fact, it is detrimental to science.

The book has two major goals: to explain what a hypothesis is and what it does, and to show why its critics are wrong.

Ordinary citizens and scientists need to be able to evaluate the claims of science about global climate change, the safety of vaccines, the risks of cancer, or any of the many other science-related challenges that keep confronting us. To assess competing claims, we need to know how to distinguish stronger, more solid, and reliable claims from weaker and less reliable ones. The claims that science makes, no matter what they're called, are hypotheses—putative explanations for phenomena—and to be able to assess the claims we need to know how to evaluate hypotheses. It is not difficult to do once you know how. But it is not an obvious or easy thing to do, especially in the face of the criticisms and contradictory information that we're bombarded with. The purpose of this book is to make reasoning with and about the hypothesis and understanding science easier.

Why is the book so long? I could use a loose analogy with what the Nobel Prize- winning physicist, Richard Feynman, once said. He claimed that a seemingly simple idea—"the atomic hypothesis . . . that all things are made of atoms—little particles that move around in perpetual motion, attracting each other when they are a little distance apart but repelling when being squeezed into one another"—contains "an enormous amount of information about the world, . . . if just a little imagination and thinking are applied."[1] The hypothesis is a seemingly simple idea that contains an enormous amount of information about science if you just apply a little imagination and thinking to it.

To give your imagination something to work with, I wanted to go into the details of scientific thinking that your teacher never told you about, probably because she had never been taught them herself. What I found was that the subject is a lot deeper—and so the book is a lot heavier—than anticipated.

In the beginning, I didn't set out to write a book of any kind, though. Actually, the project was started by a simple question:"What is my hypothesis?"

Although the graduate student appeared earnest, I wondered at first if she was joking. She was very bright and was well advanced on the research phase of her PhD thesis in a colleague's laboratory. Her project focused on a novel aspect of the neuroscience of rat behavior that might be related to human mental illness. She had done many experiments and gotten masses of high-quality data. Her experiments were focused on specific topics; this was not an open-ended "discovery science" approach. Surely, I thought, she must have a good idea of why she is doing each experiment, what the likely outcomes are, how the series of experiments hangs together—in short, an idea of what her hypothesis was? As we discussed her project, it seemed that the different pieces of the investigation were running along in parallel; near each other but not really intersecting.

"What is the working title of your dissertation?" I asked, because a well-chosen title usually announces its major theme.

"I'm not sure at this point."

We took a step back and started talking about why she was interested in the topic, what the central unanswered questions were, and what she thought that her results were leading to. But her question had touched a nerve. A clearly formulated hypothesis would have helped her organize her thinking, yet she wasn't sure she had one, even though, like all students in our program, she had learned about hypotheses in her first-year graduate Proseminar course. The course reviews basic aspects of scientific practice and reasoning to ensure that everyone starts off on the same page. She had done extremely well in Proseminar, had practiced working with hypotheses, and had written a mock, hypothesis-based grant proposal. Now, 3 years later, she was at sea when it came to her own work; the lessons from the classroom hadn't stuck with her.

Unfortunately, this vignette was not unique. Even very good students often had trouble integrating the concept of a hypothesis into their thinking. We, the faculty, seemed to be failing them somehow. We were doing a good job of getting across the mechanics of laboratory research. The students became adept at putting together specific experiments, carrying out procedures, and collecting and analyzing data; however, they often missed a broader sense of what was going on in an investigation, whether it was their own or someone else's. What exactly was the problem?

I got one clue when a senior colleague and I began co-teaching a section on scientific thinking in Proseminar. He conducted an exercise in which the students had to pick out true hypotheses from a list that included scientific statements that were not hypotheses. It quickly became apparent that he and I disagreed on the answers. Although we soon papered over the problem, I found this incident profoundly disturbing. We were both strong advocates of hypothesis-based scientific thinking, and I had naturally assumed that we had a common understanding of what a hypothesis was. The experience taught me otherwise. Evidently the concept was not as universally well understood as I had thought, and if we senior scientists did not agree on what a hypothesis was, then what on earth were we teaching? No wonder the students were struggling!

Another clue came in the form of a blog post[2] entitled, "Hypothesis overdrive?" by Jon Lorsch, Director of the National Institute for General Medical Sciences at the National Institutes of Health and an important biomedical science thought leader. To stimulate a discussion within the biomedical research community, Lorsch shares his thoughts on "the hazards of *overly hypothesis-driven* research" [emphasis added], advancing an anti-hypothesis position as a framework: "It is too easy for us to become enamored with our hypothesis . . . ," "[a] novel hypothesis will appear in a high-impact journal and lead to recognition in the field," and "focusing on a single hypothesis also produces tunnel vision, making it harder to see alternative interpretations of the data." He makes a case for an alternative—*question-driven science*—where "the focus is on answering questions: How does this system work? What does this protein do? Why does this mutation produce this phenotype?" He wonders if asking questions would be more productive than testing hypotheses and concludes with the provocative query, "Is it time to stop talking about hypothesis-driven science and focus instead on question-driven science?" The post sparked vigorous feedback from eminent biomedical scientists (including a Nobel Prize winner), provided a glimpse into community concerns about scientific thinking, and underscored a number of the questions I had. Many issues, it appeared, remained unsettled.

The blog set-up implied that a hypothesis was a simple thing. The online discussion cast doubt on this implication: one respondent said "My claim is that . . . hypotheses are generated from data . . . and then refined/eliminated with further data," while another said "you have to have some ideas worth testing, . . . to organize your research. . . . These are working hypotheses." Finally, one says that, "First, the 'hypothesis' . . . is often not hypothetical at all, but only phenomenology dressed up as hypothesis; a hypothesis is a universal statement that cannot be directly verified by observation." This was confusing. Is the experiment-stimulating function related to working hypotheses? How do working hypotheses compare to universal statements, are there other kinds of hypothesis, and how does phenomenology fit in?

The actual meaning of the word "hypothesis" was not the only sticking point. For many respondents, a connection between hypothesis and "bias" was self-evident. One respondent asked that rhetorical question "why should I have to express a bias [by having a hypothesis] toward one of the possible outcomes?" And another felt that "a hypothesis is simply the favored answer to a scientific question and by forming a hypothesis, you bias yourself." There were dissenting voices ("Most scientists I believe would not bias their interpretation of results to confirm their hypothesis"), but these were in the minority. Some writers felt the problems lay in an imperative to test the hypothesis. One said that "If we can get rid of the 'support or disprove the hypothesis mode' we will ALL get ahead" and considered hypothesis-driven research "dangerous."

Respondents were also anxious about what happens if one's "overarching hypothesis" was disproved. Was it true that the only thing that you'd have learned was that the hypothesis was wrong? One scientist acknowledged that "Knowing how a biological system *doesn't* work is certainly useful, but most basic research study sections expect that a grant will tell us more about how biological systems *do* work, regardless of the outcomes of the proposed experiments." In contrast, one commentator counsels students "that if you do propose a new 'model' then . . . try to shoot [it] down . . . before someone else does." Both writers were alluding to the concept of *falsification* promoted by the philosopher Karl Popper, which states that tests of a hypothesis can never prove its truth but can, in principle, prove its falsehood. The principle of falsification is widely recognized throughout science, and so it was disconcerting to see so much disagreement about its function and worth.

Several comments touched on philosophical matters that the hypothesis involves—the nature of scientific answers; of certainty of knowledge; the writings of Karl Popper, John Platt, and David Hume—that many of us scientists don't know very much about. And practical as well as theoretical concerns arose: the National Institutes of Health (NIH) grant application reviewing process, the attitudes of reviewers, and the "Reproducibility Crisis" in science, among others.

Many respondents took it for granted that the whole conception of hypothesis-driven research is ill-advised. Lorsch's premise was that "By putting questions ahead of hypotheses, getting the answer becomes the goal rather than 'proving' a particular idea. . . . [putting] questions first and [including] multiple models to explain our observations offers significant benefits." A number of respondents liked this proposal; however, no one explained how making the switch would bring the clarity to scientific thinking that it was supposed to. The unstated assumption running through the comments was that non–hypothesis-related approaches are free from the drawbacks of the hypothesis. Lorsch's questions suggest that there are sharp divisions between asking questions and testing hypotheses or between models and hypotheses. But is this true? What is the evidence that science can answer questions more reliably than it can test hypotheses?

Furthermore, while the post focused on "question-driven science," respondents also mentioned "discovery science," "exploratory science," "fishing expeditions," "curiosity-driven science," and "big science" as alternatives to the hypothesis-driven approach. Although these alternatives were not defined and the contrasts were not spelled out, the conclusion was clear: these alternatives and the hypothesis were mutually incompatible and a scientist would have to choose one or the other.

In short, the blog and its follow-up revealed that (1) the hypothesis is a genuinely hot topic and the scientific community has pervasive concerns about it, (2) there is no consensus on what the concerns are or how to tackle them, (3) differences in interpretations of a hypothesis contribute to the confusion, (4) much antagonism toward the hypothesis arises from nebulous worries regarding bias or other problematical behaviors, and, finally, (5) to a large extent frustration with the hypothesis is prompted by anxieties about how others, particularly reviewers of scientific grant applications, interpret it and is only tangentially about the hypothesis itself. To be sure, this was a self-selected, nonrandom sample of respondents, and we can't be sure how broadly representative their views are. Still, the variety and passionate tone of the responses made me suspect that there is a problem regarding what many people would have said is one of the bedrock tenets of science. The main message that I took away was that I needed to know more about the hypothesis—its divergent roles and attributes—and about the relationships between hypotheses, questions, exploratory science, etc.

I was beginning to get a good sense of why our graduate students had trouble with the hypothesis: the scientific community itself had trouble with it. A final piece of the puzzle was discovering that, in addition to facing neglect and misunderstanding, the hypothesis also faces active opposition. It turns out that scientists have written entire books that feature extensive attacks on the hypothesis. Surely, I thought, the hypothesis has served as the bulwark of scientific thinking for hundreds of years, hypothesis-based science has produced formerly unimaginable advances in knowledge—things can't be all that bad!

I.B Overview

Initially, I wanted to get to the bottom of why many of us have difficulty with the hypothesis, and I felt that I should try to defend it as an irreplaceable part of the scientific enterprise. A lot that I learned in writing the book surprised me: Who would have guessed, for example, that a majority of working scientists have had almost no formal instruction about scientific thinking or that word "hypothesis" has been virtually expunged from much of the pre-college science curriculum? I want to share stories like that as well.

There are many books on the general topic of *critical thinking* (For Dummies[3]; For Kids![4]) that will teach you to make better decisions and solve problems, help you tell honest science apart from *pseudoscience*, or improve your life in countless other ways. The goal of this book is more modest: it seeks to explore and explain the concept of the scientific hypothesis in enough depth to reveal its complexity and its value while, at the same time, demystifying it. Above all, I hope to show why it is such a useful tool for the craft of science. Acquiring expertise in using the hypothesis does not come naturally to most of us—I'll inquire into the reasons that it doesn't and suggest ways of enhancing your skills.

Although it assumes no special experience or background, the book was written with science students, teachers, mentors, and administrators in mind. Yet most of the issues regarding scientific thinking are broadly applicable, and I believe that general readers who are interested in the intersection of science and philosophy, or who want better insight into what science is up to, may gain new perspectives as well. Many of the examples that I use are generic and nonscientific. My own background is in biomedical sciences, specifically neuroscience, and I am familiar with how the hypothesis is viewed in this area when it comes to experimental design, grant applications, journal articles, teaching, etc. As a public that is constantly bombarded with science news, its triumphs and peccadillos, we need to know how to think about what's going on in science.

The book is divided into three parts. Part I (Chapters 1–9) lays out what the modern scientific hypothesis is, what it can and can't do, addresses misconceptions surrounding it, and reports the results of two original surveys that I conducted to find out how scientists think about and use the hypothesis. When I asked several hundred scientists about the amount of formal instruction in the scientific method, the hypothesis, hypothesis-testing, and so on, that they had received, I found that their experience was like mine: the instruction was minimal when there was any at all. Nevertheless, we're apparently supposed to know all about these things. Part II (Chapters 10–12) covers problems—actual and alleged—with hypothesis-based science, and evaluates overt opposition, and suggested alternative approaches, to using hypotheses. I'll also explore cognitive factors, including biases, that help explain why working with the hypothesis is so unnatural for most of us. Finally, in Part III (Chapters 13–15), I make suggestions for sharpening your skills in scientific thinking, especially in using the hypothesis, as well as suggest a few policy changes that might bring some clarity to national issues. In Chapter 15, I'll take up the specific challenge that Big Data poses for conventional scientific thinking.

The influence of Karl Popper's thought on the practice of science is inescapable, albeit controversial. It permeates this book, and you can't truly follow the debate about the hypothesis in science without knowing what he actually said—and what he said is frequently misrepresented. It is important to correct the record. We also need to learn about Popper's thought to see how it meshes with other modern modes of doing science and how it fits into the grand scheme of things. Hypothesis-testing is not the only way of doing science (Chapter 4); however, this fact does not mean that you have to choose among them. By the end of the book, I hope to have sorted out the advantages and disadvantages of the various modes.

Although it is centered on the hypothesis, the book is not a monolith or a novel, and you don't have to read it straight through. For instance, students and teachers might want to focus on Chapters 1–5, 8, and 14; the philosophically minded might choose Chapters 1–3, 5–12, and 15; scientists perhaps Chapters 1–9 and 14; policy makers could add Chapter 13; and so on.

If this sounds interesting, then you may want to go to Chapter 1 if you'd like details. A precis of each chapter follows.

I.C Chapter Contents

Chapter 1 is for readers with little or no background in philosophy or for those who would benefit by a short refresher in basic topics such as *fallibilism, objectivity, deduction, induction*, etc. I also discuss philosophy-of-science concepts such as *objectivity* and the nature of *explanation, uncertainty*, and *levels of organization of nature* as background that will be called on from time to time in the remainder of the book.

Chapter 2 describes what a modern scientific hypothesis is, distinguishes hypothesis from *prediction*, and explains their relationships. We need a rich and precise language for discussing the hypothesis, and Chapter 2 also discusses science and *truth* and emphasizes the flexible, recursive nature of the *Scientific Method*. I will define associated concepts, such as *direct* and *indirect* evidence, using both scientific and nonscientific examples. What makes for a good hypothesis? I cover *non-obvious predictions* (riskiness), *parsimony*, scientific significance, specificity, and *constraint*. Not all hypotheses are explicit, and there is the crucial, although underappreciated, influence of background assumptions that are nonetheless *implicit* hypotheses.

If science had a philosophical patron saint, it would be Karl Popper, an extremely influential, though often maligned, thinker. Chapter 2 goes into the programs of Popper and John Platt in greater depth than you find in science courses. Almost everyone has heard about Popper's doctrine of *falsification*, although his critics distort it so much that you may wonder why you heard about it. I'll try to clear things up. And I'll go over other elements of Popper's philosophy such as "tested-and-not-falsified" hypotheses, *corroboration* and how it compares with *confirmation*, and other issues that scientists are rarely taught. Platt's practical experimental program of *strong inference* augments Popper's philosophy in a couple of significant ways, and I'll review it briefly as well.

Philosophers of science would no doubt disapprove of my take on Popper's philosophy if they happened to hear about it. "It's so, I don't know, so 1970s," they'd say. "We've come so far since then." I plead guilty to a friendly reading of Popper, although I'll point out some issues that he glosses over. On the other hand, I'll also criticize his critics and argue that, while the philosophers have unquestionably moved away from Popper, they disagree on where they are now, which gives scientists little to go on.

Chapter 3 is somewhat more advanced than Chapters 1 and 2. It is intended for readers who want to dig a little into the nuances of Popper's and Platt's programs or who have questions about them. I provide answers to several common objections that philosophers raise, for example the problem of *holism* and *negative data*.

Disputes can arise when critics point to various non–hypothesis-based modes of science and declare that, therefore, hypotheses are unnecessary. The conclusion doesn't follow. Chapter 4 distinguishes among several major "kinds" of science, including "confirmatory versus exploratory," "natural versus social," Big Data versus Little Data, and makes the case that they all fit comfortably within the realm of modern science.

Besides the *scientific* hypothesis scientists also use the *statistical* hypothesis, which, despite its name, has little in common with the scientific hypothesis. Chapter 5 compares the two classes of hypothesis by highlighting their conceptual similarities and differences, such as how they incorporate empirical versus numerical content and how they relate to the real world. While familiarity with basic statistics terminology is helpful, expertise is not required because I focus on fundamental issues in general terms. In addition to exposing hidden complexities of the typical *null hypothesis* statistics methods, we'll look at how possible replacements or supplements to *p-valued* testing, including *confidence intervals, effect sizes*, and *errors*, affect the way in which we conceive of the statistical hypothesis.

The statistical discussion continues in Chapter 6. Whereas conventional *frequentist* statistical ideas are the focus of Chapter 5, we also need to be acquainted with *Bayesian* statistics, so I'll cover the basics of Bayesianism in this chapter. I believe that relatively few bioscientists learn about Bayesian statistics, but Bayesian methods have an increasingly prominent role in science, and the Bayesian philosophy of science is very different from the one we're used to. For example, Bayesians and traditionalists disagree on what *probability* means, as well as on the purpose of scientific research. I'll compare them and illustrate how the Bayesian approach can be incorporated into a standard hypothesis testing procedure.

In Chapter 7, I'll dive into the controversy surrounding the Reproducibility Crisis, which centers on the fraught question: Is science reliable? The acid test of scientific reliability has been *reproducibility*—when scientists in their own laboratories can duplicate the results reported by scientists in different laboratories. But what does "reproducibility" really mean, and do reported difficulties in reproducibility constitute a crisis, a problem, or just alarmist talk? Rather than a comprehensive review, this chapter goes through evidence for and against the notion that a crisis exists, and it highlights roles that hypothesis-based science can play in easing the stresses that do exist. *The Reproducibility Project: Psychology* and the statistical critiques of John Ioannidis and colleagues are at the center of the chapter.

Chapter 8 makes the case in favor of the scientific hypothesis from two distinctly different points of view: statistical and cognitive. To demonstrate the statistical advantages of the hypothesis, I pick up the discussion of the Reproducibility Crisis from where we left it in Chapter 7. I explore reasoning that implies that hypothesis-based research in general will be more reliable than, for example, open-ended gene searches. Turning to the cognitive advantages of the hypothesis, I start from the consensus among cognitive scientists that the human mind is an organ inherently driven to try to understand the world. This drive shapes our thoughts and perceptions, and the hypothesis is a natural way of channeling it into science. The hypothesis is thus a natural organizational tool that creates blueprints for our investigations and our scientific thinking and communication. Finally, I'll argue that the hypothesis specifically aids in protecting against bias.

Pronouncements about how scientists perceive and use the hypothesis are rarely accompanied by actual evidence so, in Chapter 9, I report the results of a SurveyMonkey poll of several hundred members of scientific societies that I conducted to find out how scientists themselves think about the hypothesis and related matters. The poll covered topics from the extent of the respondents' formal instruction (70% had ≤1 hour of instruction) in scientific thinking as well as how they've used it at different stages of their careers. I also screened more than 150 neuroscience research reports in top journals to get a sense of how scientists' publications are influenced by the hypothesis. Only about 33% of the papers explicitly stated the hypothesis (or model) that they were testing, although another approximately 45% were based on implicit hypotheses. I discuss how the data relate to the place of the hypothesis in scientific education and practice.

As I mentioned earlier, the hypothesis is not merely neglected, it is actively opposed, and, in Chapter 10, I review the arguments and conclusions of three vocal critics of the hypothesis: Stuart Firestein, David J. Glass, and David Deutsch. These scientists want to replace hypothesis-based procedures with other strategies, *questioning and model-building*, *curiosity-driven science*, and *conjecture and criticism*, respectively. Although they all agree that the traditional scientific hypothesis has outlived its usefulness, each approach differs greatly from the others. I disagree with their arguments; nevertheless we can learn a lot by examining their ideas.

To a substantial extent, you can trace difficulties in scientific thinking to human cognitive foibles. Chapters 11 and 12 look at issues surrounding our mental quirks and how they affect our ability and readiness to use the hypothesis. Critics tend to blame the hypothesis itself for difficulties that scientists have in scientific thinking and reasoning, but the critics often base their arguments on overly simplistic interpretations of how the mind works. In Chapter 11, I'll review fundamental concepts that challenge these interpretations, including the fact that much of our "consciousness" and the source of hypotheses reasoning is actually unconscious and genetically programmed into us. In Chapter 12, I consider representative biases (e.g., *confirmation bias, publication* bias, and others) in light of cognitive theories that account for biases as products of *heuristic* ("rule of thumb") *thinking*. There are two main competing schools of thought when it comes to heuristics and biases: the *heuristics and biases* program initiated by Daniel Kahneman and Amos Tversky, and the *fast-and-frugal* program championed by Gerd Gigerenzer and his colleagues that is inspired by evolutionary psychology. The programs differ chiefly in their view of heuristics as either sources of error (Kahneman) or as quick, generally efficient, adaptive cognitive strategies (Gigerenzer). Their divergent points of view give rise to divergent explanations of how we think and, in particular, how our biases affect our hypothesis-based thinking.

Chapter 13 continues the investigation of the problems that those wishing to become adept at scientific reasoning have to contend with. In this chapter, I review a range of educational materials dealing with scientific thinking for individual scientists at various stages of their careers (i.e., from pre-college science courses to materials on the NIH websites). A thorough review of science teaching materials would be far beyond the scope of this book, so I selected sources, including information available from a large national educational organization, the National Science Teachers' Association (NSTA), and an educational initiative that NSTA backs, the Next-Generation Science Standards (NGSS). I also reviewed the scientific thinking components of two commercially available, pre-college science teaching programs. At the college level, I briefly sampled several science textbooks and focused greater attention on a widely used textbook devoted to scientific thinking. Finally, I searched the NIH and National Science Foundation websites and found that they have almost no information about the Scientific Method or the hypothesis that could be a source for practicing scientists. I make suggestions for improving science education, and I look at some policy issues (e.g., *preregistration* of studies) to see how applicable they are to hypothesis-based science.

The message of Chapter 14 is that rigorous scientific thinking is a skill that you can acquire. Effective scientific thinking does not demand innate genius (although having genius doesn't hurt), but it does require conscientious practice if you want to become good at it. Like learning to drive a car, at first you need focused, conscious attention to execute the simplest maneuvers before they become second nature. The question is how to coax people to take seriously the task of learning to think scientifically. This chapter offers suggestions that will probably be unfamiliar to many science students (and their teachers). Two exercises are to *find* and *diagram* the hypothesis in published papers. Given that the majority of the hypothesis-based papers don't state their hypothesis (Chapter 9), these exercises may be helpful to many readers. Other suggestions include an eclectic mix of recommendations borrowed from nonbiological fields that emphasize the virtues of clear thinking. The unifying theme is that these suggestions are designed to get you to think about your own thinking, a form of *meta-cognition.*

Finally, in Chapter 15, I'll take up the subject of Big Data and the hypothesis. If the hypothesis represents precision and logical elegance in scientific thinking in its ultimate instantiation, the Big Data Mindset would represent the opposite: it is imprecise, deriving ungainly advantage from brute force process. It manipulates massive quantities of messy data yet is capable of uncannily accurate predictions. Big Data gets its predictive power through correlations, not the reasons that the correlations exist. How does the Big Data Mindset deal with the hypothesis? By dispensing with it. Eventually, when there is no longer a need to understand, there will be no reason to explain. A very different vision of the future of science emerges when we conceive of a unification of Big Data capabilities and a technology called the "Robot Scientist."

In a brief Epilogue, I consider a merger of the best aspects of Big Data and advanced artificial intelligence (AI), the Centaur Scientist, that could interact with human scientists to help advance our knowledge by testing hypotheses for centuries to come.

Notes

1. Richard P. Feynman, *Six Easy Pieces: Essentials of Physics Explained by Its Most Brilliant Teacher* (Reading: Perseus Books; 1995, p. 4).
2. https://loop.nigms.nih.gov/2014/03/hypothesis-overdrive/
3. M. Cohen, *Critical Thinking Skills for Dummies* (Chichester, UK: John Wiley & Sons; 2015).
4. Evan-Moor Corp., *Critical Thinking, Grade 1: Connecting School and Home (Skill Sharpeners)* (Monterey, CA: Evan-Moor Educational Publishers; 2017).

PART I
FUNDAMENTALS

1
Philosophical Background Matters

1.A Introduction

"Too much philosophy," was my colleague's crisp appraisal of a draft of this chapter. I think he was being diplomatic. Technically, his comment left open the possibility that a bit of philosophy would be okay, but his tone said that none would be better. He is not alone: typical working scientists do not have much use for philosophy, making their way more or less satisfactorily while ignoring it. Indeed, many philosophical issues are so arcane that we can safely skip them. Why not skip all of them? There are several reasons. Philosophers have thought a great deal about topics that scientists, whether we like it or not, must confront because these topics directly affect how we think about and do science. Critics of hypothesis-based science, including the scientist-critics whom we'll meet in Chapter 10, ground their arguments firmly in philosophical concepts. Finally, the scientific hypothesis, the primary topic of this book, is enmeshed in a web of philosophical associations acquired long before science emerged from philosophy. We'll need to be familiar with these associations to understand the hypothesis and the arguments surrounding it.

Despite their current lack of interest in philosophy, scientists themselves used to take part in pragmatic philosophical discussion. The Nobel Prize-winning Spanish neuroanatomist, Santiago Ramón y Cajal, begins his witty and still informative *Advice for a Young Investigator*,[1] written in 1898, with this: "I shall assume that the reader's general education and background in philosophy are sufficient to understand that the major sources of knowledge include, observation, experiment and reasoning by induction and deduction."

Because modern undergraduate and graduate science curricula are jam-packed with science courses and other requirements, it is doubtful that all scientists nowadays have the background that Ramón y Cajal had in mind. We'll also have to confront the estrangement that exists between science and philosophy; except on occasion, they don't talk. Like my colleague, most scientists either brush philosophy aside as irrelevant or are openly hostile to it. Nonetheless, we do need to know a few key concepts, as even die-hard, anti-philosophy scientists agree.[2]

1.B Science Versus Philosophy?

In a chapter entitled "Against Philosophy,"[3] physics Nobel Laureate Steven Weinberg posits that, apart from protecting physicists from the misconceptions of philosophers, philosophy is generally harmful to science. Its rare good ideas outlive their usefulness and hold back progress. In the 1800s, for example, the doctrine of *positivism* encouraged scientists to base their thinking concretely on observable entities and to shun unobservable ones. Einstein credited the positivist thinking of the physicist Ernst Mach for encouraging him to question the reality of Newtonian space and time, which led to Einstein's radical conception of space-time in the Special Theory of Relativity. But eventually positivism became a drag on the imagination, ultimately preventing Mach himself from ever accepting the reality of atoms.

What exactly is it that Weinberg rejects? Philosophy? All of it? It's a big subject. He makes the revealing comment that "In our hunt for the final theory, physicists are more like hounds than hawks; we have become good at sniffing around on the ground for traces of the beauty we expect in the laws of nature, but we do not seem to be able to see the path to the truth from the heights of philosophy."

Setting aside the fascinating issue of natural laws having beauty (not to mention an odor), Weinberg's main complaint is that the lofty philosophical outlook is disconnected from the gritty world of science. He is bemused by those who "think that there are things that can be said about space and time on the basis of pure thought." Weinberg has read a lot of philosophy and, in mulling over the "unreasonable effectiveness of mathematics in the natural sciences,"[4] he gets in a dig about the "unreasonable ineffectiveness of *philosophy*" for science. Richard Feynman, another Nobel Prize-winning physicist, was more succinct and less subtle: "Philosophers are always on the outside making stupid remarks."[5] Feynman has no patience with the grand but endless "why" questions that philosophers tend to ask. In the grand scheme of things, the only answer is: nobody knows. So let's forget about why and get on with figuring nature out.

On the other hand, neither Weinberg nor Feynman is entirely innocent of philosophy. Weinberg admits to a "rough and ready *realism*" (a belief that an objective reality exists outside of our brains) and says we need a "tentative worldview to make progress." He accepts that science "must ultimately test its theories against observation"; all of these are bona fide philosophical topics. And Feynman took to heart the significance of a number of philosophical problems and wrestled publicly with them: for example, what is the nature of explanation.[6] It appears that there are philosophical threads that these scientists tacitly admit are important and beneficial.

For their part, philosophers have not always helped build bridges. The philosopher W. V. O. Quine once declared that "for scientific or philosophical purposes, the best we can do is to give up on the notion of knowledge as a bad job."[7] Yet scientists are in the business of finding (or creating) knowledge that, despite its uncertainties, mostly works well enough—sometimes spectacularly well. And, indeed, some philosophers see the interactions between philosophy and science in a way that scientists would find more to their liking. According to philosopher of science Larry Laudan,[8] "New or innovative methodological ideas have generally not emerged, nor have old ones been abandoned, as the result of internal, dialectical counterpoint between rival philosophical positions or schools. . . . Rather, it is shifting scientific beliefs which have been chiefly responsible for the major doctrinal shifts within the philosophy of science." In other words, science leads philosophy, not the other way around. In any case, although science grew out of philosophy, there is now a rift between them.

1.C Philosophy and the Philosophy of Science

Let's start with philosophy itself; obviously a vast enterprise, but what is it all about, really? Philosopher of science Peter Godfrey-Smith attributes "The best one-sentence summary of what philosophy is up to"[9] to the philosopher Wilfrid Sellars, who had said that philosophy is concerned with "how things in the broadest possible sense of that term hang together in the broadest possible sense of that term."

What, we wonder, would the "broadest possible sense of things" be? Its an all-encompassing vision that covers, literally, anything you can name. Within it, we find the *philosophy of science*, which studies the activity and products of science and is distinct from the *philosophy of nature*, which tries to work out what the data provided by science are telling us about the world. Sellars wants to find out how all the things "hang together," which implies that things do hang together somehow. That there is a sort of unity to philosophy. A nonphilosopher may find the suggestion of unity (or maybe unities) of philosophical knowledge to be surprising since philosophers specialize in analytical dissections of terms and concepts that, judging by the erudite summary essays in the *Stanford Encyclopedia of Philosophy*,[10] for example, reveal more evidence of things hanging apart than hanging together. In fact, the philosopher's skill in analysis is probably the one that can benefit scientists the most.

The aerial view of science, via the works of Thomas Kuhn's *The Structure of Scientific Revolutions*,[11] with its "paradigm shifts" and "normal science," and the responses to Kuhn made by philosophers Imre Lakatos,[12] Larry Laudan, and Paul Feyerabend, might enrich your life, but it doesn't hold much in the way of practical utility for us. The philosophers generally want to account for "rational theory choice" and explain the scientist's state of mind; that is, why we adopt some ideas and resist others. But Kuhn thinks that theory choice is not a purely rational process; it depends on several factors, including social influences. While social and other irrational forces are important, we don't know much about how they affect scientific decisions. Later, we'll touch on how our psychology affects our use of the hypothesis, although this is not what philosophers concern themselves with. In fact, where the philosopher's quest is heading is not very obvious. Indeed, one philosopher of science asks, "What *kind* of theory should the philosopher of science develop?"[13] Perhaps scientists will benefit more from philosophy of science when this question is answered.

To keep from fighting old battles, I suggest that we focus on a few philosophical ideas that might help us out, even if our treatment of them would not pass philosophical muster. We need a small, workaday collection of concepts and nuggets of philosophical wisdom, rather than the hyper-abstract perspective that philosophy of science tends to offer.

For instance, most scientists would accept Weinberg's common-sense notion of *realism*. The world ("world" means the entire physical universe, or "nature"; the three are synonymous here) does exist independently of our experience, although our perceptions (cognitive as well as sensory) can shape our experience and, therefore, our understanding of it. If you're a scientific realist, you accept the principles of *empiricism*: scientific knowledge is acquired by observations and experimental tests of the real world. The philosopher allows that it is reasonable for science to aim for "accurate descriptions (and other representations) of what reality is actually like."[14] However, scientists aspire to more than mere descriptions, which is what a tourist visiting a new city does for the folks back home. Scientists want to understand reality by finding and testing explanations, including theoretical mechanisms, hypotheses, predictions, and other explanatory devices.

1.D Modern Science

How do we distinguish science from other activities? Is there a "unity of science"? Scientific fields could have common features that nonscientific fields don't have, or there might be connections among the scientific fields that directly link them together, or both. A group of activities might all warrant being labeled "scientific" without sharing subject matter, methods, techniques, conceptual underpinnings, etc. Think about another activity say, fishing, and ask, "Is there is a unity of fishing?" You can go about fishing with an expensive fly-casting rod in a remote wooded stream, huge trawler nets on ocean-going factory ships, a string and a baited hook at the end of a pier, a spear while diving on a coral reef, etc. And it doesn't matter whether or not you are successful in catching anything;

trying is enough because fishing is defined by its objective—to catch fish—not by the equipment or techniques used, background knowledge available, or immediate purposes (sport, profit, food acquisition, etc.).

Is there an analogous unifying theme in science? We don't have to go through numerous possible answers to this question to find a practical framework. To put my cards on the table, I favor the answer that has dominated the Western intellectual tradition since the Enlightenment: science is a systematic study of nature having the core values of *rationality, objectivity, testability*, and *reproducibility*. It aims to discover universal natural Truth while accepting that attaining that goal is impossible. Science is rooted in empiricism, the Scientific Method, and freedom from authoritarian dogma. It rests on the belief that nature is quantifiably regular and understandable. And real. Science admits of no supernatural causes or effects.

Philosophers of science find almost every word in this description debatable, and we'll get to their criticisms shortly. Furthermore, because of its historical philosophical roots and its association with the industrialized West, modern science is frequently referred to as "modern Western science," although this term ignores the increasingly voluminous contributions made by scientists from non-Western countries.[15] "Modern science" captures the affiliation with modern industrialized societies without the cultural baggage (we'll discuss an alternative, *Indigenous science*, in Chapter 4). The next section has a selection of philosophical concepts that all scientists should know.

1.E Fallibilism

Philosophers of many stripes and scientists—who disagree about much else—agree at least on this: we cannot attain absolutely certain knowledge about the world, a position so profound that it has a name. It is called *fallibilism*, and if you agree with it, you are a *fallibilist*. Deep down all scientists are fallibilists, even if we don't use the term. We don't believe that achieving complete, unambiguously correct knowledge (capital "T" Truth) is possible, and we accept that our most definite and cherished facts may be wrong. For better or worse, scientists are in a perpetual state of doubt at some level (see sections 1.F–1.H), though it's true that we don't always act that way. Doubt, however, is not a bad thing; in fact, gaining our freedom to doubt and to express doubts publicly was a major achievement in intellectual history. Until the Enlightenment, belief and acceptance of principles laid down by ancient authority were the dominant intellectual virtues. Doubt was not welcome in the ages when *dogma* ruled.

Although it is an unavoidable consequence of our human mental, temporal, and spatial limitations, accepting the ultimately insecure basis of our knowledge can create a sense of queasiness. Other systems, especially religions, provide faith-based certainty that relieves the anxieties that accompany uncertainty. Science offers no such comforts. On the bright side, if we put our minds to it, getting accustomed to fallibilism shouldn't be much more difficult than getting accustomed to the fact that there is no Santa Claus. Nonetheless, if you are like many people, you can commit to the principle of fallibilism intellectually but maybe not quite whole-heartedly. We'll encounter the problem several times throughout this book, and I'll suggest solutions. Fallibilism affects the question of how science knows what it knows. One answer is that science relies on *objectivity*, which we turn to next.

1.F Objectivity

"The authority of science relies on its objectivity."[16] This may seem so obvious, even trite, that you'll be surprised to learn that objectivity has not always been a core characteristic of science. Indeed, Lorraine Daston and Peter Galison argue[17] that recognition of the importance of objectivity is relatively new to science, with its origins dating to the mid-late nineteenth century. It emerged as scientists gradually realized that, in order to communicate and carry on science as a cumulative, collective activity, there had to be ways of establishing what the facts in any given field were. In those pre-statistical days, scientists dealt with random variability mainly by ignoring it, which they justified essentially as follows: if you're interested in finding the true regularities that you know must underlie natural phenomena, the minor details are unimportant. They trusted their own powers of observation and imagination to guide them to the truth. Unfortunately, as they discovered, not everybody's observations of the same phenomenon led to the same truth about it. Worse, they found that their own observations could be unreliable. There is a touching tale[18] of a physicist in the late 1800s who wanted to investigate the splash that occurred at the instant that a falling drop of liquid contacted a surface. He had a stroboscopic setup that let him visualize the splash during a brief flash of light and, with no way of recording what happened, painstakingly hand-drew each visual afterimage as it transiently imprinted itself on his retinas. He sketched thousands of beautifully symmetrical splashes, until, after 20 years of this, he succeeded in making a camera that could capture the images photographically. He was crestfallen to discover that his meticulous drawings were largely figments of his imagination; reality was not nearly as orderly as he had perceived.

Eventually, it became clear that facts had to be independent of the idiosyncrasies of the individual scientist. "To be objective is to aspire to knowledge that bears no trace of the knower—knowledge unmarked by prejudice or skill, fantasy or judgement, wishing or striving," according to Daston and Galison. The authors use the development of scientific *atlases*, compendia of pictures of birds, clouds, images of brain scans, etc., to examine how standards of objectivity emerged. The atlases were the common working objects that made it possible for communal science, for *collective empiricism*, to arise and engage scientists from all over the world. Sadly, high hopes that atlases and other products of instrumental measurements would be the cure-all for the ills of subjectivity were doomed to disappointment, too. Instruments such as cameras recorded too much detail—experts had to decide what was crucial so the user could develop her own judgmental skills. Thus objectivity standards passed from the pre-objective *truth-in-nature* (seeking ideal natural forms; what the droplet-drawing physicist was doing) through stages of *mechanical objectivity* (recording information instrumentally, warts and all; what the camera enabled him to do) to *mechanical objectivity* supplemented by "trained judgment." Objectivity, it seemed, had an ineradicable element of subjectivity to it.

Scientists today take it for granted that we can observe the world in reasonably accurate ways. However, imperfections in our sensory and instrumental apparatus, in our ability to remain detached from our theoretical constructions or the influence of our social surroundings, all conspire to hinder attainment of a pure state of scientific objectivity. Some philosophers view objectivity and subjectivity as opposites that "define each other"[19] and that, therefore, in order to achieve objectivity, we must be aggressive in dealing with subjectivity, striving for "suppression of some aspect of the self" or admitting that objectivity is a hopeless dream.

Luckily, summoning up steely, self-abnegating willpower is not required for the day-to-day demands of experimental science. Instead, modern science tends to equate the necessary objectivity with "intersubjective"[20] reliability: an observation is objective if anybody can, in principle, make it. No special personal fortitude required. People who are subjectively biased can make observations that corroborate the findings of other equally biased individuals and thereby produce objectively reliable results. Indeed, science counts on its finding to be replicated in different independent laboratories as a way of telling whether they're correct. Yet science is not "value free" and cannot give one "tidy account" of nature.[21] Judgment does come into play, but in the form of general agreement or consensus among scientists that determines what they accept as true.[22] Objectivity, therefore, arises as a community virtue and doesn't refer only to the behavior of individual scientists. Concern about reproducibility in science (Chapter 7) is mainly a concern about community-wide objectivity.

In some quarters, "objectivity" has acquired offensive overtones.[23] It has become a surrogate for "superior" or "true." The word has been used to sanctify certain kinds of research. Feminist philosopher of science Sandra Harding is not primarily bothered by the research lab–delimited view of objectivity, which she calls "weak objectivity." She believes that research should be "fair to the evidence, . . . to one's critics, and fair to the most severe criticisms that one can imagine." For her, "the question is how to go about doing research that simultaneously advances the comprehensiveness and reliability of its results and also produces resources for answering the kinds of questions that are most important to an oppressed group."

Social issues, in other words, especially issues of social justice, are interwoven with this notion of objectivity. "The shape and purpose of the research project . . . are at issue," and Harding sees truly "value-free" research as a myth, an ideal so mistaken as to be worse than useless. The driving forces behind the research—the identity and motivations of the research funders, the cultural factors that determine the problems considered worth investigating—all necessarily influence research from beginning to end. To counter the negative influences, Harding proposes a program of "strong objectivity" that encompasses weak objectivity and then goes beyond to advocate for a vision of a deliberately inclusive, socially aware, and responsive kind of science.[24]

Is science *value-free*? It aspires to be but, if value-free means a perfectly detached, Olympian "view from Nowhere"[25] that is disconnected from and unsullied by ordinary human foibles, biases, limitations, and emotions, then it can't be. Scientists are human beings who are immersed in complex social systems, and it would be a miracle if we were not affected by them. However, acknowledging that science is not totally removed from the concerns of society and culture ("weak cultural constructivism"[26]) is not the same as agreeing with "postmodernist" critics that science is no more than a cultural construction ("strong cultural constructivism"). Science is much more than the expression of the biases of one group or another. Though incomplete and imperfect, the understanding of nature provided by modern science is vastly greater than that achieved in earlier ages. Scientists aim to get their knowledge of the world right, and they are aware (or are reminded) of the possibility that they could be wrong.

Philosophers, though not fully comfortable with this state of affairs, agree that we ought not to "throw out the valuable baby with the bathwater"[27] when it comes to assessing objectivity. They thus concede that scientists cannot wait until all the philosophical brain teasers have been solved, that scientists have to take action, and that an imperfect principle of objectivity is better than none.

1.G Reductionism

Nature exists on an immense range of physical scales and degrees of complexity. *Reductionism* is the doctrine that sciences of complex systems rest on sciences of simpler systems. Psychology rests on biology, which rests on chemistry, which rests on physics. The scientist's naïve view that the world is made up of entities and processes that are within the realm of physics is sometimes called *physicalism*. Physicalism is a tame form of reductionism, little more than saying that there are no supernatural causes or effects in science.

A strong form of *reductionism* says that everything will eventually be explained by physics alone. Probably very few scientists, and no philosophers, buy into this version. The philosopher Daniel Dennett calls the tame form "weak reductionism" and the extreme form "greedy" reductionionism.[28] The limit of greedy reductionism is *determinism* (or *causal determinism* since, of course, there are many kinds of determinism), which is the concept that, if we knew everything about all of the particles and forces everywhere in the universe, then we could use a super-duper computer to predict all past and future events. Weak reductionism denies that we'll be able to fully explain the complex aspects of nature in terms of the simple ones.

The physicist's grail of a "Theory of Everything"[29] (ToE) illustrates the distinction: ToE is the catchy name given to the hoped-for theory that will unite the major known forces of nature; namely, gravity, electromagnetism, and the strong and weak forces at play in the nucleus of the atom. If it could be constructed, a ToE would represent an amazing intellectual feat, but it will not explain, literally, everything. A ToE would explain the basic forces, fields, and particles in the universe; detailed quantitative descriptions of their properties and interactions; and, no doubt, jaw-dropping explanations of phenomena right out of science fiction—"time-travel!," "worm-holes!" Yet, although people and oil paints do obey the laws of physics, a ToE will not be able to predict the existence of either people or oil paints, let alone how a portrait will look before the artist paints it. There will always be far more things left unexplained than explained by physical theories.

Reductionism cannot mean that all sciences are ultimately derivable from a more basic science. This raises another question, however. If strong reductionism is invalid, then how should we think about the relationships among the different levels of analysis of science?

1.H Levels of Organization of Nature

The biologists Ernst Mayr[30] and Nobel Prize winner Niko Tinbergen[31] discuss explanation and the levels of analysis or levels of organization in biology in similar ways. They argue that a full understanding of a biological phenomenon requires probing the phenomenon at the species (phylogenetic), individual developmental (ontogenetic), physiological (mechanistic), and immediately practical (functional) levels. Let's take an example from Mayr: Why, he asks, does the wren outside his house in the northeast corner of the United States suddenly decide to migrate south in the winter?

At one level, we note that these birds eat insects and that, during the winter in that part of the country, outdoor insects are scarce. Therefore, wrens migrate south to where the insects are because they would starve otherwise (the functional level of explanation). And, because migration south in the colder months has solved the food problem for thousands of generations of wrens, the tendency to migrate is encoded in the wren's genes (the phylogenetic level of explanation). Both functional and genetic levels speak to the *ultimate cause* of migration, but ultimate causes do not explain everything. We could know all there is to know about its diet and genome and still not understand why a given wren suddenly decides to fly south one autumn day. It turns out that the electrical activity of cells in the wren's nervous system responds to day length, and, as the hours of daylight decrease, the neuronal activity primes the birds to migrate. Even so, the birds don't go until the temperature drops sufficiently; when it does, they take to the air. Shorter day length and cool temperature are physiological (intrinsic and extrinsic, respectively) *proximate causes*. Thus, the answer to the question, "Why did the wren migrate?" is complex: the bird was genetically inclined to do so, food was becoming scarce, the days were getting shorter, and, finally, the weather got cold enough to trigger migration. Inasmuch as all of these elements are necessary for a complete explanation of the phenomenon, they are all on the same explanatory plane. In some sense, they are of co-equal importance. Not all explanations are like this.

We should probably start by asking, "what exactly is an explanation?" An intuitively straightforward answer is that an explanation is a statement of the (presumed) cause of something's being the way it is. An explanation often answers the question "why?" We are uncertain about something to the extent that we can't explain it adequately. You'll often hear of scientists wanting to know the *mechanism of action* of a phenomenon. The mechanism of action is often an explanation that is based on a slightly lower level of organization than the one you're directly working on. If you want to know why one muscle in your arm contracts, you'll need to know which nerves activate it and why they are activated. If you want to know about why nerves cause muscles to contract, you'll need to know about the biochemical reactions that let nerves communicate with muscles. Other properties of explanation include its ability to unify aspects of knowledge by showing the interconnectedness of phenomena.[32] Sometimes entirely different kinds of explanation are needed for different levels of natural complexity.

1.I Explanation, Uncertainty, and Levels of Organization

Let me try to clarify something pertaining to issues of explanations and levels of organization, that will have bothered some readers. A few pages back, I suggested that, like many people, you might harbor reservations about the notion of fallibilism; that, despite intellectually going along with it, you might not completely buy into the notion that you can't be 100% certain of anything in science. Now, let's acknowledge that, on the face of it, fallibilism seems nonsensical. We must know some scientific facts with absolute certainty, mustn't we? Can there be any doubt, for example, that the earth goes around the sun? How can we square such blatantly obvious truths like this with the dictum that science can't achieve certain truth?

The reason is that the provisional truth of scientific statements, their explanatory adequacy for a given problem, is bound up in the concept of *levels of organization of nature*. At one level, represented by artists' graphic illustrations of the solar system, the earth does go around the sun in a smooth, elliptical orbit. From a much larger cosmological perspective, this is not exactly what happens. Both earth and sun revolve around the center of mass of the solar system (technically, the "barycenter"), and the location of the center of mass is constantly changing as the planets orbit the sun. Zoom out farther, and you'll see that the solar system moves through the Milky Way galaxy; the Milky Way moves within the local cluster of galaxies; the local cluster moves within the universe; and the universe expands. In summary, it is convenient, and true enough for most purposes, to think of the earth as traveling in an elliptical orbit on a flat plane around the sun in spite of the fact that a complete explanation of its motion is way more elaborate than that.

A homey allegory borrowed from Richard Feynman teaches a similar lesson in a different way.[33] Suppose, he says, you are told that your dear old Aunt Minnie is suddenly laid up in the hospital because "she slipped on some ice and fell and broke her hip." At first this strikes you as a perfectly satisfactory answer, and normally you'd head to the store for a sympathy card. But suppose that this time you want a thorough explanation, and you ask why she slipped on the ice. You are told that her stepping on the ice caused a thin layer of liquid water to form on the surface, the water reduced the friction between her boot and the solid ice, and the reduced friction caused her to slip. This seems odd; when you step on a wood floor, the wood does not liquefy, so you ask about the situation with the ice; after all, if you are really going to understand why the poor old soul is in the hospital, you need to know that! At this point, the answers become increasingly tricky.[34] Water expands when it freezes, and scientists used to think that the slipperiness came about when the pressure and friction caused by, say, somebody's boot heel, melted a bit of ice. However, explanations based solely on melting predict that if no melting could occur, ice would not be slippery, and so you ask about that and learn that at temperatures near absolute zero, where ice can't melt, it is still slippery. And nobody knows exactly why. The present thinking is that, at the surface of a sheet of ice, water molecules never truly freeze in place because, at the surface, which is essentially two-dimensional, there are fewer stabilizing atomic bonds among the molecules than there are around the molecules in the three-dimensional depths of the ice. The interactions between the weakly bound surface molecules and the deeper lying ones mean that the surface of ice is always in a quasi-liquid state that accounts for its slipperiness. The apparently simple question about why Aunt Minnie is in the hospital takes us to levels of nature where we don't have complete answers. All the same, it would be idiotic to say that, because we don't understand ice at its most basic physical levels, we don't know why she is in the hospital. We do. She slipped on the ice!

The moral of the story is that uncertainty, the concept of fallibilism, is related to the level of organization of nature, or its complexity, that you're interested in. An adequate explanation depends on what you need to know, the level of complexity of the system you are working with, and the kind of question you are asking. An explanation may only suffice at a particular level. This is why we can feel sure of some things and, at the same time, accept that all of our scientific knowledge is uncertain.

Science has different standards for different purposes. For basic (i.e., "pure") research, nothing short of perfect knowledge will do. So pure research keeps striving with no end in sight. For applied research and technology, it's another matter entirely; for applied purposes, we have to take action based on the best information we have. We know enough about ice and elderly people to put down sand or salt on ice on a sidewalk without knowing everything there is to know about ice. We understand enough about physics and the movements of planets to launch a space probe and put it into orbit around Jupiter 5 years later[35] without having a ToE.

1.J Why Try to Attain Truth If We Can Never Reach It?

Consideration of fallibilism gives rise to another conundrum for anyone trying to understand science: If *attaining* Truth is the goal, but reaching the goal is impossible, then why bother? Or, if for some reason we were to bother, how should we proceed?

The answer to the first question goes roughly along these lines: we have to. Humans are innately driven to try to understand; the urge is present even in children who have not yet learned to talk.[36] Our primate ancestors evidently found it useful to comprehend and control nature as much as they could, and they produced more offspring than those who couldn't comprehend and control as well, and those offspring produced more offspring, etc.—fast-forward a few million years, and here we are.[37,38] The drive to understand nature enabled our ancestors to find food, mates, and shelter and stave off death long enough to procreate. As time went on, the drive became generalized, was *exapted*, and people figured out how to control fire, develop agriculture, forge metals, form civilizations, and, eventually, invent cell phones. Whatever the precise details are, the fundamental need to comprehend nature is built into the genome of *Homo sapiens*, a conclusion with numerous implications for science and philosophy.

If we assume that understanding for its own sake is now an intrinsic drive, then it is apparent that the success of our species has not depended on a perfect grasp of nature's mysteries. It hasn't mattered that we haven't achieved Truth, because we have gotten countless practical benefits from discovering partial, provisional, small "t" truths. Thus, there are at least two answers to the question of why we try to achieve the unreachable ideal of Truth: "we have to" and "incomplete knowledge is still useful."

The second question is thornier than the first: How to proceed rationally to search for Truth if we're not sure where to find it? For millennia two answers have commanded the most attention in the Western philosophic tradition: *deductive, reasoning* and *inductive reasoning—deduction* and *induction*, for short. Both of these concepts are key to understanding scientific reasoning, especially the hypothesis and its proposed alternatives, and so I want to go into them in greater depth than science students ordinarily encounter.

1.K Deduction, Matters of Reason, and Matters of Fact

Philosophers sometimes divide concepts into *matters of reason* and *matters of fact* (these categories have other names as well). We can unambiguously determine the truths of matters of reason because they have to do with the meanings of words and their relationships in sentences. In *deductive* reasoning, if the starting assumptions, the *premises*, of an argument are true, then the conclusions derived from these premises *must be True* because the premises logically imply the conclusions. A philosopher would say that the premises *entail* the conclusion.

A common form of deductive reasoning involves the *syllogism*, an argument having a general statement, a specific statement, and a conclusion. The general statement is the major premise; the specific statement is the minor premise. Sometimes you see this with the major premise designated "P," the minor one "Q," and the conclusion "R." The syllogism says that "if P and Q are true, R *must* be true." If you agree that *no rocks are alive* (P) and that *this object is a rock* (Q), then you must agree that *this object is not alive* (R). And you must agree, not because the logic police could arrest you if you don't agree, but because you are a rational person who understands the meanings of the words and who expects meanings to apply consistently.

The syllogism is one form of a deductively valid argument. As a rule, if a true conclusion necessarily follows from true premises, then the argument is a valid deductive argument.[39] Mathematical truths are in this category. "Two plus two equals four" is true because the meanings of "two," "plus," "equals," and "four" are defined so that this statement has to be true. In the artificial environments of mathematics and symbolic logic, so-called *closed systems*, you can prove the Truth of most, though not all, statements by deductive reasoning. Deductively valid statements in these systems tell you nothing about the nonmathematical, empirical world. In the closed systems that they are part of you can tell whether or not a sentence is true by examining the words that make it up, as we did when concluding that the object must not be alive. The truth of the statement that "two chairs plus two chairs equals four chairs" is guaranteed by the mathematical terms and is not related to any property of "chair." Associating real-world objects with numbers allows us to take advantage of the power of mathematical reasoning to try to understand nature.

Another form of nonmathematical argument that is logically true is a *tautology*.

> "Why is Fido eating that smelly cat food?"
> "Because he likes it."
> "How do you know he likes it?"
> "Because he's eating it."

That's a tautology; it essentially equates eating and liking. The conclusion that Fido likes smelly cat food is based on the fact that he's eating it, together with the implicit assumption that eating it means he likes it. The tautology is logically valid though empirically meaningless; it says nothing about the real motivation for Fido's food choice. Maybe he doesn't actually like cat food of any kind—he wouldn't choose it if he had options—but now he is starving and will eat anything he can gag down.

Deductive reasoning is safe and secure and, for nearly two thousand years, from the time of Aristotle until the 1500s, philosophers thought that they could accumulate knowledge about the world through deductive reasoning alone. It might work like this: Suppose we assume that "people who earn more than $250,000 annually are rich," and, since Pat earns more than $250,000 annually, we conclude that Pat is rich. It might appear that we have gotten some new knowledge about Pat, but, once again, we are only working with word meanings. The argument essentially defines "rich" to mean "is earning more than $250,000 annually." However, being rich in some social strata might mean earning more than $1,000,000 annually, and, in that case, Pat would merely be well-off, not rich. Eventually it became clear that, by itself, this deductive procedure was not getting anywhere. The problem was that the fundamental notions that philosophers had about the world, which they had taken to be as certain as Euclid's axioms about geometry, were actually uncertain and that deductive reasoning from them could not lead to indubitably true conclusions. So deduction could not guarantee the truth of scientific conclusions.

The question is not whether deductive reasoning alone is *sufficient* for making scientific progress—it is not; the question is whether deductive reasoning has a place in science or scientific reasoning, and it does. As we'll see in Chapter 2, hypotheses and predictions are connected logically; we deduce predictions from hypotheses. Before exploring that relationship, however, we need to examine the major alternative to deductive reasoning.

1.L Inductive Reasoning

If it is impossible to make scientific progress by deductive reasoning alone, how can we discover scientific truths? Traditionally, the most popular answer, which has innumerable variants, is induction; the idea that that we go from observed regularities in nature to general rules that we then take to be true. Probably no single issue has caused more misery for the analysis of scientific thinking than inductive reasoning.[40] There are certainly intrinsically interesting and important aspects of induction, although nowadays induction may be more interesting from a cognitive science point of view than from a philosophical one. We'll see how the hypothesis can rescue science after we inquire into the cause of the misery.

1.L.1 Enumerative Induction

Learning about the natural world scientifically involves careful observation, data collection, and analysis. When this began to be done systematically, experimental science came into being. An early advocate of the empirical approach was Francis Bacon, whose major contribution to the philosophy of science was to emphasize the importance of experiment and inductive, rather than deductive, reasoning as a way of making scientific advances. Every day the earth turns and the sun rises in the sky. Supposedly, we reason by induction that the sun will reappear tomorrow. This sort of argument is known as *enumerative induction* because the general rule follows from the enumeration of individual cases. Larry Laudan classifies enumerative induction as a form of low-class, ordinary induction—"plebian induction"[41]—and distinguishes it from high-class "aristocratic induction" that tries to assess the validity of a theory or hypothesis from a number of confirmatory tests of predictions that it makes. First, the low-class form.

1.L.2 The Problem of Induction

From Bacon onward until the nineteenth century, many thinkers considered that there was "no significant element of uncertainty or doubt attached to the conclusions of so-called inductive inference."[42] This opinion was anchored by the widespread belief that Isaac Newton had arrived at his famous laws of motion via induction. Their stunning success apparently affirmed the power of inductive reasoning to lead to Truth; Newton himself thought that he had discovered God's own plans for the workings of the universe in this manner.[43] Still, doubt about whether induction was generally reliable crept in and even about whether Newton employed inductive reasoning at all. (He didn't; at least not what we're calling induction here[44,45]; see also Box 10.1.) What shook philosophic and scientific faith in its power?

If you take the claims of enumerative inductive reasoning too literally, you get the impression that experimental observations themselves lead to general rules. This is not true, of course. Generalizations do not arise from data; they are created by human minds thinking about data. Moreover, when the minds invent the generalizations, predictions are often the limit of what you get with induction; patterns, not deep understanding. Richard Feynman notes that the ancient Mayans made repeated astronomical observations of lunar eclipses that allowed them to *predict* future lunar eclipses accurately but not to grasp the celestial mechanics that would have *explained* eclipses.[46] Their reproducible observations could only take their astronomical knowledge so far.

Any serious analysis of induction has to start with the philosopher David Hume[47] who exposed its shortcomings by showing that its validity rests on the huge unproven assumption that Nature—the external, objective universe—has always been the way it is now and will always be as it is. The validity of induction depends on Nature's being *uniform* in this sense, and the idea is called the *Uniformity of Nature* (UN) assumption. If the UN assumption were true, then you could legitimately argue by induction that the rule you generalized from past observations would hold true in the future. Therefore, to determine if induction is a valid way to argue, we need to know if the UN assumption is true. This is the famous *Problem of Induction.*[48]

We could try to prove the truth of the UN assumption using either deductive or inductive reasoning. We begin by asking, "Is the UN assumption a *matter of reason*; does it logically follow from premises that we know to be absolutely true?" No, because the only statements like that are parts of closed, self-contained, artificial systems, such as mathematics or symbolic logic, or are tautologies. In contrast, the UN assumption is explicitly about the world outside of these man-made systems; hence, deductive logic cannot prove it. Thousands of years of failed effort to discover logically certain statements about nature was why Bacon turned to inductive reasoning in the first place.

If the UN assumption is not true as a *matter of reason*, could it be true as a *matter of fact*? Can we demonstrate its validity via empirical information gotten through experience, observations, experiments, etc.? One way to find out would be to look to the past and ask if Nature was always the same as it is now. If so, then, reasoning inductively, we could infer that Nature will remain the same in the future. Voila! We would have proved that the UN assumption is true. There is, of course, one snag. We said that the validity of inductive reasoning depends on the validity of the UN assumption, and now we're saying that the validity of the UN assumption depends on the validity of inductive reasoning. This is a completely circular argument, and there is no way out. We can't prove the UN assumption is true as a *matter of fact*, either.

1.L.3 Proposed Solutions to the Problem of Induction

The Problem of Induction is a tough one. Philosophers tried to develop a workaround in the form of a *Principle of Induction* to shore up inductive reasoning. By adhering to the Principle, the truths of inductive reasoning could, it was hoped, become as secure as the truths of deductive reasoning. Every scheme like this was utterly dependent on the particular Principle it rested on; if the Principle wasn't true, the scheme couldn't work. The question shifted and became, "How can we be sure that the Principle of Induction is true?" Trying to solve this problem led the philosophers back to where they started: they'd have to rely on deductive or inductive reasoning or call upon an another, pre-existing Principle of Induction known to be true to guarantee the truth of the latest Principle of Induction, and so on, into an infinitude of prior Principles. For the philosopher Alfred Whitehead "The theory of Induction is the despair of philosophy. . . ."[49] However, philosophers are resilient types, not easily put off by a thousand years of shattered dreams.

If we can't *prove* that a Principle of Induction is true, then why not simply assert that such a principle *must* be true? Just declare that it has to be true a priori—that is, true without proof—and go from there. Indeed, this strategy neatly escapes the problem of proving Principles of Induction, but if your goal is to justify the truths of scientific facts about the rough and changeable world, it isn't fair to call in reinforcements from an otherworldly realm of perfect, though unprovable, ideas. Only testable ones are allowed in science, so a priorism was out, too.

In the end, all of this became pretty frustrating. At one point the philosopher Bertrand Russell threw up his hands and declared that "induction is an independent logical principle, incapable of being inferred either from experience or from other logical principles, and that without this principle science is impossible."[50] In other words, for Russell, induction was simultaneously *indefensible* and *indispensable*. An astonishing and revealing conclusion.

The nub of the problem of induction for the philosophy of science is the conviction that scientific experimentation as a rational endeavor *must* rely on the predictability of prior experience to proceed.[51] Is this true? And, whether it is or not, why do so many people seem so convinced that it must be true?

To start with, we must admit that the world does appear to be pretty predictable, and if the world were totally chaotic, with fundamental physical constants changing daily, science could not proceed. On the other hand, if the values of the fundamental physical constants were in flux, there would be no galaxies, no planets, no life, etc., and it would be correct to say that science, as a rational endeavor, could not exist. This is probably not the profound insight that fans of inductive reasoning have in mind, though. The absolute requirement for physical reality as we know it to have a reasonable degree of regularity is one of those awkward points at which philosophers and nonphilosophic scientists usually part company. As nonphilosophers, scientists wonder what they gain by worrying about a condition that, if *it* existed, would mean that *they* would not exist? We can't prove the existence of objective reality, either: "Fascinating," muses the scientist as she heads to the lab to study objective reality—we can't take the problem too seriously.

Nonetheless, for the sake of the argument, let us grant enough regularity that scientists and philosophers of science can evolve. How would science progress without trusting in the validity of inductive reasoning from past experience to develop solid generalizations about the future? Defenders of induction invite us to "Imagine a universe in which 'computers might explode for no reason!' "[52] It is ridiculous, a *reductio ad absurdum*, to conceive of successful science in such a place, they say.

We can only hope that none of these critics was handling a laptop or a cell phone when it suddenly burst into flames.[53] Extensive experience with laptops and cell phones, coupled with the exercise of inductive reasoning, should have guaranteed that nothing of the kind could possibly happen. Still, it has happened many times. (Indeed, you might be wondering if it's time to draw the inductive inference that these devices are dangerous and, hence, that users should, at least wear oven-mittens when using them.)

What went wrong in the case of the exploding computers is what can always go wrong with inductive reasoning: unexpected things happen. This doesn't mean that the universe is random, as the arguments posit, but only that we are not omniscient. Electronic devices do not burst into flames "for no reason"; when they catch fire, they do so for perfectly good reasons that we weren't aware of beforehand: laptop batteries can be faulty and liable to overheat. The fact that we sensibly expect, without being able to prove, that the universe is governed by regular, orderly laws does not mean that we can safely rely on every observed regularity to continue indefinitely into the future.

1.L.4 Other Forms of Induction

Given that philosophers gradually embraced the view that the problem of enumerative induction "is insoluble," [54] that the attempt to solve it is a "dead end," you might think that they would simply give up on it altogether—but that would underestimate the power of the conviction that we must, simply must, rely on something like induction to carry out science and the resourcefulness of philosophers who devised an entirely new approach to the problem.

Before going on, I want to digress to bring out a point that, although it becomes central to our story, does not get the emphasis it deserves. It is a matter of definition: What does "confirm" (or "verify") mean? There are two related senses of these words: one is "to establish the indubitable truth of a statement," and the other is "to strengthen or support a statement that might or might not be true." Here's an example. You might be asked, say before testifying at a trial, to confirm your name. You know your name beyond doubt—you've got your birth certificate with your tiny footprints on it, assurances from your parents, testimonials from others who've known you forever, official papers and records documenting who you are, etc.—it's your name. Period. You attest to the certainty of that fact. But "confirm" is sometimes used as an indicator of support for a statement having questionable truth value. In this case, evidence is said to make the statement more likely to be true without guaranteeing that it is true. For a philosopher, "a statement can be highly confirmed and still false."[55] The ambiguity is why its best for scientists to avoid "confirm." You might mean it in the soft sense of support for your hypothesis, but if a reviewer thinks you mean it in the hard sense—as proving your hypothesis—then he or she may wonder if you know what you're doing.

Keeping the two senses of "confirm" in mind makes it easier to understand the reinterpretations of induction that we'll turn to next. Until now, the discussion has centered on the first meaning: induction as a way of establishing scientific Truths beyond doubt. When that project failed, the goalposts shifted. Instead of trying to guarantee the logical validity of statements, proponents of induction argued that induction could increase the *probability* that statements would be true. From there, they moved on to trying to justify our *confidence* in the findings of science.

1.L.5 Induction and Probability as Criteria for Success

The strategy of the new attack was to supplement pure induction with probabilistic reasoning. By incorporating the idea of probability, inductive reasoning from true premises could lead to conclusions that would *probably* be true rather than *certainly* be true. Inductive logic would from this perspective let you assess how close competing scientific theories and hypotheses were to the truth. [56] But without knowing the final Truth in advance, how can you say exactly how close you are to it? And you can't compute the probability of the Truth of competing theories the way you can compute the chance of rolling a seven with two dice (we'll review probability concepts in Chapters 5 and 6). When advocates of inductive reasoning say that the truth of scientific conclusions can be made more probable with further experiments, they are talking about *subjective probability* (also known as *Bayesian* probability, see Chapter 6). An outcome is probable to the extent that they believe it is.

This subjective way of assessing probability also makes it possible to distinguish among degrees of certainty, which creates an aura of precision around inductive conclusions. For example, reasoning from the statements that "a recent tally shows that 50.82% of Americans are female" and "Pat is an American," we could conclude that "Pat is probably a female." Since most Americans are female, this must be true (assuming that the categories of male, female, and American are unambiguous). Still, the difference is very slight—49.18% of Americans would be male—and Pat could well be a male. You would not want to base any vitally important decision, such as whether to assign Pat to the "Boys" or the "Girls" gym locker room on the first day of junior high school, on such reasoning. Arguments like these are sometimes classed as "weak inferences" because the conclusions are not very secure.

Reasoning from the information that "97% of American workers earn less than $250,000 per year" and "Pat is an American worker" generates the inductive inference: Pat probably earns less than $250,000 per year. This would be an example of "strong inference" because the probability of correctly guessing Pat's income range is high. Unfortunately, if we're thinking like basic scientists we don't care about that; we want the Truth. There is a 3% chance that our guess is wrong and that Pat makes a pile of money.

1.L.6 Does Science Seek Probable Truth?

People who put a lot of stock in probabilistic inferences argue that, say, a 97% certainty is an appropriate goal for science; it is realistic, and we should be satisfied with some such standard. This may be true to for an applied scientist, where 97% certainty might be enough to justify taking action. On the contrary, basic scientists are not satisfied with probabilistic standards. Nor would you be if the stakes were high.

Imagine a scenario in which a healthcare official offers you a choice between two vaccines against a horrible, often fatal disease such as Ebola. Both vaccines have been carefully tested in large, well-designed and conducted clinical trials. The vaccines are identical in cost, availability, and ease of administration, but one is 90% effective and the other is 97% effective. Obviously, you are eager to get the 97% effective one, when you learn that a 99% effective vaccine has just become available. Now which vaccine do you choose? It is a no-brainer: When compared to 99% protection, 97% is not acceptable, and you would switch with equal eagerness at the chance for 99% safety. Effectively, you make no distinction between 90% and 97% protection. And if a 99.9% effective vaccine came along, you'd have the same reaction—after all, in a country of over 325,000,000 people, an increase of 0.9% would represent approximately ~ 3,000,000 people who didn't get Ebola, and you would want to be among them. And this is the way it is in basic science, where scientists do not seek probable truth any more than they seek partial protection against disease.

1.L.7 Inference not Induction

Conclusions based on reasoning with probabilities do not represent enumerative induction. They are not reached by extrapolating a general rule from particular instances. Still, some philosophers would classify probabilistic conclusions as coming from inductive reasoning because they are not certainly (i.e., deductively) True. However, this stretches the definition of "induction" so far that it becomes meaningless.

Many philosophers have dispensed with the term "induction," preferring various forms of "inference" instead. In addition to encompassing a wider variety of logical constructions, "inference" is also less burdened by the disappointing history of "induction." Hence, we now have *inductive inference, explanatory inference*, and *inference to the best explanation*[57]; for example, explanatory inference is "inference from a set of data to a hypothesis about a structure or process." From the variety of names, you might suppose that each kind of inference represents a distinct cognitive process. On the contrary, it appears that the kinds of inference are defined more by the behavioral contexts used to identify them than by the mental operations at work. To illustrate: In one situation, you're shown a series of particular events and, after a pause, you say "Aha!" and state the generalization that predicts the next event. This would be *inductive inference*. In another situation, you're given some clues to a puzzle, and, after a pause, you say "Aha!" and solve the puzzle. This could be classified as *explanatory inference*. And so forth. The philosophical project here is to identify the conditions that call for using one term or the other, while the origins of the "Aha!" moments remain equally mysterious products of the mind. It is hard to see how the distinctions among kinds of inference, whatever they are, will be much help in trying to understand scientific thinking, especially hypothesis-based thinking, which is what we're interested in.

1.L.8 Affirming the Consequent

At beginning of this chapter, I noted that Larry Laudan distinguished his own two kinds of induction: plebian and aristocratic. So far, we have been discussing his ordinary, plebian kinds and have not directly considered the high-class, aristocratic type of induction. While we won't go into the topic in detail, there is a related philosophical principle we should know about.

The concept of fallibilism assures us that we can't prove the truth of a scientific hypothesis or theory. We're going to take up the question of the hypothesis in earnest in Chapter 2, but an intuitive feel for it will do at the moment. The problem is that aristocratic induction wants to describe the relationship between empirical evidence and the Truth of theories. The philosopher Peter Godfrey-Smith frames the issue like this: "What connection between an observation and a theory makes that observation *evidence for* the theory?" and he considers it to be possibly "*the* fundamental problem in the last hundred years of the philosophy of science."[58] Your theory makes a prediction that you test, and the results are consistent with it. Now what? How many tests would it take before you could reasonably conclude that a theory is correct, or at least better than another one? As with other problems with induction, the rigorous answer is that such conclusions are *never* logically justified, and, in fact, concluding otherwise would constitute a *logical fallacy*.

There are two related problems for aristocratic induction. The first is that no one theory is uniquely capable of accounting for a real-world phenomenon because it is always possible that other theories might do as well.

To appreciate the difficulty, we can consider a car that wouldn't start as a stand-in for a real scientific problem. Your car mechanic takes a look and makes a diagnosis: "The car won't start because the battery is dead." This is his hypothesis, and it makes a prediction: if the battery is dead, the headlights won't come on when you flip the light switch. So he flips the switch and lights don't come on. Does this *prove* that the hypothesis was correct? Obviously not; many other possible explanations (broken or corroded wires, stolen battery, etc.) also predict that the lights will not turn on. The information he got by flipping the switch was not sufficient to restrict the conclusion to "the battery is dead." To give it a name, this problem is called *under-determinism*, and it means that the data cannot be uniquely accounted for by one hypothesis; another one would do as well. It is a serious matter that is not confined to situations involving inert automobiles. When Einstein noted[59] that, for all of the successes of the extremely complex and precise theory of quantum mechanics, there might be an infinite number of alternative theories that could do as good a job, he was saying that the quantum theory was under-determined by the data.

Under-determinism is clearly a hindrance, but where is the *logical fallacy* in aristocratic induction? This is the second problem. To illustrate it, we can go back to the discussion of deduction, where true conclusions necessarily follow from true premises. It turns out that we cannot reverse the order and go backward. You can't start from a true conclusion and argue that the premises that led to it *must* be true—they might or might not be. Let's look at the car problem again and reformulate it like this: *if the car battery were dead, then the headlights would not light* (premise), and *the headlights do not light* (premise). Would the conclusion—*the battery must be dead*—follow logically (i.e., would it be indisputably true)? No. One of the other explanations could explain the unlit headlights even if the battery were fine. The conclusion in this case could be false even though the premises were true because the conclusion was *not entailed* by the premises: understanding the terms in the premises did not automatically lead you to the conclusion, and thinking it does is the logical fallacy, which is known (for reasons that don't matter here) as *the fallacy of affirming the consequent*. We commit this fallacy when we try to argue that getting a predicted result means that the hypothesis that made the prediction must be true.

The desire to avoid the fallacy of affirming the consequent is why, when scientists get a result that agrees with their hypothesis, they say that the results are "consistent" with the hypothesis, rather than that the results prove its truth, or "confirm" it.

1.L.9 Is Induction a Philosophical Problem?

If it is not possible to integrate induction into the philosophy of science, maybe it's time to take a different tack. Maybe induction is a psychological, neuroscientific, and cognitive phenomenon that can be best understood in the contexts of those fields. David Hume[60] was, as usual, ahead of his time with this insight. He reasoned that, if induction from experience does not take us to an "effect from an impression of its cause" by true understanding or reasoning, then "the imagination"—the largely unconscious, nonrational part of the mind—must be responsible. We can stretch the insight a bit further with an example.

With his canine intellect, your puppy, Snowball, infers from his past experience that the sounds of a can of dog food being opened mean that dinner will soon appear and comes bounding into the kitchen. If we call what Snowball does *induction*, then the erudite term loses its luster, and yet we'd need a good reason not to call Snowball's cognitive feat a form of induction, especially since we don't understand our own thought processes any better than we understand his. Both dogs and people observe, learn about, and generalize from regularities in our environments and act accordingly. Naturally, we *experience* our minds differently, but we cannot explain where our ideas come from any more than we can explain where his come from.

If induction is not an exclusive product of the human mind, why should it remain within the purview of philosophy? The study of planetary motions was initially carried out within philosophy but eventually left philosophy and transmogrified into physics, astronomy, and cosmology. One of philosophy's functions is to serve as an "incubator" of ideas,[61] nurturing and fostering them until they can be expressed as testable hypotheses and segue into other fields. Surely the incubation period for induction must be up by now, and this problem should move into the neurocognitive sciences?

1.M How Does Science Acquire Information?

Evidently neither deduction nor induction can account for the advancement of scientific knowledge. How, then, does it happen? The answer, given in the next chapter, is that we make observations and try to explain them. We stick with explanations that work, for as long as they work, and discard them when they don't. It is a trial-and-error process that is remarkably effective, though never complete. It is called the Scientific Method and its chief tool is the hypothesis.

1.N Coda

This chapter reviewed a few fundamental philosophical principles that are involved in scientific thinking. We considered the relationship between philosophy and science, and we discussed several philosophical concepts, including fallibilism, objectivity, reductionism, and affirming the consequent. We touched on the difficult question of the relationships among explanation, uncertainty, and levels of organization of nature that runs through the book. And we explored two major forms of reasoning—deduction and induction—and showed that, while they are inadequate in themselves to account for the growth of scientific knowledge, they have roles in science that we will begin to explore in the next chapter, where we take up the central topic: the modern scientific hypothesis.

Notes

1. Santiago Ramón y Cajal, *Advice for a Young Investigator*, translated by N. Swanson and L. Swanson (Cambridge: MIT Press; 1999).
2. This book is not about the philosophy of science, and I use philosophical terms from the point of view of a lay consumer of philosophy. For an in-depth treatment of this material's online sources, the *Stanford Encyclopedia of Philosophy* (plato.stanford.edu) and the *Internet Encyclopedia of Philosophy* offer authoritative essays. Peter Godfrey-Smith's *Theory and Reality* (Chicago: University of Chicago Press; 2003) is an excellent introduction to the philosophy of science, and the *Oxford Very Short Introduction* series, including Samir Okasha's *Philosophy of Science* (New York: Oxford University Press; 2016) and Graham Priest's *Logic* (New York: Oxford University Press; 2017) provide compact resources.
3. Steven Weinberg, "Against Philosophy," in *Dreams of a Final Theory* (New York: Pantheon; 1993), pp. 166–190.
4. James Gleick, *Genius* (London: Little Brown; 1992, Richard Feynman quote, p. 13).
5. Eugene Wigner, "The Unreasonable Effectiveness of Mathematics in the Natural Sciences," *Communications in Pure and Applied Mathematics* 13: 1, 1960.
6. Richard Feynman, *The Meaning of It All: Thoughts of a Citizen-Scientist* (New York: Perseus Books; 1998).
7. Willard V. O. Quine, quoted in Gleick, *Genius*, p. 371.
8. Larry Laudan, *Science and Hypothesis: Historical Essays on Scientific Methodology* (Boston: D. Reidel; 1981).
9. Peter Godfrey-Smith, *Philosophy of Biology* (Princeton: Princeton University Press; 2014).
10. Jordi Cat, "The Unity of Science," in *The Stanford Encyclopedia of Philosophy* (Winter 2014 Edition), Edward N. Zalta (ed.). http: plato.stanford.edu, archives, win2014, entries, scientific-unity.
11. Thomas Kuhn, *The Structure of Scientific Revolutions* (Chicago: University of Chicago Press; 1970).
12. Godfrey-Smith, *Philosophy*. Godfrey-Smith provides a concise overview of Lakatos's, Laudan's, and Feyerabend's thought vis-à-vis Kuhn in his Chapter 7.
13. Ibid., p. 149.
14. Ibid., p. 176.
15. Nobel Laureates from non-Western countries; visit https:en.wikipedia.org, wiki, List_of_Nobel_laureates.

16. Julian Reiss, Jan Sprenger, "Scientific Objectivity," in *The Stanford Encyclopedia of Philosophy* (Summer 2016 Edition), Edward N. Zalta (ed.). http:plato.stanford.edu, archives, sum2016, entries, scientific-objectivity.
17. Lorraine Daston, Peter Galison, *Objectivity* (New York: Zone Books; 2010, prologue and chapter 1).
18. ibid., pp. 11–13.
19. ibid., p. 36.
20. Karl Popper, *The Logic of Scientific Discovery* (New York: Routledge Classics; 2002, p. 22).
21. Sandra Harding, *Objectivity and Diversity: Another Logic of Scientific Research* (Chicago: University of Chicago Press; 2015, "value-free," p. 1 and following; "tidy account," 116). Concern about the view of science as one "tidy account" owes much to philosophies of science, such as Harding's, that advocate for a pluralism of sciences, especially as these relate to interactions at the borderlands between science and social issues.
22. Naomi Oreskes, Erik Conway, *Merchants of Doubt: How a Handful of Scientists Obscured the Truth on Issues from Tobacco Smoke to Global Warming* (New York: Bloomsbury; 2015).
23. Harding, *Objectivity*. Defining and analyzing "objectivity" is Harding's central concern.
24. Sandra Harding, "Gender, Democracy, and Philosophy of Science," http:www.pantaneto.co.uk, issue38, harding.htm.
25. Harding, *Objectivity,* pp. 34–35.
26. Paul R. Gross, Norman Levitt, *Higher Superstition: The Academic Left and Its Quarrels with Science* (Baltimore: Johns Hopkins University Press; 1994).
27. Reiss, Sprenger, *Scientific Objectivity,* ibid.
28. Daniel Dennett, *Darwin's Dangerous Idea* (New York: Simon & Schuster; 1995).
29. Stephen W. Hawking, *Theory of Everything: The Origin and Fate of the Universe* (Rutland: Phoenix Books; 2006).
30. Ernst Mayr, *Towards a New Philosophy of Biology: Observations of an Evolutionist* (Cambridge: Harvard University Press; 1988).
31. https:en.wikipedia.org, wiki, Tinbergen%27s_four_questions
32. Godfrey-Smith, *Philosophy*, chapter 13; Okasha, *Philosophy*, chapter 3; James Woodward, "Scientific Explanation," in *The Stanford Encyclopedia of Philosophy* (Winter 2014 Edition), Edward N. Zalta (ed.), http:plato.stanford.edu, archives, win2014, entries, scientific-explanation, for introductions to the concept of "explanation."
33. Gleick, *Genius*, p. 370.
34. http:dujs.dartmouth.edu, 2013, 04, what-causes-ice-to-be-slippery, #.V_pJZ-UrIdU
35. https:www.nasa.gov, mission_pages, juno, main, index.html
36. Allison Gopnik, "Scientific Thinking in Young Children: Theoretical Advances, Empirical Research, and Policy Implications," *Science* 337: 1623–1627, 2012.
37. John Tooby, Leda Cosmides, "The Physiological Foundations of Culture," in *The Adapted Mind* (New York: Oxford University Press; 1992), pp. 19–136.
38. Steven Pinker, *How the Mind Works* (New York: W. W. Norton; 2009).

39. Technically, a *valid* logical argument can be made starting from false premises, provided that the conclusion *must* be true. For instance, from the premises "Martians are real" and "the moon is made of cheese," we can logically conclude that "two plus two equals four," because that conclusion is true no matter what. Nevertheless, although this argument is *valid*, it is not *sound* because the premises are false.
40. Leah Henderson, "The Problem of Induction," in *The Stanford Encyclopedia of Philosophy* (Summer 2018 Edition), Edward N. Zalta (ed.), https:plato.stanford.edu, archives, sum2018, entries, induction-problem, >.
41. Laudan, *Science and Hypothesis.*
42. Ibid.
43. Edward Dolnick, *The Clockwork Universe: Isaac Newton, the Royal Society, and the Birth of the Modern World* (New York: HarperCollins; 2011).
44. Laudan, *Science and Hypothesis.*
45. Richard Westfall, *Never at Rest: A Biography of Isaac Newton* (Cambridge: Cambridge University Press; 1980). The astronomer and mathematician Johannes Kepler, working from the precise astronomical measurements of Tycho Brahe, had figured out, among other things, that the planets traveled around the sun in elliptical orbits, but he had no idea why they would do that. Newton's contribution was to postulate a force acting on all of the bodies in the universe—gravity—that fell off as the inverse square of the distance from a body. Interestingly, his laws were just a mathematical description of what was happening; Newton refused to speculate on the nature of the force. Newton proved that his law of gravity predicted elliptical planetary orbits; however, his reasoning did not use enumerative induction.
46. Gleick, *Genius*, p. 367; Richard Feynman, *The Very Best of the Feynman Lectures* (Pasadena: California Institute of Technology; 2005). See also A. Loeb, "Good Data Are Not Enough," *Nature 59*: 2–25, 2016, http:www.nature.com, news, good-data-are-not-enough-1.20906, for interesting details of the Mayans' accomplishments. This comment makes strong arguments in favor of smaller research groups, multiple competing points of view, and hypothesis testing in research.
47. David Hume, *A Treatise of Human Nature*, reprinted with an Introduction by D. G. C. Macnabb (New York: Meridian Books; 1740, 1969).
48. Ibid.
49. "The despair of philosophy": Alfred Whitehead, quoted in http:,philsci-archive.pitt.edu, 13057, 1, USM_Induction2017.pdf.
50. Bertrand Russell, *A History of Western Philosophy* (New York: Simon & Schuster; 1945).
51. Okasha, *Philosophy*, chapter 2.
52. Ibid., pp. 24–29.
53. https:www.youtube.com, watch?v=pizFsY0yjss; https:consumerist.com, 2016, 06, 27, hp-and-sony-recall-laptop-batteries-due-to-possible-overheating-and-fires.
54. Henderson, "Problem of Induction."
55. Godfrey-Smith, *Philosophy*, chapter 3.
56. https:www.youtube.com, watch?v=pizFsY0yjss; https:consumerist.com, 2016, 06, 27, hp-and-sony-recall-laptop-batteries-due-to-possible-overheating-and-fires.

57. Okasha, *Philosophy*, pp. 24–29.
58. Godfrey-Smith, *Philosophy*, chapter 3.
59. Ronald W. Clark, *Einstein: The Life and Times* (New York: Avon Books; 1971).
60. Hume, *A Treatise of Human Nature*.
61. Godfrey-Smith, Philosophy, chapter 3.

2

The Scientific Hypothesis Today

2.A Introduction

"*Sony Says PlayStation Hacker Got Personal Data*" announced the headline on the front page of the *New York Times* (April 27, 2011).[1] Let's suppose that you do not immediately get the gist of what happened and reach for your handy *Webster's New Third International Dictionary* (circa 1968) to find out. You learn that a "hacker" is "one who hacks," "a hand implement or hooked fork for grubbing out roots," "one who is inexperienced or unskillful in a sport," or a "cab-driver," but that "to hack" means "annoy or vex" or "to cut with repeated irregular blows." You wonder which definition applies here. Would you guess that a clumsy novice video-game player had uncovered some personal data in a console after breaking into it with repeated blows of a hooked root-digging tool? Probably not. You'd keep looking until you came across the modern definition of a hacker as "a person who secretly gets access to a computer system in order to get information, cause damage, etc.,"[2] which seem more sensible. The context would help; the older interpretations seem unlikely to lead to a front-page headline. In fact, the definitions in the old dictionary turn out to be useless because it predated the internet and the Computer Age and could not have included the most relevant meaning. Word meaning typically changes with time, retaining some of the original sense but stretching to cover new situations. (Hackers sometimes gain unauthorized access to computers by an unskillful, brute-force method, such as trying thousands of potential passwords.)

Like "hacker," the word "hypothesis" has undergone significant shifts in meaning over the centuries. We can trace part of the controversy surrounding the hypothesis (see the Introduction) to the outdated definitions that its critics use.[3]

To keep from getting bogged down in semantics later, I'll begin by going into the hypothesis and related topics in greater detail than most science courses do. I'll also go into the thinking of Karl Popper and John Platt because they've had the greatest philosophical influence on scientific thinking, which is not to say that their programs are the only ways of doing science, as I'll explain in Chapter 4. The ideas of Popper and Platt do clarify the hypothesis and its role in science, especially for those who, like me, did not get much formal exposure to these topics in our science classes.

2.B Science and the Scientific Method

Science aims to give a complete, comprehensive, accurate accounting of nature: how things are, how they got to be that way, and how they will be in the future. And it seeks this information for all things and events, everywhere in the universe, and for all times—past, present, and future. In short, science wants to know everything about nature, and it wants to be absolutely certain of its knowledge: this is Truth with a capital *T*—the shining ideal. Of course, achieving 100% certain Truth is just not going to happen; both logic and physical laws prevent it. To one extent or another, we must be eternally uncertain. Much of what we'll have to say about the hypothesis has to do with the paradox created by the search for Truth in the context of uncertainty.

Almost everyone in the discussion of the hypothesis agrees that, whatever it is, a hypothesis is part of something called the *Scientific Method*, a reality-based, evidence-driven approach to studying nature. However, the Method does not actually exist as a thing or even a process, and it is not defined by a single list of steps. It is a collection of guidelines or principles, a mindset, that forms the basis of scientific inquiry. The *Oxford Dictionary*[4] says that the Scientific Method is "a method of procedure that has characterized natural science since the 17th century, consisting in systematic observation, measurement, and experiment, and the formulation, testing, and modification of hypotheses." This is not a set recipe: the types of observation, measurement, experiment and the formulation, testing, and modification of hypotheses vary across investigations. Furthermore, the steps don't have to happen in any fixed order and rarely occur in a linear sequence; they repeat or loop back on each other, as Figure 2.1 suggests. So, knowing that the hypothesis is a fundamental component of the Scientific Method and is useful for the rational investigation of nature doesn't tell us much about it.

2.C What Is a Hypothesis?

Given that there is a rough consensus about what the Scientific Method is and that the hypothesis is part of it, you would think that scientists must have agreed on what a hypothesis is. Not so; there has been a wide range of interpretations of "hypothesis" (from the Greek, a "base, basis of an argument, supposition," literally "a placing under," from *hypo-* "under" + *thesis* "a placing, proposition") and a brief list would include the following: (1) a scientific axiom[5] intended to provide a foundation for deductive reasoning about nature, though not itself subject to being tested; (2) a necessary motivating force behind an investigation, a mandatory starting point for one[6]; (3) a postulate about fictional entities that is useful

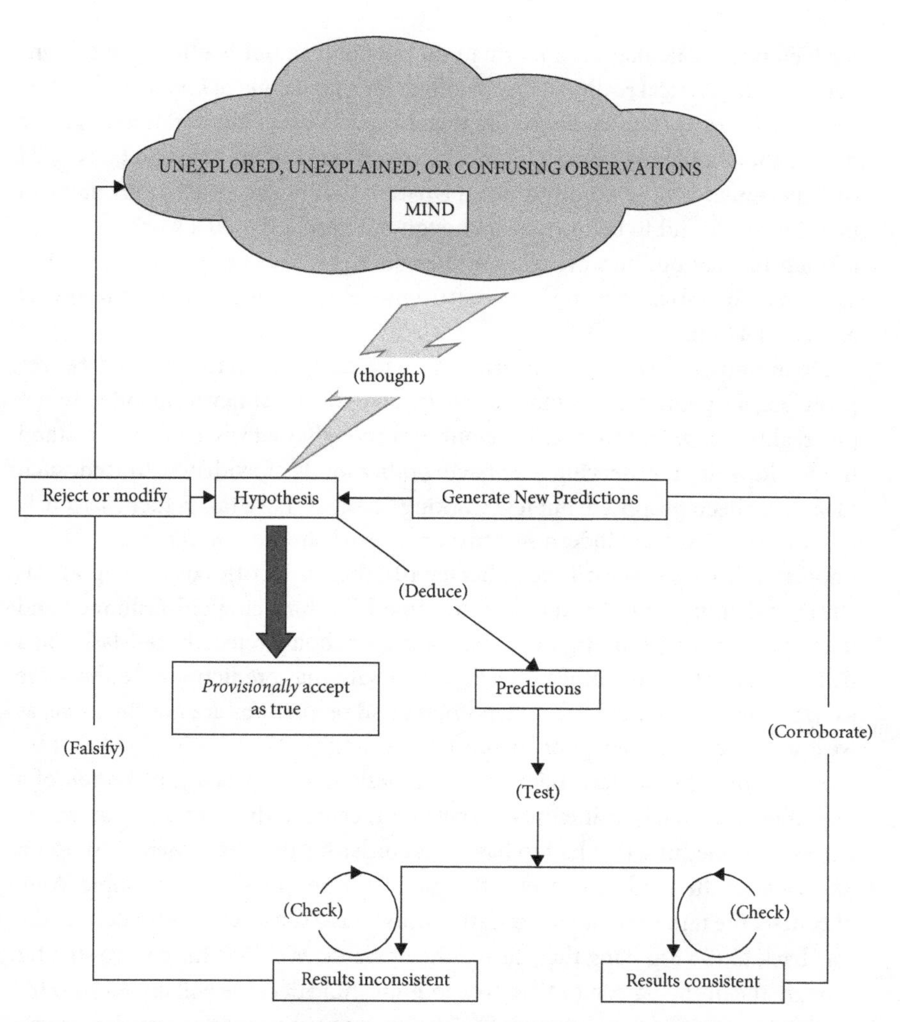

Figure 2.1 Schematic diagram of the recursive hypothesis testing cycle according to the Scientific Method and Karl Popper's standard of falsification. The boxes represent decisions or intentions, arrows represent processes and relationships between states, and the parenthetical labels identify the processes. The cycle begins with a thought, perhaps in response to an observation that you don't understand. You think of a testable explanation for your observation: this is your hypothesis. You deduce predictions from the hypothesis and test them. The test results are either consistent or inconsistent with the predictions. If they are consistent with the predictions, your hypothesis is provisionally "corroborated." If they are inconsistent, your hypothesis is provisionally falsified. If your hypothesis is falsified, you must reject it in its present form, meaning either reject it entirely or modify it. If your hypothesis is corroborated, you can go back and derive new predictions, etc. At some point you decide to accept the hypothesis. The cycle is recursive in toto or in part. For example, you should always check tests of predictions that either agree or disagree with the hypothesis.

for facilitating calculations about natural phenomena but is otherwise disconnected from physical reality[7]; (4) a synonym for a *prediction of how a given experiment will turn out*[8]; (5) a conjecture that explains observable and unobservable details about a phenomenon[9]; (6) a conjectural explanation that can be *verified* by experiment[10]; (7) a conjectural explanation that makes predictions that can be *falsified*—found to be untrue—although not verified[11]; (8) a working plan[12], a blueprint that outlines the relationships among ideas and implies the logical steps in an investigation; and, finally, (9) a model that summarizes the results of an investigation.

Definitions 1, 2, and 3 characterize the hypothesis, respectively, as a data-free postulate, a rigid requirement, or an imaginary construct never intended to depict reality. You might wonder if anyone ever thought that you could understand the world without observing it or having other kinds of evidence. Indeed, such ideas have been proposed, but few working scientists have taken them seriously for at least 100 years. Modern scientist are overwhelmingly *empiricists.*[13] This is important because, as we'll see (Chapter 10), there are critics of the hypothesis who use definitions 1–3 and attack the hypothesis for being old-fashioned and inadequate. They get the argument backward; we should reject the old-fashioned definitions, not the hypothesis. Definition 4 says that predictions are the same as hypotheses, whereas, in fact, hypotheses and predictions are not the same, as we'll soon see, so we can ignore definition 4 as well.

At its core, the modern scientific hypothesis is a proposed *explanation* of a phenomenon. That is, it attempts to explain the cause of the phenomenon, where a *cause* is something that had to happen in order for the phenomenon to occur. The cause of the road being wet is that rain fell from the sky, for example. And, of course, the rain itself has a cause, that cause has a cause, etc., ultimately all the way back to the Big Bang that started the universe. We don't have to go that far though; the immediate, intuitively obvious meaning of cause will do. As your investigation proceeds, your search for "the" cause may change or broaden as you learn more.

Definitions 5–7 agree that hypotheses are explanations—we call them "conjectures" ("guesses") to express our ignorance of the mental activity that produces them. In one way or another, they "happen" in our minds, and we don't know how. Although hypotheses are guesses in this sense, they are not random, purposeless stabs in dark, and, conversely, all guesses are not hypotheses. Hypotheses can be tested to see whether they're likely to be correct or not, and they have other properties we'll talk about. Definitions 6 and 7 disagree on the issue of whether hypotheses can be proved to be true, *verified*, which is a crucial distinction and the source of an enormous and often acrimonious struggle. We'll devote a lot of attention to the principle that we can't verify our hypotheses but only find out if they're incorrect; that is, we can *falsify* them. Finally, there seems to be good agreement that hypotheses are useful as working plans and models that guide and summarize investigations—definitions 8 and 9.

This may all seem confusing, so to anticipate the conclusion of this chapter: a modern hypothesis is a conjectural explanation for some aspect of nature that can be tested and potentially falsified and that serves as both a blueprint and a summary of an investigation. The hypothesis captures what is known about a phenomenon and frequently goes beyond appearances to identify unobservable causes for the phenomenon.

2.D Hypotheses Are Not Predictions

Although hypotheses and predictions are often linked, they are different concepts, and failing to distinguish between them creates needless confusion. A *hypothesis* is a putative explanation for actual observations. It makes, or *entails*, predictions, but is not itself a prediction. We can logically deduce predictions from a hypothesis. Therefore, if a hypothesis is true, then the predictions that follow from it must be true. The logical relationship tells you which predictions you should test and which ones you can omit because they don't follow from the hypothesis. And the fact that predictions follow with deductive certainty from hypotheses is what allows you to test and reject a hypothesis. If its predictions are false, then the hypothesis *must* be false. Logic facilitates scientific thinking and communication by making it clear what a hypothesis actually says, and the link between hypotheses and predictions is crucial for experimental, hypothesis-testing science.

A *prediction* is a forecast of how some future event will play out. We test predictions *directly* with experiments that show whether they are true or false. We test hypotheses *indirectly* by testing their predictions. By making predictions, the hypothesis reveals exactly what it does and does not mean, establishes a rationale for the conduct of investigations, and specifies the relationships among relevant variables. You might hypothesize that your new puppy, Snowball, jumps around hysterically when you return home because he has never learned to act in a more civilized manner. One prediction of your hypothesis would be that Snowball would behave himself better if he were properly trained. The hypothesis itself is not directly testable (you cannot measure the state of learning in the relevant parts of Snowball's nervous system); you can test it indirectly, however, by sending Snowball to Puppy School to be trained.

Because predictions are logical deductions from hypotheses, if a hypothesis makes a prediction, then the prediction must be true. It does not work the other way, though: true predictions do not allow you to conclude that your hypothesis must be true. If Snowball's welcoming activity decreased after training it might mean that he simply got used to your returning home; it was no longer an amazing event, and the training itself did nothing. Or the problem might not have been his lack of self-control. You, his owner, might have been reinforcing his over-the-top greeting by getting down and rolling around on the floor with him. Maybe the training just got *you* to calm down. Your prediction (he'll behave better after training) would be correct, while your hypothesis to explain it (the dog is the problem) could still be wrong.

People sometimes say they "hypothesize" about what the outcome of a measurement will be. What they mean is that, while they haven't done the measurement and don't know how it will turn out, they do have a guess that they call a hypothesis. However, their guess does not explain anything; it is not a hypothesis and merely indicates that there is something they want to measure. For example, you might wonder if your tap water is acidic and guess that it might be. There is no hypothesis involved; you get a pH meter, take a sample of the water, and determine its acidity. You get an answer: the pH is or is not more acidic than 7.0. Predictions like this are not related to hypotheses. And it is not a question of being able to do the test right away (maybe you don't happen to own a pH meter and can't test the water); what is fundamentally a prediction does not turn into a hypothesis merely because the measurement can't be carried out.

The difference between hypothesis and prediction depends on why you're doing a measurement, not the measurement itself. Identical measurements might test hypotheses or predictions. One day you might think, "Dang, my tap water sure tastes funny; I'll bet it's because of that acid rain I've been reading about." In this case your hypothesis is explanatory—the tap water's taste is different because acid rain has tainted it—and extremely complex. You can't test it directly. Rather, you'd have to do many tests before you could be reasonably confident that there was anything to the hypothesis. Maybe the funny taste comes from the rusty pipes in your apartment or the effluvia spilling into the river from the chemical plant a few miles upstream. Maybe age is taking its toll on your taste buds, or the side effects from your latest fad diet are kicking in and there is nothing wrong with the water at all.

Whatever the final answer, the acid rain hypothesis would *predict* at least that the pH of your tap water would be acidic and you could test this prediction with the pH meter. (Even "pure" rain has a slightly acidic pH; nevertheless acid rain, at approximately pH 4.0[14] is more acidic still.) If the pH of your water is not acidic, then its taste probably wasn't caused by acid rain, and you should entertain other hypotheses: the chemical plant upstream, etc. Likewise, if the pH of your tap water was acidic, that would be consistent with your hypothesis but not prove it was correct. You might be encouraged to test other predictions (e.g., that the source of your water—lake, reservoir, water tower on top of your building—is acidic and that it is acidic because of the rain). The difference between prediction and hypothesis has nothing to do with *what* measurement is made, but with *why* it is made and what conclusions it allows.

A hypothesis is an explanation for a phenomenon, and I've suggested that it is often expressed in terms of properties or causes that we can't directly observe or measure. The reason is that, if we could directly measure or observe them, then that's what we'd do—no reason to beat around the bush with hypotheses. But what do the concepts "direct" and "indirect" mean? I've been using them in a logical context—we test hypotheses *indirectly* via their predictions, and predictions are statements about the world that we can test *directly* by making measurements. There is also a practical sense in which scientists refer to these terms.

2.E Direct and Indirect Measurements

A more-direct experimental test involves fewer assumptions about what the test measurements show, fewer and simpler transductions between the physical properties of what is measured and the output of the measuring device, and fewer transformations of the data needed to interpret the test.

Here's an example: Nobel Prize-winning neuroscientists David Hubel and Torsten Wiesel at first studied the mammalian visual system by recording the electrical activity of individual neurons in the brain while shining a light into (anesthetized) cats' or monkeys' eyes and mapping the locations of the neurons that were activated by each eye.[15] After years of tedious effort, analyzing hundreds of cells one at a time and collating the results, they identified patterns (the "ocular dominance columns") in the arrangement of the responsive neurons in the visual cortex. Discovery of these columns was a breakthrough in our understanding of how the brain processes visual information. Somewhat later, using a dye that could travel through the nerve fibers connecting the eye to the brain, they captured in a single histological image the same patterns, now formed by millions of dye-filled neurons, that they had struggled to decipher from the electrical measurements of single neurons. "Seeing is believing," said Hubel as he displayed the picture at a large international conference, and the audience agreed: compared with the indirect information painstakingly assembled from reams of electrophysiological data, the pictorial evidence was impressively direct.

From a loftier perspective, essentially every test a scientist makes is indirect because the things a scientist is usually interested in are invisible or inaccessible. When scientists talk about doing direct tests, they are using "direct" as a relative term—it simply means more direct than other tests. To see this, let's take measuring the pH of a solution as an example of a "direct" experimental test. A pH meter is supposed to determine the concentration of hydrogen ions in a solution, but it does not do that by counting ions, which are minute electrically charged particles. The meter responds to a difference in ionic charge separation, a gradient in the density of charged particles between the inside and the outside of the specialized glass sensor at the tip of the pH probe when the probe is in a watery solution. The charge separation, which is related to the hydrogen ion gradient, sets up a potential difference, a voltage, in the sensor that is measured with respect to a reference electrode in the testing solution. The voltage drives an electric current in the circuitry of the meter which translates the amount of current into a number on the digital display. Whew! But the complications do not end there: other ions and the temperature of the solution also influence the sensor. The measurement is so delicate that tiny changes in the numbers of ions coating the glass of the sensor or reference electrodes can throw things off. To make an accurate pH measurement, you must always start by adjusting the meter to ensure that it gives the correct value for the pH of known solutions. This is why we say that even apparently straightforward (i.e., "direct") scientific measurements are, in truth, indirect.

Our hypothesis regarding acid rain was actually about the hydrogen ion concentration in rain water, and one prediction that it made depended on the output of a pH meter. The convoluted process of testing this prediction is something to keep in mind when considering that modern science commonly uses measuring devices that are enormously more complicated than a pH meter.

With these philosophical notions in mind, we can turn to the philosopher who had the greatest influence on twentieth-century science, Karl Popper, and the philosophically minded scientist, John Platt, who added practical dimensions to Popper's ideas.

2.G Karl Popper and John Platt

The middle of the twentieth century saw the development of two programs that have had lasting effects on practicing scientists: *Critical Rationalism*, a program most famously associated with Karl Popper[16] and *Strong Inference*, by John Platt.[17] Popper was a philosopher who was outside of the philosophical mainstream of his day. His friends considered him prickly, combative, and generally "difficult"[18]; his enemies really didn't like him. Once, while giving a lecture at Cambridge University, he was purportedly threatened with a hot fireplace poker by the philosopher Ludwig Wittgenstein.[19] John Platt was a biophysicist who became interested in the philosophy of science and published significant papers on it. His article in the journal *Science* in 1964 on scientific thinking and practice affected generations of scientists without, however, interesting philosophers much. Popper's and Platt's programs overlap considerably, with Popper largely providing the underpinning for Platt's system for testing hypotheses.

Although philosophers of science acknowledge Popper's sway, by and large, they reject his ideas. In fact, his continued appeal to working scientists frankly puzzles them.[20] Hence, philosophers, who mostly ignore scientists' dabblings in philosophy or treat them with polite indulgence, would not be sympathetic to my stance on the hypothesis. As noted in the Introduction, I am a Popperian who strays from strict Popperian orthodoxy from time to time. I think that Popper got the key issues right and that there are many good reasons for scientists to learn about his thinking despite the nuances and objections that philosophers fixate on (Chapter 3 lays out and responds to some of the objections). Popper's program is not as straightforward as scientists often assume it is. I'll bring out a few aspects of his thinking that tend to be overlooked and suggest how to understand them.

The positions of Popper and Platt are especially congenial to doing science, as opposed to contemplating science. For me, the following, which I offer as a fundamental tenet, essentially an axiom, gets to the heart of the matter.

> *Scientists will* never *accept a scientific statement as true if they can think of a feasible test that might show that it is not true.*

If you believe this, then you, too, may be a Popperian.

2.G.1 Critical Rationalism (Conjectures and Refutations)

The philosophical school that Popper founded is called Critical Rationalism, although Popper referred to his program as *Conjectures and Refutations*, which describes its essential principles.[21] Conjectures and Refutations is a form of trial-and-error reasoning that encapsulated the fundamental procedure of science: you *conjecture* a hypothesis to explain an aspect of the world; try to *refute*, or falsify, it; and, if you can't refute it, then you may tentatively conclude that you've discovered a *fact*.[22] While we need to be able to tell if a hypothesis is false, there is no such requirement to tell if it is true. To a nonscientist, this may seem like a glaring omission: If the hypothesis is an important element of scientific reasoning, and if science seeks the Truth, why not do something to prove that a hypothesis is True?

Popper's best-known example, slightly embellished, of how his program works is this: one day, while observing a group of swans in the park, you notice that they are all white, and you wonder why. The hypothesis occurs to you that they are all white because "all swans are white." (It is a stretch to call this a genuine hypothesis, but it is classical and serves the immediate purpose: I'll have more to say about it in Chapter 3.C.2.d.) You realize that you can't test this hypothesis to see if it's true by finding more white swans because, no matter how many you find, you won't ever know if *all* swans are white unless you observe "all swans" (i.e., all things that were, are, or ever will be, swans). You can't use inductive reasoning to generalize from a group of swans that all swans are white for the same reason. However, says Popper, if you find one bird that is unquestionably a swan and is black, then you know that your hypothesis is *false*. Thus, falsification, not verification, is the true test for science.

In short, rather than trying to solve the problem of induction, Popper revolted against the very concept of philosophical induction. He was impressed that scientists were able to arrive at powerful "laws" that sustained practical actions and made further research advances possible, even though the present laws might eventually need to be replaced by improved laws. And Popper stressed that inductive observations themselves could not "lead to" knowledge; knowledge comes from people applying their minds to understand the world, not from the world itself. In glossing over the role of the mind, inductivists missed the main story.

Popper considered the question of how knowledge could increase in the absence of induction. The Logical Positivist philosophers had argued that science should only admit or include *statements that can be shown to be true*,[23] they sought justification (or *confirmation* or *verification* or *validation*) for scientific statements. While at first glance their objective seems eminently worthwhile, the impossibility of *guaranteeing* that scientific statements are True (i.e., of demonstrating their validity beyond any conceivable doubt) is exactly the fundamental philosophical problem of science: we remain uncertain that our "truths" are True. Strict justificationist programs are doomed to disappointment, and Popper worked to find a way of explaining the growth of knowledge without them.

2.G.2 Elimination of Induction

One of Popper's chief objectives was to "eliminate induction from science," but his goal is frequently mischaracterized.[24] He was not so much concerned with induction as a means of generating scientific hypotheses.[25] The origin of concepts was a problem for psychologists—or, nowadays, cognitive scientists—to solve, not philosophers. On the other hand, generations of thinkers had failed to find a way of verifying scientific statements as True, and Popper did want to do away with the notion that induction could fulfill that role.

Popper was a committed fallibilist who took the concept that we are eternally uncertain seriously, even though he recognized that it can be awkward to do so. For instance, many of us have trouble discussing scientific findings that we "know" to be true (global warming is real!), together with the uncertain state of our knowledge. (The discussion of levels of explanation and Aunt Minnie and the ice in Chapter 1 show one way to resolve this paradox. We'll come back to it in Chapter 3.) Nevertheless, Popper wanted to ensure to the greatest extent possible that the *contents of science* (i.e., *the facts as we know them*) were true.[26] If you seek Truth, but are dead set against induction, what do you do?

2.G.3 The Truth About Falsification

Popper's solution was radical: if we can't guarantee the Truth of scientific knowledge, then we should stop trying to guarantee it and stop worrying that we can't guarantee it. A valid hypothesis must predict at least one experimental outcome that, if it actually occurred, would mean that the hypothesis was wrong (i.e., it would be *falsified*). Falsifiability, not provability, is the standard that we live by. Testing is not easy though; it must be *severe*: we have to demand that our hypothesis demonstrates its "mettle" by surviving the tests.[27] In Chapter 3, we'll discuss fall-back positions (e.g., "probabilistic truth") and see why, although it has its place in science, it is unsatisfactory as a final standard for basic science.

We must, says Popper, forget trying to *prove* that a hypothesis is True and accept that the best we can hope for is to *avoid accepting hypotheses that are false*. We have to be satisfied if a hypothesis is *intended* to be True, is falsifiable, and withstands attempts to falsify it. After the testing, we must *decide* whether to accept the test results (i.e., what they mean, whether they are reliable and unambiguous). The empirical results of the experiment provide the grounds for coming to a conclusion, but the interpretive decision is what moves science forward. The hypotheses that are not false are the ones we're after because they are the closest we can get to Truth.

For Popper, observations must be robust and repeatable (or "reproducible"). Repeatability is crucial for two reasons. You can't test a hypothesis about a one-time observation, such as "a miracle," because you don't know if it will ever occur again, let alone subject it to tests. And repeatability is important because it allows for *intersubjective* testing—that is, more than one person can make the observation—and "intersubjectivity" is another definition of "objectivity." If science is to be objective, then its observations must be repeatable.

Popper is also clear that observations of all kinds are influenced by our existing hypotheses (they are "theory laden"[28]). When we examine tissue under the microscope, we do so on the assumption that its microscopic structure holds important secrets. We also assume that we know how microscopes work, that the theories that underlie the microscopes are correct and, therefore, that what they reveal is correct. Finally, we interpret what we see through the microscope in light of what we already know and what we expect to see. And, of course, if we detect an anomaly at any stage of the process—something quite unexpected happens—then we become alert because it may signal that an opportunity to make a big discovery is lurking nearby. Or that we're making a mistake.

Some philosophers ask why being able to find out whether a hypothesis is wrong is such a good thing?[29,30] Philosophy aside, falsifiability is not as counterintuitive as it is sometimes made to seem. Remember our car mechanic—call him Bob—uses the same logic in diagnosing the reason that your car engine did not make a sound when you pushed the Start button (or turned the ignition key if it's an older car). He might first evaluate the hypothesis that "the battery is dead," which predicts that the car lights and radio would not work either. If he tests the prediction—turns on the switches for the radio and the lights—and finds that they work perfectly, then he would conclude that the hypothesis is wrong (the battery is not dead), it was falsified by the tests, and he would come up with a new explanation.[31] Falsification is how we eliminate worse ideas so that we can find better ones; it is part of the method of science.

Falsification also solved the "demarcation problem" of the philosophy of science. The question is how can we tell scientific statements apart from nonscientific ones or, more generally, how distinguish science from nonscience? According to Popper, you apply the *falsifiability* standard. A scientific statement can, in principle, be falsified, it is subject to the *possibility* of disproof. Therefore, statements that are not disprovable, such as religious statements, are not scientific. We'll explore the role of falsifiability and the demarcation problem in the context of the *unity of science*, which we'll go into in Chapter 4.

The falsifiability criterion removes ambiguity, preventing the intellectual hocus-pocus that allows astrologers or cult leaders to continue to claim that their belief systems are correct despite massive evidence to the contrary. Because they do not set out unambiguous predictions that could falsify their hypotheses, charlatans can rescue their claims no matter what happens. (Unless they give up: when noted end-of-the-world oracle Harold Camping missed in his prediction of a global cataclysm on September 6, 1994, he changed the date, first to September 29, 1994; then to October 2, 1994; March 31, 1995; May 21, 2011, and finally, October 21, 2011. After suffering an unexpected brain stroke, Camping admitted that he couldn't get it right and quit the predicting business.[32])

2.G.4 Revising Versus Rejecting Hypotheses

If a hypothesis makes a false prediction, then the hypothesis is falsified and must be rejected. For many people, "rejection" seems harsh and insensitive. They'd like a milder, more nurturing way of treating a clever idea that hasn't lived up to its promise, and they have difficulty in accepting Popper's program because of it. Their attitude may stem from overly negative connotations of rejection. An issue that we'll return to is whether one test is sufficient to falsify a hypothesis—it isn't—and we'll see why later.

In any case, "rejecting" a falsified hypothesis does not mean that you have to purge your brain of all vestiges of it. Assume that your testing was appropriately severe and repeatable and that you falsified the hypothesis. This only means that, *as it was literally stated*, the hypothesis was wrong; hence, you reject it and need to find a fresh one. The fresh one may be a little or a lot different from the old one; you may start from scratch or build on elements of the falsified hypothesis. If for example, the data can be explained by a revised hypothesis, then it is okay revise it, to derive new predictions, and test them. If Bob tested your battery and found that it was not strictly speaking "dead" (i.e., no longer rechargeable) but merely had a low voltage, then he'd revise his hypothesis to "discharged battery," recharge it, and begin trying to discover why it ran down. Bob's revision would be legitimate because it would explain the original and the new data in a natural way, and, in fact, Bob could have advanced it originally as an alternative hypothesis.

However, you can't legitimately revise a falsified hypothesis simply by doing an ad hoc patch to rescue it. You can't introduce extraneous elements that were not suggested by any of the data. A good way to avoid making ad hoc patches is to ensure that your revised hypothesis makes predictions that the falsified one did not make. Bob's revised hypothesis predicts that he'll be able to recharge the battery, but he wouldn't be able to recharge a truly dead battery. This procedure is the opposite of what the defenders of the Ptolemaic planetary system were doing by tweaking their epicycle hypothesis with ad hoc fixes, trying to catch up with and explain new anomalies as they cropped up. That is, they retained and embellished their original hypothesis with new details to "rescue" rather than reject it.

It's true that you have to be careful in revising hypotheses and be aware of the slippery slope: if Snowball's biweekly sessions at Puppy School did not help, this would argue that your hypothesis about his hyperkinetic behavior was false. You might then guess that Snowball's problem is that his training did not take place in his normal environment (i.e., your apartment). Technically, this would be a revised hypothesis; it makes a new prediction, and you could go ahead and spend the money to have Snowball's trainer come over and do the training on site. Obviously, you could continue down this road indefinitely. If Snowball's at-home training regimen were unsuccessful, you might be tempted to revise the training deficit hypothesis again: maybe he needs to be trained at home, but only between the hours of 5 and 7 p.m., which is when you usually get home from work. There is, literally, no limit to the number of hypotheses that you can generate. Poor revisions of hypotheses are narrower, less streamlined, and have more ties to a particular case than good revisions. In the end, however, there are only guidelines and no hard and fast rules for deciding when to stick with a falsified hypothesis and revise it or toss it out and start over from scratch.

2.G.4 Tested-and-Not-Falsified? (True, as Far as We Know)

A peculiar sense of dismay seems to grip almost everyone who first encounters Popper's philosophy. How are we supposed to think about hypotheses that have been severely tested and not falsified? Did passing the severe test affect its validity? For Popper, if a hypotheses is tested and not rejected, nothing special happens to it. "Wait a minute!," you may object, "it *seems* stronger somehow," and almost everyone feels the same. Despite our official allegiance to the principle of eternal uncertainty, we expect tested hypotheses to be truer. We'll take up the complicated idea of "probable truth" in the next chapter; for the moment let's remember that the ideal goal of science is the Truth, not "probable truth," even if we knew what that meant. In the meantime, if we look closely at Popper's analysis, we'll see that the situation is not so complicated.

Popper says that, while you can't *know* if a tested hypothesis actually *is* closer to the Truth, it is entirely reasonable to *act* as if it is. Again, *scientific facts (the "contents of science") are hypotheses that have been severely tested but not falsified.*[33] They represent the best information we have.

Putting the matter this way seems awkward, though. Another Critical Rationalist philosopher, David Miller,[34] chides Popper for being "squeamish" and not coming right out and saying that we can *classify* our "tested-and-not-falsified" hypotheses as actually "true." It would be much easier if he had, and saying so would make sense. Why would this be justified? Miller points out that there is a big difference between "classifying" a hypothesis as "true as far as we know" and "certifying" that it is actually True. We can classify, but not certify. Recall that when you proposed a hypotheses, you were proposing that it is a true explanation; you provisionally classified it as being true before you tested it, and the testing didn't make it false. Ergo, to the best of your knowledge, the hypothesis is true.

2.G.5 Corroboration Versus Confidence

Now another confusing point. We can't say that tested-and-not-rejected hypotheses are "confirmed" or "verified" (Chapter 1 L.4.), but we do need a way to talk about them. Popper presses a different word into service; he says that tested-and-not-rejected hypotheses have been *corroborated.* "Corroborated" is not a synonym for "confirmed" or "verified." We don't know if a corroborated hypothesis is true; we only know that, at this point, it's not untrue. Corroborated thus merely means "tested severely and not falsified."

Corroboration is not a trivial concept. We know more about a corroborated hypothesis than we do an untested hypothesis. Popper toys with the notion that we could rank order hypotheses according to the degrees to which they have been corroborated, although nothing much comes of this.[35] In any case, corroboration refers only to our present state of knowledge; it does not, in a sneaky inductivist way, hint anything about the future fitness of the hypothesis, which is, of course, what "confirmation" and the like do. A well-corroborated hypothesis may turn out to be falsified tomorrow. We don't know. In the meantime, we treat it as being provisionally true.

While we might—and I think should—accept this reasoning intellectually, there is no denying that it goes much against our instincts. Intuitively, we feel that a tested-and-not-falsified-hypothesis is better somehow. And we feel an irrepressible surge of triumph ("Yes! [fist pump], we did it!") when our nifty hypothetical explanation passes a tough experimental test with flying colors. There is nothing wrong with these emotions, as long as we don't mix them up with our objective knowledge. We are not actually any more sure of the Truth of the hypothesis today after the experiment than we were yesterday, before the experiment.

Popper's thinking provides the basis of the modern hypothesis. Before fleshing it out more, we should cover the basics of John Platt's program, which adds practical elements to Popper's abstract program.

2.H Strong Inference (John Platt)

Strong Inference is a step-by-step scheme for carrying out scientific investigations that Platt argued is the most efficient way to make progress.[36] He claims that the major difference between rapidly advancing, quantitatively rigorous sciences (e.g., physics and chemistry) and sciences that progress slowly and more haltingly (e.g., social sciences) is that the former followed the principles of Strong Inference. His paper describing the program attracted enormous attention (it has been cited more than 3,500 times[37]), not least because of Platt's frank, colloquial, not to say in-your-face style. Platt felt that a firm conceptual basis was essential for working with hypotheses and for progress; he opposed excessive mathematizing in favor of close logical reasoning and advocated applying the following steps to every scientific problem.

2.H.1 The Program

1. Conceive of multiple alternative hypotheses to account for the phenomenon you're studying (Platt credits T. C. Chamberlin [1897][38] for the original proposal of multiple hypotheses).
2. Devise crucial experiments to exclude one or more of the alternatives.
3. Carry out the experiments and interpret the data.
4. Recycle the procedure to develop subhypotheses and sequential hypotheses to refine the results with further testing.

2.H.2 Deduction and Exclusion

Although Platt often refers to "inductive inference," he acknowledges that the crucial first steps in this program depend on conjecture to generate the hypotheses and on deduction to extract the key predictions for experimental testing. Indeed, his use of "induction" seems more colloquial than technical. Platt repeatedly emphasizes the vital importance of *exclusion* in Step 2, and he argues forcefully that merely confirmatory testing cannot lead to progress. The ability to eliminate one or more alternative hypotheses is the sign of a good test, and you continue testing and eliminating until only one hypothesis is left standing. A test that cleanly distinguishes between experiments is how Platt defines a "crucial experiment."[39]

Platt asks "How many of us write down our alternatives and crucial experiments every day, focusing on the exclusion of a hypothesis?" He is a strong proponent of formal procedures such as keeping a permanent laboratory notebook and consciously analyzing alternative hypotheses and how to test them. He took it for granted that skill in employing Strong Inference can be learned but that you have to practice consciously if you're going to become an expert. Platt cites Popper approvingly and equates exclusion with falsification.

2.H.3 Multiple Hypotheses

While Platt sees Strong Inference as a reworking of the traditional Scientific Method (observe-hypothesize-predict-test-modify), Strong Inference does emphasize two points that get little attention in the traditional accounts: the need for multiple competing hypotheses for a given problem and the iterative nature of the process. Having multiple hypotheses helps guard against becoming psychologically attached to any one of them and encourages a concentrated analytic approach to a problem. And the Strong Inference procedure is inherently iterative—conclusions drawn at the end of one investigation serve as the starting material for the next round of hypothesis testing. Platt, a scientist, aimed to establish a workable program for carrying out rigorous scientific investigations and didn't trouble himself with philosophical enigmas such as justified confidence in hypotheses.

In summary, for both Popper and Platt, the hypothesis is, at its core, an investigation, and falsification is the driving force for hypothesis testing and elimination For convenience, I'll usually lump their programs together under Popper's phrase "Conjectures and Refutations." I'd like to round out the concept of the modern hypothesis by emphasizing a few of its other characteristics and introducing other forms of hypothesis.

2.I Characteristics of a Good Hypothesis

From what we've covered up to now, you know that a hypothesis is much more than the "educated guess" that people often say it is, but we haven't looked into its properties in any detail. What makes for a good one?

2.I.1 Significance and Generality

A good hypothesis tackles a *scientifically meaningful* issue; an unsolved problem, a discrepancy in the data, an avenue of investigation made possible by some technological advance. Of the infinite number of conceivable hypotheses, only a tiny minority are significant. A good hypothesis has generality, "reach" in David Deutsch's lexicon[40]; it is not narrow and ad hoc, meaning that it tries for a comprehensive perspective that rises above particular cases. For example, a hypothesis about Snowball's frenetic welcoming behavior would be scientifically insignificant. A more significant hypothesis might try to account for the hyperactivity syndrome that affects lonely dogs who suffer from sensory deprivation. A still more significant one might be that young mammals, including young humans, develop behavioral anomalies because of the lack of social stimulation. What constitutes scientific importance is a judgment call that can be tricky, especially since the significance of an investigation is not always immediately apparent. Few people would have anticipated that the biology of a bacterium living in the sulfurous hot springs of Yellowstone National Park could be significant, and yet an enzyme isolated from such bacteria is the basis of the most widely used technique in molecular biology,[41] the polymerase chain reaction (PCR) that is used to analyze DNA rapidly and efficiently. The example of PCR illustrates that the significance of a hypothesis may become apparent only long after it has been advanced and tested.

2.1.2 Riskiness

A good hypothesis makes novel *non-obvious* predictions (i.e., *risky* ones). Non-obvious predictions have long had special standing in science because they do not follow from current hypotheses. Indeed, the best ones predict phenomena that weren't known when the hypothesis was formulated. Since they were unknown, they can't possibly have influenced the hypothesis, which increases its generality and explanatory power.

The classic historical example of a non-obvious prediction came from Einstein's General Theory of Relativity, which predicted that the path of light from a distant star would curve significantly as it passed by a massive object such as another star. The Newtonian theories of gravity and space anticipated at most a minor bending of starlight, and Einstein's prediction seemed to come out of nowhere—but it was testable. It was also risky because so much was riding on it; if starlight did not bend a lot, Einstein's whole theory would have been in trouble. The British astrophysicist, Arthur Eddington, sailed to the island of Principe off the west coast of Africa to observe the light from distant stars that would become visible near the sun during a total eclipse. Normally, the glare of sunlight hides those stars, but they can be seen during the eclipse when the moon blocks the sun. By comparing the apparent position of the stars during the eclipse when their light passed close by the sun to their position at night, when the sun was on the other side of the earth, Eddington found that starlight did indeed follow a curved path as it passed by the Sun. His report caused a sensation and persuaded physicists that Newton's model was incomplete[42]; "just wrong" according to Richard Feynman.[43]

We can contrast non-obvious predictions with *obvious* ones. For instance, Bob, the mechanic, would probably consider that, when the car wouldn't start and the radio and the lights didn't work, an obvious prediction would be that the electric power windows would not work either. Obvious predictions are not wrong; however, they're not expected to yield breakthrough information. Again, there is no rigid rule for telling obvious and non-obvious predictions apart. Tests of obvious predictions can produce surprising and powerful insights because they are unexpected. If Bob had found that the power windows worked perfectly, he'd probably be back at square one, completely rethinking the problem. Ordinarily, though, less obvious predictions are more highly prized than obvious ones.

2.1.3 Simplicity

A good hypothesis is *the simplest explanation that encompasses the facts as they are currently understood*. This is the *rule (or law) of parsimony*, also called *Occam's*

Razor, because it was stated by the Bishop of Occam (c. 1287–1347); the rule has been restated many times over the centuries.[44] The virtue of simplicity is ubiquitous and, according to the physicist and mathematician, Henri Poincare, inescapable: "People will scoff at the rule of parsimony, but still draw a smooth curve through their data points . . . the smooth curve is an expression of the rule of parsimony. It is the simplest explanation for the data."[45] Less obviously perhaps, parsimony also expresses *bias*, as we'll see in Chapter 11.

Here's an example of how parsimony can affect an investigation: when Bob found that the engine didn't start and the lights and radio didn't work, he reasoned parsimoniously that the battery was probably dead; it could explain all of his observations. However, it is conceivable that the battery and the radio were both dead and the light switch broken; or that the battery was not dead but the switches to the ignition, radio, and lights all malfunctioned; or that the starter motor was defective, the radio switch broken, and the light bulbs burnt out; and so on and on, even for this mundane case. The alternatives, though conceivable, are complex and, because several parts would have to break down at the same time, they are less likely and less parsimonious than a battery failure. You do not want to be paying hourly labor rates to car mechanics who do not reason parsimoniously when repairing your car. Likewise, taxpayers and research funders expect scientists to act parsimoniously when approaching more involved and expensive problems.

Although you'd think it would be easy to grasp, the rule of parsimony is occasionally misunderstood or misrepresented. *The rule does not say that the world is simple or that the simplest explanation is the most likely to be correct.* It only says that, when you're trying to explain a phenomenon, you should first look at the simplest explanation for it, where "simplest" means the one with the fewest unknown variables, intermediate steps, or assumptions.

Probably most scientists nowadays prefer simplicity for practical or aesthetic, rather than philosophical, reasons. There are an infinite number of potential explanations for any phenomenon, and, if you have a fertile imagination, you can conjure up alternatives at every step and quickly be overwhelmed with elaborate possibilities. The rule of parsimony says you should formulate hypotheses in an orderly way, going from simple to complex, to minimize fruitless effort. Starting with the simplest explanation that can explain all of the relevant data, you only reluctantly resort to more complicated explanations when new results leave you no choice.

Historically, people have not always interpreted parsimony in this way. A devout conviction that the universe was governed by discoverable mathematical laws ordained by an all-powerful God convinced scientists such as Newton that the simplest explanation for a phenomenon would, in fact, be the True one. They believed that "God was a mathematician"[46] and thought that mathematical simplicity was a sure sign of the eternal Truth of the physical laws that they developed.

In contrast, Karl Popper likes simple hypotheses because they are the most open to falsification through testing. According to Popper, the more possibilities that a hypothesis forbids, the more powerful it is; and the simpler it is, the more it forbids.[47] Popper's example of the white swans was a model of simplicity: it forbids swans from having any color but white; therefore any non-white swan would have falsified it. Newton's laws of planetary motion were simple and strong, Ptolemy's system was complicated and weak.

There is an important caveat to the concept of parsimony. A hypothesis must explain *all* of the pertinent information. The key word is "pertinent," and what information is pertinent depends on how you define your scientific question. Scientists are free to establish the limits of the problems that they are working on. They are not obliged to account for every scrap of data that they stumble across. Bob would be free to ignore a broken turn-signal lever, for instance, as a relationship between it and the engine's failure to start would be hard to imagine. It is also perfectly acceptable for scientists to break a big problem into manageable bits. If necessary they can set aside complexities that they can't deal with as long as that they are totally transparent about what they are doing. The scientific community can review their reasoning and criticize it if need be. However, *selectively eliminating data without being open about it is not allowed.*

In neurophysiology, the Nobel Prize-winning physiologist Bernard Katz and his colleague, Paul Fatt, made a fundamental discovery concerning the operation of chemical synapses where a presynaptic cell meets a postsynaptic cell.[48] Initially, Fatt and Katz knew that a neurotransmitter was released from the presynaptic cell in tiny spritzes that occurred in distinct temporal patterns: either steady though irregular streams or sudden high-frequency bursts. Taken together, the two patterns formed a very confusing picture, so Fatt and Katz announced that they would focus on the steady though irregular streams and temporarily ignore the bursts. With this simplification, they were able to concentrate on individual spritzes and eventually found that each one took place when one tiny presynaptic sac, a vesicle filled with neurotransmitter, suddenly dumped its entire contents into the synaptic space. And, sure enough, the bursts of release turned out to represent an entirely distinct process. To this day, the discoveries of Fatt and Katz remain the backbone of our knowledge of how chemical synapses work. Carefully defining and simplifying a problem as they did does not violate parsimony because what you believe is one problem may in fact be a conglomeration of more than one.

Indeed, if the boundaries of a problem are blurred you may face insurmountable barriers to solving it. For example, the ancient Greek natural philosophers had great difficulty in thinking about motion, in part because their idea of "motion" was so expansive: besides physical translocation of an object from one place to another, motion for them included the growth of trees, the aging of men, and the rusting of iron.[49] Since their notion of motion was for us something more like "change" in a very broad sense, it is no wonder they couldn't make headway in understanding it. By the time of Newton, physicists had dispensed with the extraneous meanings and could think about motion as we normally do.

2.I.4 Specificity

A good hypothesis is *specific*, or restricted, meaning that it is exclusionary; it not only explains a phenomenon, but it also rules out other explanations, which makes it precise and informative. Suppose a biochemist hypothesizes that a particular enzyme, say a mutant form of protein kinase C, causes a tumor to grow, and she tests the hypothesis by altering the amount of the mutant enzyme while measuring tumor growth. If the amount of enzyme affected tumor growth as her hypothesis predicted, she could reasonably conclude that the mutant enzyme was a factor. A stronger hypothesis would be that, not only is mutant protein kinase C activity enzyme a factor, but that other similar protein kinases are not. A still stronger hypothesis would be that mutant protein kinase C is a *necessary and sufficient condition* for growth of the tumor: that you have to have the mutant for the tumor to grow and that having the mutant guarantees that it will grow. Specific and powerful hypotheses make it possible to do rigorous tests that exclude numerous alternative hypotheses. The conclusions that you can draw from narrow hypotheses are not necessarily wrong but are wishy-washy when compared to the ones you can get from testing restricted hypotheses.

It is worth noting that, as a practical matter, it is often best to begin an investigation with a coarser, less powerful hypothesis so as not to miss making a discovery because your focus is too narrow. In the tumor biology example, you might start with the general hypothesis that a protein kinase of some sort is involved in tumor growth and do a nonspecific test, perhaps with a treatment that would affect all such enzymes. If you find no evidence that a protein kinase of any kind was important, your specific candidate, mutant protein kinase C, almost certainly would not be. Hence, you might want to start with a "dirty" drug (i.e., a nonselective one) that will block all kinases. Initially, casting a wide net with a nonselective test can be helpful because it can effectively eliminate many hypotheses at once.

2.1.5 Constraint

Finally, according to David Deutsch,[50] a good hypothesis is "hard to vary." All of the details of a good explanation are vital; change one of them and the explanation no longer explains what it was supposed to. With a bad explanation, however, you can change many details and still be able to concoct an explanation that will work. Deutsch's favorite illustration involves the accounts of the annual earthly seasons given by ancient mythologies and by modern astronomy. In the myths, the actions of a god or gods cause the transition from summer to fall. You can freely substitute one god for another, alter his or her circumstances, motivations, powers, or objectives and generate an explanation for the seasonal progression that is as good as the one you started with. In contrast, change any detail of the modern theory, say the degree of tilt of the earth's axis of rotation with respect to its orbital plane around the sun, and the astronomical explanation is ruined. It is a good explanation that is hard to vary, and the myths are bad ones. Of course, the modern astronomer has a lot more data to work with than the Greek mythologists did, and indeed the "hard to vary" criterion implies that a good explanation is more tightly constrained by data than a poor one.

2.1.6 Falsifiability in Practice

If a proposition is not falsifiable, then it is not a hypothesis at all. Still, it is hard to judge a hypothesis that you can only test in principle to be as good as one that you can test in practice. There may be both theoretical and practical barriers in the way of actually testing a hypothesis. If your hypothetical explanation of Snowball's hyperactive behavior is that he is deprived of canine society during the day, then it is certainly falsifiable; but if you will be evicted if you get another dog, then it may well remain testable only in principle. Like other criteria for good hypotheses, this one is not an absolute requirement. Provocative, though presently untestable, hypotheses inhabit fields from psychology to theoretical physics. A few will lead to major new developments, so we shouldn't write off apparently untestable hypotheses altogether. Still, it's probably best to aim for achievable results, and that means you want practically testable hypotheses.

Up to now, I've been discussing the hypothesis as if it were a public intellectual product, an object with specific properties that we could evaluate. Yet, scientists don't always make their hypotheses overt. Their hypotheses exist in an implied state and we, readers and science-consumers, have to work out what they are. This tendency toward leaving hypotheses unstated makes scientific communication more difficult than it needs to be. Implicit hypotheses are pervasive, and they are significant for deeper reasons than communication alone.

2.J Implicit Hypotheses

"No one searches without a plan," said Santiago Ramón y Cajal,[51] and you can't have a plan without a hypothesis about where and how you should search and what you'll encounter. A related concept, voiced by Popper and other philosophers, is that all observations are "theory-laden."[52] You can't make a scientific observation without having a theoretical framework for it. The Critical Rationalist philosopher Bryan Magee gets this point across to beginning students by instructing them to "Observe" and record their observations.[53] The students, naturally, are baffled—what should they observe; what should they record? The world has much too much information to observe it all at once. "Observation is always selective . . . it needs an object." You have to break down the world into manageable categories—birds, colors, temperatures—and for that you need hypotheses: what constitutes a bird, a color, a temperature? And you need to know how to observe, what instruments you need, and what hypotheses the instruments depend on, etc. Observations, he concludes, "are always *interpretations of the facts observed; that they are interpretations in the light of [hypotheses].*" But the hypotheses that shape the observations are rarely laid out explicitly; they are implicit.

At first the notion of implicit hypotheses seems paradoxical: hypothesis formation and testing demand focused, conscious thought, so how could a hypothesis not be explicit? Implicit hypotheses exist in two forms: as unstated experimental frameworks and as deep assumptions (deep implicit hypotheses).

2.J.1 Implicit Experimental Hypotheses

Authors of scientific papers frequently do not state their hypothesis (Chapter 9) even when they have one. Teasing out an implicit experimental hypothesis buried within a paper requires careful reading. A common modern tactic is to note in the Abstract or Introduction to a report that a certain phenomenon is "poorly understood." (I regretfully confess to having used this formula.) Saying something is poorly understood is almost entirely vacuous. The fundamentally uncertain state of all scientific knowledge guarantees that all phenomenon have poorly understood features. In making the remark, the experimenters are usually just announcing the topic of the paper and, perhaps, hinting at their implicit hypothesis and its predictions. The authors ordinarily go on to lay out their experiments logically and give a rationale for each critical test. Here and there, they may sprinkle in "predictions," even if the predictions aren't part of a coherent argument. The Discussion section integrates the major findings into a coherent whole, puts them into a larger context, and occasionally proposes a summary hypothesis (or model) that suggests what the experimental hypothesis had been all along. With some effort you can find the implicit hypothesis, and, indeed, trying to find them can be an educational exercise (Chapter 14).

We could, for example, deduce our mechanic's implicit hypothesis when he was trying to diagnose why the car wouldn't start. By reading his report afterward; ("Because the reason for the car's not starting was poorly understood . . ."), he pressed the Start button, heard no sound, and tried the lights and the radio; then, when still nothing happened, he opened the hood and checked the battery terminals, cables, etc. Without being told what his reasoning was, we could figure it out, assuming, of course, that we already knew how cars worked. Otherwise, we might be puzzled about why he tried the lights and radio (did he really think that *they* had anything to do with starting the engine?). Reading a report based on an implicit hypothesis requires extra effort and presumes that the reader can see that the conclusions do follow from the results. While such a report is easier to understand than one without any apparent plan, it is generally harder to follow than one with an explicit hypothesis, especially for readers who are not experts.

2.J.2 Deep Implicit Hypotheses

Implicit hypotheses are also part of the *background knowledge* that we bring to any project. Since all scientific knowledge is ultimately uncertain, a "fact" is essentially a hypothesis—"A fact is where the investigation rests." As with other tested-and-not-falsified hypotheses, we assume that they're true, but, unlike implicit experimental hypotheses, we are not currently investigating them. They are *deep implicit hypotheses.*

When neuroscientists study the brain, one such deep implicit hypothesis was based on decades of accumulated information about what cells are and what they do. The brain processes information at chemical synapses between cells, as we discussed earlier. There are two fundamentally different kinds of cells in the brain, neurons and *glial cells* (aka *glia*). For nearly 100 years neuroscientists interested in information processing focused on neurons, the highly electrically excitable signaling cells that control everything from muscle contractions to visual perception. Although there are about as many glia as neurons, the glia were mostly ignored; their electrical excitability was too sluggish to keep up with the rapid signaling capabilities of neurons. The deep implicit hypothesis was that glia were the go-fers and support staff of the brain, cleaning up the local microenvironment and guiding the slow process of brain development, but having no role in the glamourous job of information processing. Ignoring glia allowed neuroscientists to make enormous strides in understanding neuronal communication.

Still, the assumption that neurons alone carry out information processing was an implicit hypothesis, and hypotheses, even deep implicit ones, exist to be tested. When the neurons-only hypothesis was made explicit and tested, it failed miserably. It turns out that if you disrupt glial cells so they can't function, you prevent neuronal synapses from operating normally. In fact, in recent years, we've learned of so many ways in which glia influence neuronal signaling that some neuroscientists believe that the key communication structure in the brain is the "tripartite synapse,"[54] a microstructure made up of pre- and postsynaptic neurons together with their nearby glial partners.

You'll often hear that you've got to have a completely open mind to do a scientific investigation. It is an admirable objective that you won't be able to achieve. We can't escape the constraints of our minds, which are constantly at work constructing our understanding of the world by spinning webs of deep implicit hypotheses. A tell-tale sign of this activity comes in the form of a surprise or an unexpected result. "Expectation" is another name for a prediction made by a deep implicit hypothesis, a bit of background knowledge that you took for granted. Surprise is what happens when a prediction of a deep implicit hypothesis is falsified.

Besides forming background knowledge, the deep implicit hypothesis also shows up in the choice of the research technique. "To a man with a hammer, everything looks like a nail" fits when it comes to scientific experiments; your choice of measuring instruments shows which measurements you think are worth making. Microscopists believe that secrets lie in cellular structure, while electrophysiologists measure electrical potentials because they believe that cellular activity is the key feature of biological function. Science takes place within a web of deep implicit hypotheses.

Now that we've surveyed some properties of hypotheses, in the next chapter I'll return to the main one, falsification, and address a few of the questions or objections that philosophers have raised and that you might have about it. Popper's falsification program was novel and completely at odds with the philosophical mainstream of his day, so it is not surprising that it ran into a lot of opposition.

2.K Coda

In this chapter, we reviewed the hypothesis and its properties. The modern hypothesis is an empirically testable explanation for an aspect of nature. Hypotheses are not predictions, but they imply predictions; you deduce predictions from hypotheses. The ultimate uncertainty of knowledge causes the most persistent problem for scientific thinking, and the programs associated with Karl Popper and John Platt are robust scientific methods for dealing with it. They both rely on the ability of experimental tests to falsify or eliminate incorrect hypotheses;

the factual content of science is the collection of tested and not falsified hypotheses. Inductive reasoning does not provide a sound basis for establishing scientific truths. Hypotheses that are falsified must be rejected but can be revised and retested. Popper's refusal to grant any special status to a hypothesis that is tested-and-not-falsified cuts across the grain of our intuition. He says that tested and not rejected hypotheses have been corroborated but not confirmed and that corroborated hypotheses can be rationally used as the basis for action. We behave as if they are true while remaining aware that they may be false.

Notes

1. *The New York Times*, April 2, 2011, Technology, p. B1
2. Merriam-Webster online dictionary; visit http://www.merriam-webster.com/dictionary/hacker.
3. While hypotheses, theories, and laws are sometimes distinguished on quantitative grounds—theories and laws are said to have more experimental support than hypotheses, or to be more mathematical—the boundaries between them are flexible (*Merriam-Webster's Dictionary of Synonyms*) and imprecise. Because all three are conjectural, falsifiable, and provisional, I won't try to make any distinctions among them. The book is about the scientific hypothesis, not the hypothesis of formal logic, which refers to the antecedent of a proposition (e.g., in the argument, "*If P, then Q*," *P* is the antecedent, the hypothesis, or the premise and *Q* is the consequent). Finally, a peculiarity of colloquial American English is that "hypothetical," which should suggest only that the statement is not indisputably true, can imply that a statement is so doubtful as to be—in fact probably is—false. "Hypothetically" in this sense, often accompanied by hand gestures, "air quotes," suggests "Sure, maybe it could be that way, but it isn't." That is not what I mean.
4. Oxford Dictionary online; visit http://www.oxforddictionaries.com/us/definition/american_english/scientific-method.
5. Henri Poincare, *Science and Hypothesis* (originally New York: Walter Scott; reprinted CreateSpace Independent Publishing Platform; 2013), Preface.
6. Stuart Firestein, *Ignorance: How It Drives Science* (New York: Oxford University Press; 2012), p. 77: "the hypothesis is supposed to be the starting point for all experiments."
7. For example, the nineteenth-century physicist and positivist philosopher Ernst Mach did not accept the existence of atoms as actual entities, but he did allow that the atomic hypothesis was useful because it made certain calculations possible; see Laurens Laudan, *Science and Hypothesis* (Dordrecht: D. Eridel; 1981), chapter 13. In earlier ages, the Catholic Church had had no problems with Galileo's description of a heliocentric solar system as long as he acknowledged that the idea of the sun being at the center of the solar system was merely a hypothesis, a useful tool for simplifying astronomical calculations. When Galileo asserted that his model accurately described reality, the Church's patience came to an end and he was forced to recant the idea.

8. Representative sources that confound hypothesis and prediction: https://answers.yahoo.com/question/index?qid=20070922185430AAFFd6G; http://www.differencebetween.net/science/difference-between-hypothesis-and-prediction/; http://www.differencebetween.info/difference-between-hypothesis-and-prediction; http://madaboutscience.weebly.com/prediction-vs-hypothesis.html;
9. David Deutsch, *The Beginning of Infinity: Explanations that Transform the World* (New York: Penguin Books, 2011) discusses the hypothesis as a conjecture about unobservables.
10. The Logical Positivist philosophers (see, e.g., A. J. Ayer, *Language Truth and Logic* [New York: Dover Publications; 1946]), argued that verifiability was the main criterion for a genuine *statement of fact* (i.e., a hypothesis). Verifiability implies the ability to make observations that would show the statement was either true or false. Ayer admitted that, while conclusive verifiability was not attainable (i.e., that the truth of a hypothesis could not be proved absolutely), we could achieve *verifiability in principle* if the appropriate experiments or observations were conceivable or if the truth of a statement could be made probable (*weak verifiability*), rather than certain.
11. Karl Popper, *The Logic of Scientific Discovery* (New York: Routledge Classics; 2002). Popper rejected the concept of verifiability because hypotheses could never be conclusively shown to be True. Hypotheses could be found to be false, however, and science could advance by rigorously testing and rejecting the bad ones. Popper's ideas are considered in detail in Chapter 3.
12. Santiago Ramón y Cajal, *Advice to a Young Investigator.* Translated by N. Swanson and L. Swanson (Cambridge, MA: MIT Press; 1999).
13. A few theoretical physicists, such as David Deutsch, are proposing theories that go beyond empiricism, and I'll touch on his ideas in Chapter 10.
14. pH of acid rain: http://www.epa.gov/acidrain/education/site_students/phscale.html
15. David Hubel and Torsten Wiesel, "Early Explorations of the Visual Cortex," *Neuron* 20:401–412, 1998.
16. Not all scientists admire Popper, and I'll review the work of three critics in Chapter 10.
17. John R. Platt, "Strong Inference," *Science* 146:347–353, 1964. Platt's Strong Inference is unrelated to the probabilistic sort of inferential reasoning called "strong inference" that was discussed in Chapter 1.
18. Malachi H. Hacohen, *Karl Popper: The Formative Years 1902–1945* (Cambridge, MA: Cambridge University Press, 2000).
19. David Edmonds and John Eidinow, *Wittgenstein's Poker* (New York: HarperCollins; 2 001).
20. Peter Godfrey-Smith, *Theory and Reality* (Chicago: University of Chicago Press; 2003). Godfrey-Smith is a philosopher of science whose lucid descriptions of Popper's work I will rely on throughout this chapter.
21. Karl Popper, *Conjectures and Refutations* (New York: Routledge, 2002).
22. See David Wooton, *The Invention of Science* (New York: Harper Perennial; 2015), chapter 6, for a discussion of the Renaissance invention of the concept of a *fact* as a "peculiar blend of reality and thought" that distinguishes modern from ancient science.

23. Justificationism. The school of Logical Positivism (see Note 10) distinguished the "context of justification" (i.e., the aspect of scientific thinking that involved public and hence, philosophically accessible, aspects in the conduct of science; (e.g., Okasha, *Philosophy of Science*, Note 24, p. 79) from the "context of discovery," which included those aspects of thinking that were entirely internal and hidden, even from the thinker (i.e., the psychological parts). Logical Positivists thought that it was important to be able to justify scientific beliefs. Thus, even the weak form of probable verifiability or justification still depends on the validity of inductive reasoning for its truth value. See D. Miller, *Critical Rationalism*, Note 33, for a comprehensive discussion of justificationism and Popper's views.
24. Samir Okasha, *Philosophy of Science: A Very Short Introduction* (Oxford: Oxford University Press; 2002). I do not believe that Okasha accurately represents Popper's thought.
25. What is induction? See discussion in Chapter 1.
26. Popper, *Logic*.
27. The philosopher and statistician Deborah G. Mayo, *Statistical Inference as Severe Testing: How to Get Beyond the Statistics Wars* (New York: Cambridge University Press; 2018), faults Popper for not defining "severe" and proposes that a severe test is one that would "probably" have detected a flaw in the hypothesis had it existed. She describes in technical detail how statistical tests fail the severity criterion. Mayo goes considerably beyond what Popper would have endorsed, however, arguing that, if a hypothesis does pass a severe test, its passing counts as evidence in favor of the hypothesis, contrary to what Popper believed, as I'll show.
28. All hypotheses, observations, and, in fact, scientific reasoning of many kinds are said to be "theory laden"; that is, they are shaped by our prior understanding and assumptions about the world; for example, visit https://en.wikipedia.org/wiki/Theory-ladenness.
29. Okasha, *Philosophy of Science*.
30. Godfrey-Smith, *Theory*, chapter 4.
31. The analogy between the automotive problem and science breaks down because Bob could eventually get all relevant information regarding the car battery and manipulate all the variables directly; scientists rarely have access to all the variables or the ability to manipulate them. The point here is that Bob's strategy in testing and eliminating alternatives is analogous to the kinds of reasoning from hypotheses that scientists use.
32. Howard Camping tribute https://www.christianpost.com/news/tribute-to-harold-camping-on-family-radio-network-leaves-out-any-mention-of-his-end-times-prophecies.html, *Christian Post*, 2011; Howard Camping, Wikipedia.org https://en.wikipedia.org/wiki/Harold_Camping, Howard Camping Dies, Huffington post https://www.huffpost.com/entry/harold-camping-dead-dies_n_4459716. Over the years, Camping's organization, *Family Radio*, amassed hundreds of millions of dollars, many from donors, and was notably successful with the campaign associated with his March 11, 2011, end-of-the-world prediction. When that didn't happen,

Camping saw that his attempts at forecasting were "incorrect and sinful" and stopped trying.

33. Popper, *Logic*, p. 17; "the system that represents our world of experience... has been submitted to tests and has stood up to tests."
34. David Miller, *Critical Rationalism: A Restatement and Defense.* (Peru, IL: Open Court; 1994). Miller is a critical rationalist philosopher who, in his "restatement" of Popper's positions, presents an in-depth critique rebuttal of many of Popper's critics, as well as an accessible explication of Popper's own ideas. Miller's book was a major resource for my discussion of Popper.
35. "Degrees of corroboration" This is one of the gaps in Popper's program that Mayo (see Note 27) seeks to fill, seeing passing a severe test as positive evidence in favor of a hypothesis.
36. Platt, ibid. Again, Platt's program is primarily hypothetico-deductive, as Popper's is, not inductive.
37. Number of citations for Platt's article from Google Scholar search (May 31, 2018).
38. T. C. Chamberlin, "The Method of Multiple Working Hypotheses: With This Method the Dangers of Parental Affection for a Favorite Theory Can Be Circumvented," *Journal of Geology* 1897 (reprinted in *Science* 148:754–759, 1965.)
39. Crucial experiment. Francis Bacon emphasized the concept, but some thinkers, for example. Pierre Duhem, "Physical Theory and Experiment, in Sandra G. Harding (Ed.), *Can Theories Be Refuted? Essays on the Duhem-Quine Thesis* (Boston: D. Reidel; 1976), deny that any such thing exists. Again, science does what it can. Perhaps we can say that some experiments are more crucial than others.
40. Ayer, *Language Truth and Logic*. See Note 10.
41. See https://en.wikipedia.org/wiki/Polymerase_chain_reaction.
42. Bertrand Russell, *The ABC of Relativity*, Revised ed. (New York: The New American Library, 1959).
43. Richard Feynman, *The Very Best of the Feynman Lectures* (New York: Basic Books; 1961), audio CD.
44. Richard Westfall, *Never at Rest: A Biography of Isaac Newton* (New York: Cambridge University Press; 1979); Isaac Newton's first rule—"No more causes are to be admitted than those which are both true and sufficient to explain the appearances." Similarly, A. Comte's "Law One" (cited by Laudan, *Science*, p. 154) was "the rule that we should in all cases form the simplest hypothesis consistent with the whole of the facts to be presented." Or "Everything should be as simple as possible, but not simpler." Dubiously attributed to Albert Einstein (math.ucr.edu/home/baez/physics/general/occam.html).
45. Henri Poincare, *Science and Hypothesis*.
46. Edward Dolnick, *The Clockwork Universe* (New York: HarperCollins; 2011).
47. "Not for nothing do we call the laws of nature 'laws': the more they prohibit, the more they say." Popper, *Logic*, p. 19.
48. Paul Fatt and Bernard Katz, "Spontaneous Subthreshold Activity at Motor Nerve Endings," *Journal of Physiology (London)* 117:109–128, 1952. Fatt and Katz studied

the chemical synapses between nerve and muscle cells, but the principles they found apply to all synapses, such as those between nerve cells in the brain.

49. Bertrand Russell, *A History of Western Philosophy* (New York: Simon & Schuster; 1945).
50. Deutsch, *The Beginning of Infinity*.
51. Ramón y Cajal, *Advice*, p. 117.
52. Mayo, *Statistical Inference as Severe Testing*.
53. Bryan Magee, *Philosophy and the Real World: An Introduction to Karl Popper* (La Salle, IL: Open Court; 1994), p. 30.
54. Alfonso Araque, Vladimir Parpura, R. P. Sanzgiri, and Phillip G. Haydon, "Tripartite Synapses: Glia, the Unacknowledged Partner," *Trends in Neuroscience* 22:208–215, 1999. An early review by a few of the pioneers in this field.

3
Critical Rationalism
Common Questions Asked and Answered

3.A Introduction

The preceding chapter reviewed the basic features of the modern scientific hypothesis, but you may have some lingering questions about it or have heard criticisms of it. As I noted in the Introduction, objections to the hypothesis or hypothesis-based science come from a number of different directions and are driven by themes that recur throughout the book. This chapter takes up one small collection of criticisms that mainly have to do with a key property of the hypothesis: namely, that it is falsifiable in principle. These criticisms of Karl Popper's and John Platt's programs, collectively called "Conjectures and Refutations," are common complaints of philosophers and one or more of them may have occurred to you. I've put a few somewhat technical topics in an optional section at the end of the chapter, which you can skip if you're so inclined.

3.B Objections to Falsification

To reduce the voluminous philosophical commentary to a manageable size, I'll take the positions of the philosophers Carol Cleland,[1] Peter Godfrey-Smith,[2] Samir Okasha,[3] and Massimo Pigliucci[4] to be broadly representative of the field of Popper/Platt critics, and I've extracted most of the criticisms from their work. Despite having different agendas, all agree that many of the main principles of Conjectures and Refutations are probably misunderstood by scientists and, in any case, are fundamentally wrong. I won't always single out the individual critic's comments because the concerns overlap considerably. In addition, these writers all appear to be puzzled by a question which I will summarize as "Why, apart from sheer ignorance of philosophy, have scientists been so slow to recognize the manifest failings of Popper and Platt?" I am not sure I can say why scientists have not latched onto the abstract teachings of philosophy more vigorously, but scientists do seem to have a practical appreciation of Conjectures and Refutations that philosophers do not share, and I'll try to explain the scientists' perspective here.

Actually, when I say that philosophers don't know why Popper holds such allure for scientists, I can't include Godfrey-Smith, who does know why. As he explains[5]: Popper "is the only philosopher . . . who is regarded as a hero by many scientists." Popper conveys a "noble and heroic" vision of science, and his ideas are simple and clear (presumably a prerequisite so that scientists can grasp them). In short, Popper holds up a mirror that allows scientists see themselves as being both imaginative and creative, even artistic (they invent theories), and yet rugged and tough-minded (they ruthlessly test and reject those theories). Think, says Godfrey-Smith, of a "hard-headed cowboy out on the range with a Stradivarius violin in his saddle-bag."

No wonder scientists love Popper!

Needless to say, neither Godfrey-Smith nor the other philosophers think that Popper's flattering treatment is sufficient to justify the scientists' admiration.

3.B.1 Falsification Is Never Final

If our understanding of the world is inevitably uncertain at some level, then it follows that the tests of hypothesis may themselves be mistaken or that new data may reveal that the test was not conclusive. This is just a restatement of the principle of fallibilism that we encountered in Chapter 1 that apparently all philosophers and all scientists accept. Hence, you can't be 100% sure that your hypothesis is wrong, any more than you can be 100% sure it is right. Doesn't this make a mockery of Popper's whole program?

Critics also accuse Popper of holding to a naïve concept of falsification; that it can be simply and cleanly accomplished with "a single piece of contrary evidence."[6] They[7] claim that he thought falsification was not only simple but "decisive" and would take place "instantly" when a falsifying test occurred. In contrast, a careful reading of *The Logic of Scientific Discovery* reveals that Popper was not naïve; remaining perpetually open to the possibility of change was a hallmark of his philosophy. He repeatedly stresses that the results of hypothesis testing can never be conclusive. He acknowledges that your test results might be erroneous and that you might eventually have to reject a well-corroborated (we'll revisit "corroborated" in a while) hypothesis. The more important question that we're left with is, "so what?" What are the consequences of inconclusive falsification for science?

While it's true that we can't decisively and permanently classify a statement as either true or false, it only means that the state of our knowledge, like the preservation of our liberty, demands eternal vigilance, not that science is impossible. Fallibilism poses a challenge for our simplistic view of the world, but it is not a threat to science. Perhaps we're not used to thinking about scientific truths as tentative, but that's probably because we weren't taught how to think about them. We need to expand our minds and give up the comforts of certainty while continuing to act decisively.

3.B.2 Falsification Is Pointless

Popper's critics are unconvinced that his solution for approaching Truth via falsification makes sense. The program, says one, "fails," because "scientists are not only interested in showing that certain theories are false."[8] To which a Popperian would say "Of course they're not! They are interested in finding true ones." Unfortunately, the best they can do is to weed out the false ones. Another critic thinks that hypotheses are created "in order to be shown to be false."[9] Wrong again. The *purpose* of the falsification criterion is not to *show* a hypothesis is false; rather, the purpose of testing it severely is to find our whether it is false or not.

An illustration might help. Consider the philosophically minded engineers at the Transcendent Epistemology Safety Tire Company (TESTCo) in their quest to make the perfect tire for the family car. By varying materials and design, the engineers produced hundreds of prototypes, put them on various vehicles, and ran them through grueling tests, trying to find what made them fail. Although the overwhelming majority of prototypes did fail, the engineers finally succeeded in creating one that survived all of the realistic ordeals that they could dream up, so they turned it over to the manufacturing and marketing divisions. At the press conference announcing the new tire, the TESTCo CEO deflected the question of whether it was the *perfect* tire, pointing out that it met the most rigorous standards that had been devised to date and had his highest confidence. He was proud to say that his own family (a slide of an attractive woman and cute 11-year-old twins appeared on a screen behind him) rode in vehicles exclusively fitted with the new tire. He did say that TESTCo engineers would never rest until they produced the perfect tire, however.

An onlooker, motioning toward the mountains of ruptured tires dotting the fields around the proving grounds, nudged his companion, "I guess that TESTCo's aim is to produce shards of rubber, since that's what they mostly do!" His friend dissented, saying that TESTCo just wanted "to make tires in order to destroy them.". They both chuckled.

These folks mistook TESTCo's methods for its goal; its procedures for its actual accomplishments. Disintegrated rubber was not the company's primary product: the knowledge gained by finding out what didn't work was, because that was what allowed the engineers to come up with better tire designs. Likewise, although the TESTCo program did destroy countless prototypes, it did not make them *in order to* destroy them, but to learn how to make improvements.

In getting distracted by superficial details, the onlookers missed TESTCo's overarching purpose. Some of Popper's critics may be making a similar error.

3.B.3 Without Rules, the Decision to Reject a Hypothesis Is Arbitrary

If falsification is never complete, then rejection of a hypothesis becomes a matter of judgment: we must *decide* how to interpret the experimental results and when to declare that a hypothesis has been falsified. To philosophers, this looks haphazard and irrational. If we can't specify the precise logical connection between the data and the decision, then why not cut out the middle man? Skip the data gathering step and decide about the hypotheses without doing any experiments at all!

This is nonsense, of course. No one is suggesting that we actually do that, but it is the sort of muddle that you get into if you insist on logically airtight deductive reasoning about empirical evidence in an arena—human decision-making—in which the basic rules are not airtight or fully understood. Every day we make practical, rational decisions even without a theory of practical rational decision-making (see Chapter 12). It would be great if we did have such a theory and maybe eventually we will, but not having one is no reason to disparage Conjectures and Refutations now.

Whether or not scientists accept a falsifying test of a hypothesis depends on many psychological and sociological factors, as well as on the claims that the hypothesis makes. The magician and phoney-science debunker James Randi uses a vivid analogy regarding skeptical thinking: if you are told that a man keeps a pony in his backyard, a phone call to the man's neighbor should be enough to convince you whether he does or not. If you are told that a man keeps a unicorn in his backyard, you would insist on seeing it yourself before you'd consider believing it. Well-established theories (e.g., the Theory of Evolution, the General Theory of Relativity) would need to face multiple, serious challenges before they'd be let go. Probably your hypothesis of the neural mechanism of an eating disorder affecting lab rats could be falsified with much less work.

Moreover, Popper knew that scientists are not likely to be persuaded by the outcome of one test. Scientists may cling to a well-tested hypothesis for a while even if it fails a test. This proves, say the critics, that scientists don't truly buy into the importance of falsification. Here's a classical example: when confronted by a deviation in the orbit of the planet Uranus that appeared to be inconsistent with Newton's law of gravity, a theoretician, Urbain Le Verrier, did not immediately reject the law. By assuming that it continued to hold, he correctly predicted the existence of a previously unknown planet, Neptune, and thereby accounted for the orbital anomaly. Had he rejected the law of gravity instead, he would have made a gross error. Does this mean that scientists do not accept falsification? Or does it simply demonstrate that they are not philosophers? Scientist have a gut-level appreciation for the concept that falsification itself is never complete and that, therefore, an apparently convincing falsifying test might be wrong or misleading. It seems obvious that you shouldn't abandon well-corroborated theories, such as Newton's, at the first sign of trouble. We're convinced that it would be foolish to do otherwise, even if we can't define "foolish."

Science is a social endeavor, and, in the end, the weight of opinion in the scientific community—consensus—determines what science "knows."[10] This conclusion unsettles some philosophers; it is "a puzzling way to make decisions,"[11] says one; it represents "mob psychology,"[12] says another. What would make these commentators happy? Expecting a unanimous opinion from a large group of educated and intelligent people who come equipped with the usual complement of human foibles—competitiveness, obstinacy, irrationality, self-interest, vanity, and so on—is unrealistic. Popper was aware of these and other "nonempirical" factors that affect scientific thinking, but he saw them as problems for psychology, not philosophy. Whether the solution to the question of how beliefs form is a proper topic for philosophy or scientific psychology, as the philosopher W. V. O. Quine believed,[13] is not yet settled.

Consider that about 97% of scientists today accept that global climate change is a real threat to the world (at least to civilization as we know it) and that it is a result of human activity.[14] No one, including, we may believe, Karl Popper, would argue that we should consider that the hypothesis of global climate change is falsified because 3% of climate scientists say that they do not accept the data. They could be geniuses with uniquely brilliant insights, complete crackpots, people driven by personal or political motivations, or something else. Philosopher of science Thomas Kuhn, in his *The Structure of Scientific Revolutions*,[15] argues that sudden, large-scale changes in scientific attitudes are common in the history of science and reflect shifts in community opinions, especially those of its most vocal and persuasive members. The physicist Max Planck drily observed[16] that science advances "one funeral at time," as the stalwarts who cling to the old thinking gradually die off. We seem obliged to live with the conclusion that science is not a fully rational, philosophically pure endeavor because scientists are not fully rational, philosophically pure individuals.

3.B.4. Two Roles of Falsification in Science: Method and Contents

The concepts of falsification and demarcation have given rise to serious controversies, and an especially big one surrounds the issues of the *method* and *contents* of science. "Method" has to do with how science operates, and "contents" refers to the body of knowledge that science builds up. These are plainly separate subjects yet, surprisingly, philosophers of science—both pro- and anti-Popper—don't always highlight the distinctions between them, and this leads to trouble. This is one of the areas in which I drift away from strict Popperian orthodoxy.

Let's look how falsification applies to the method of science. The first thing a Popperian scientist does with a candidate "scientific statement" (to be concrete, we'll imagine that it is an explicit hypothesis) is to assess its potential for falsifiability. If she can think of an experimental test or observation that could demonstrate that the hypothesis is false, then she admits it, *provisionally*, into the world of science in order to test it. A hypothesis meeting the standard of falsifiability at this stage is equivalent to a job candidate's getting by a preliminary screening procedure; if he passes, then he moves on to the next, more rigorous stage of evaluation to determine if he'll get the job.

Despite the fact that the demand for a falsifiable form for a hypothesis is firm, the bar is initially set fairly low, and any conceivably testable statement passes. The initial bar is low for two reasons: first, there is little riding on the decision to accept a statement for testing since neither our scientist, nor anyone else, would base any significant action on an entirely untested hypothesis. Second, and more importantly, *an untested hypothesis is not part of the contents of science.* Popper should probably have stressed this more than he does.

The stakes go up at the next stage, when the scientist puts the hypothesis to severe tests and decides whether or not it has been falsified. If the hypothesis passes the tests, then it is *retained as part of the contents of science*; it can rationally be used as a basis for action or subjected to further tests. Scientific knowledge, remember, is just this body of *corroborated* hypotheses that we have classified as "true as far as we know." They remain falsifiable.

The question of what happens if a putative hypothesis fails the falsification tests is the one that trips up many anti-Popper critics. The answer is that a tested and falsified hypothesis is *ejected from science.*[17] The testing phase showed that it does not describe or explain an aspect of our physical, empirical world. Hence, despite having formally met the falsifiability criterion at the preliminary stage, it is now barred from joining the system of corroborated statements that describe the world. Falsifiability, in other words, is a *necessary but not sufficient* condition for inclusion in the contents of science.

Ignoring the crucial step of removing falsified statements from science leaves some philosophers wondering about what Popper would do with "nutty" theories (e.g., astrology or phrenology) that have been falsified. Do they hang around and clutter up "the pantheon of science"[18] just because they are nominally in falsifiable form? In fact, for Popper, such theories do not constitute a bother for science because they are not part of science. They're out. The process of falsification, testing, and ejection is how science protects itself against infestation by crazy theories.

If such theories want to try again, they're welcome to reapply for admission, but then they're back at square one, the initial audition step, and I believe that most scientists would agree that this time the hypothesis has to clear a higher bar before they'll be willing to reevaluate it. If you're rejected for the job at your first interview and show up in the same scruffy jeans and T-shirt for a second one, you can't expect better treatment. You've got to show some improvement if you want to be taken seriously. Scientific case in point: the theory of inheritance of acquired characteristics (sometimes called "Larmarckism"). Roughly speaking, this was the proposal that traits acquired during an organism's lifetime could be reproductively passed on to its offspring. The theory was eventually discredited and replaced by the modern gene theory in the early twentieth century. However, mounting evidence suggests that a cluster of molecular mechanisms collectively referred to as "epigenetics" can cause chemical modifications of DNA that are acquired during an individual's lifetime and that affect gene transcription. Epigenetic changes could be acquired, say following a period of prenatal stress that your mother experienced when you were in the womb.[19] These genetic changes could have untold effects on your later development, behavior, etc., though you did not, strictly speaking, inherit them from your parents. Whether or not epigenetics or other similar extragenetic factors will require a fundamental "rethink" of conventional evolutionary theory is being debated,[20] but it looks as though the theory of inheritance of acquired characteristics, all dressed up in new clothes, is reapplying for a position in science.

Popper's critics are not the only ones who've added to the confusion surrounding the key notion of falsifiability. I believe that pro-Popper writers, for example the Critical Rationalist philosopher David Miller, occasionally share the blame. For example, Miller says[21] that "a hypothesis may be admitted to the realm of scientific knowledge only if it is falsifiable by experience," but this can't be quite right, according to Popper. In the *Logic of Scientific Discovery*, section 5, Popper states that "the system that represents our world of experience [is] to be distinguished . . . by the fact that *it has been subjected to tests and stood up to tests*" (emphasis added). Scientific knowledge of the world, in other words, does not include untested or uncorroborated hypotheses, notwithstanding their formal falsifiability. Again, there is a multistep filtration process that includes proposing a falsifiable hypothesis, testing it, and, if it passes the tests and is corroborated, provisionally accepting it as part of the contents of science.

If it is true that not every falsifiable hypothesis is automatically logged into the annals of scientific knowledge, then another of Miller's remarks also muddies the waters. He says that "if [a hypothesis] passes" many tests, "then nothing happens—that is to say, it is retained" in science. But again, this cannot be entirely correct. In fact, something very significant happens to a hypothesis when it passes its initial experimental tests; it is qualitatively transformed from "falsifiable, but untested" to "falsifiable, tested, and corroborated." It's ticket has been punched, and it is now a member in good standing of the contents of science.

3.B.5 Isn't Popper a Clandestine Inductivist?

Doesn't Popper implicitly require that what is true today will be true tomorrow for his program to work? No. A Critical Rationalist may well *assume* that things will be the same tomorrow as they are today[22]; it is certainly convenient to do so, and, after all, the world as we experience it is changeable, but not kaleidoscopic or random. We expect to get the same experimental result tomorrow that we got today, though we might not, and, if we don't, we'll try to find out why. Maybe we'll make a discovery when we do. The falsification program, in other words, does not *require* that nature to be predictable in order to work, and it thrives even when nature isn't predictable. This is a far cry from programs whose very foundations are sunk into inductive reasoning and which, therefore, absolutely demand uniformity in the future for their predictions to make any sense.

Inductivists insist that a hypothesis that has passed a test is thereby strengthened, though they can't explain how this can be. If Popper also prefers a tested-and-not-falsified-hypothesis—he does—the critics conclude he must be relying on induction, and, therefore, given his firm anti-inductivist stance, that he is being inconsistent, hypocritical, or ridiculous. Let's look at a textbook problem that is used to argue the point: we want to build a bridge, and we have two designs to choose from: one has been used before and is well-tested, while the other is new and has never been tested. If we imagine that the designs are hypotheses (they aren't, but that is how the argument goes), then, since Popper refuses to grant that hypotheses are made stronger with experience, the critics infer that he should refuse to choose the tested bridge design over the untested one since the "inductive" answer is so obviously the correct one.

Popper disagrees with the reasoning. He, too, prefers the tested bridge design, but not because its past performance has conferred on it an eerie power to influence the future. The fact is that we have more information about the well-tested design: it has worked well in the past and we have no reason to think it won't work in the future. Let's pause to consider what that means: you know that, now, in this moment, one bridge design has worked in the past, and you know almost nothing about an untested design. Assume, moreover, that anything whatever could happen in the future—the laws of physics could change! Anything! You don't know. In this case, even with a maximally uncertain future, what possible reason could you give for preferring the untested design over the tested one? And if you don't have a reason to choose the untested one, then you're acting, by definition, unreasonably. In other words, the challenge for reason is not, as inductivists argue, to account for *choosing* the tested design; it is to account for *not choosing* it. This point may still seem complicated, so I explore it further in the optional Section 3.C.

Now let's take a step back. Ask yourself, once you'd decided to go with the tested design, exactly how the situation would change if you also made an *inductive inference* and added the words "and I believe the design will perform well in this case?" Apart from possibly making you feel better, what concrete effect could the words have? The reliability of the design is all you really care about, and past performance is all you have to go on.

Or, if you were asked the question, "do you think this design will work in the future?" you'd answer it by comparing the past and present circumstances, the materials used, the loads predicted, the terrain, the construction techniques, etc. You'd make a guess or venture an expert opinion about what you think will happen. In the end, however, no matter how solid your evidence, how extensive your experience, or how sage your advice, you could not guarantee with 100% certainty that the design will work in the future. "Stuff" happens. The only way to know for sure is to try it and see. You might make the wrong choice, but an inductivist could do no better.

3.B.6 Philosophy of Action

Those who complain that Popper ducks the issues raised by confirmatory evidence tend to ignore the distinctions between different kinds of science. While Popper does not think that merely confirmatory evidence strengthens the case that our theory is really True, he stresses that a severely and repeatedly tested-and-not falsified theory (i.e., one that has been well-corroborated by the data) can serve as a "basis for action"; the key word being *action*. "You have to act," he says,[23] and Popper is a philosopher of action.[24] When we have to act, we are no longer discussing hypotheses in the abstract, as general explanations for some natural phenomenon; in short, we're not talking about basic science. Instead we are in the world of applied science (Chapter 4). In applied science, we are obliged to act—to build a bridge, an airplane, or a vaccine against a deadly disease; we do not have the luxury, or burden, of indefinitely continuing a research program that seeks Truth.

Having decided to act, we express our pragmatic confidence in the products of our theories by betting our lives on them: we drive over bridges and fly in planes based on the principles laid down by our best theories. This pragmatic confidence (which may be misplaced—bridges do collapse and planes do fall out of the sky!) is motivated by the demonstrated success of the products of applied science; it does not translate into a similar confidence in the ultimate Truth of our theories themselves.

3.B.7 Popper Can't Explain Why We Feel Confident in Corroborated Hypotheses

Peter Godfrey-Smith thinks that Popper's search for Truth amounts to this: scientists wander around aimlessly selecting one theory or another,[25] holding on to one for as long as it seems to work and, when it fails, haphazardly tossing it aside and grabbing at another, hoping to stumble onto the True Theory. It is "an unusual kind of search," he notes. It is an even more unusual view of science. The fact that we do not understand enough about the mind to give an account of how reasoning works does not imply that we don't reason. The search for better theories is not aimless; we seek better ones by deliberately testing the ones we have and rejecting the failures. Godfrey-Smith continues "You will eventually die . . . without knowing whether you succeeded." Moreover, "A theory that we have failed to falsify might, in fact be true. But if so we will never know this or even have a reason to increase our confidence [in it]."

Given his grim assessment of Popperian science, you might think that you're about to hear about a better way of doing things. But no. Godfrey-Smith reports that "most philosophers" do accept fallibilism, the concept that we can never be 100% certain of truth. (Apparently, Popperian or not, we are all fated to "die without knowing whether [we] have succeeded.") It is the last part of his comment, that we will never "even have a reason to increase our confidence" in our theories, that he really wants to talk about. He thinks it will be better if scientists *believe* they are on the right track, marching steadily toward better theories, whether or not they actually are.

Even if we agree (I do) that it would be good to know the causes of scientists' beliefs, why would we think that this is a matter that philosophy can settle? Complete confidence in our hypotheses would only be justified if we knew for sure that they were True, and we don't. On the other hand, seeking confidence merely for the good feeling that it affords seems pointless. What the philosophers want is a *theory of evidence* that can provide for justified confidence; a way to tell how much a confirmed prediction strengthens the hypothesis that predicted it, which, by now, you recognize as the ghost of induction back to haunt us again. We are no longer talking about approaching scientific Truth, but about approaching a state of warranted confidence. Regrettably, neither Popper nor, to be fair, philosophy in general, offers much help at present. For Popper, partial, incomplete, perhaps misleading assurances of the validity of a theory serve no purpose; the search for Truth alone is what counts. A scientist might be more emotionally attached to the corroborated theory than to an untested one, but emotional attachment is no substitute for Truth. Given that we can't know if a corroborated theory is a True one, we've got to keep going in any case.

This reasoning drives philosophers crazy—metaphorically speaking, of course. Although scientists say that they accept the proposition that all knowledge is ultimately uncertain, they do seem to be more confident in a tested hypothesis. Philosophers want to know why.

The shift from seeking true theories to seeking *confidence* in our theories is significant for several reasons. As we've noted, confidence is a psychological phenomenon. Two people looking at the same data may well come away with different degrees of confidence about it—differing opinions make for horse races, etc. Godfrey-Smith touches lightly on cognitive issues by noting that "people" make "bad logical errors," such as in the famous "selection task" of Peter Wason (we'll review a couple of Wason's reasoning tasks in Chapter 12 if you'd like to check them out now). He doesn't draw directly applicable conclusions from this, but it opens the door to the notion that limitations in human cognition might be important in areas such as judging the reliability of scientific theories.

How much confidence to place in a corroborated hypothesis also depends on the level of organization of nature that we're talking about. Although Popper does not seem to reckon with levels of organization of nature (Chapter 1, and in this chapter Section 3.C.2.), a scientific explanation at the deepest levels, say about the subatomic structure of ice, may not have any obvious implications for action; it doesn't affect whether we're going to put sand down to prevent people like Aunt Minnie from slipping and falling. At a particular level of analysis, we can put aside our skepticism and act confidently on the basis of the best corroborated hypothesis that we have. If we are required to explain the precise molecular details of her fall at a level that will satisfy advanced physics students at a top university, we are likely be much less confident in our explanation.

Eventually, the inquiry into confidence in hypotheses veers off into purely psychological or neurobiological realms: Why is anybody confident of anything? As an emotional phenomenon, the degree of confidence you feel may come down to the amount of testosterone in your prefrontal cortex.[26] As a philosophical phenomenon, the decision as to how much confidence returns to the problem of induction or, perhaps, to Bayesian statistics, which we'll take up in Chapter 6.

Indeed, you might ask if having confidence in theories is always such a good thing for a scientist. Doesn't confidence lead to bias, and isn't bias said to be bad? Skepticism is a strong antidote to overconfidence.

Godfrey-Smith remarks that scientists are unaware of Popper's position that a theory is not made stronger by passing potentially falsifying tests. He believes that if they were aware of it they'd drop Popper like a hot potato. I suspect that he's right that many scientists do not know about this consequence of Popper's doctrine; neither I nor a number of colleagues I've asked had been taught about it. I'm not so sure that it would make much difference if we had been; we'd still regard falsification as the best way to go about testing hypotheses. Scientists are, as Godfrey-Smith points out, practical people.

3.B.8 What About Holism (the Duhem-Quine Thesis)?

The Duhem-Quine thesis is a consequence of the deep implicit hypotheses (Chapter 2) that constitute the background assumptions of science. The thesis says that because every experimental hypothesis is inextricably embedded in a network of auxiliary (implicit) hypotheses, it is impossible to test a single hypothesis in isolation. When you measure the pH of a solution, you implicitly assume that numerous hypotheses about chemistry and physics, not to mention hypotheses about the technology that went into the manufacture and operation of the meter, are true. If any of them were false, then the results of testing a pH-dependent hypothesis would be in error. This problem, called *holism*, is alleged to expose a serious weakness of Conjectures and Refutations, although the holism argument itself has been criticized.[27] Let's ignore the controversy and see where the argument goes. Scientists necessarily take many things for granted in formulating and testing hypotheses (e.g., what the measuring instruments really measure, how they work, etc.). They are what I'm calling "deep implicit hypotheses." The test of the acid rain hypothesis would fail if the pH sensor didn't respond to hydrogen ions as it should and we mistakenly rejected the hypothesis. In a way, the Duhem-Quine conundrum is just another manifestation of the uncertainty that science must always cope with.

Although the challenge presented by the Duhem-Quine is genuine, the philosophical criticism stemming from it is inconsistent. Although philosophers say holism is a danger for Conjectures and Refutations, they themselves favor of some kind of hypothesis-testing process without, however, showing how to escape the dilemma that holism creates. Pigliucci's view, presented well after he has dispatched Popper's arguments, is typical: "The common thread in all science is the ability to produce and test hypotheses based on systematically collected empirical data." No word on how the common thread deals with holism.

Likewise, Cleland gives examples from the annals of science of hypotheses that were successfully tested without explaining how the tests eluded the Duhem-Quine problem.

Godfrey-Smith suggests that the best way to make progress in understanding science is to recognize that "[t]esting in science is typically an attempt to choose between rival hypotheses about the hidden structure of the world," and that we need a theory of *explanatory inference* to understand how scientists makes their choices. He then evaluates various attempts to construct such a theory and finds them all wanting, though he holds out hope that a viable Bayesian-hybrid approach may materialize (see Chapter 6 for a discussion of Bayesian methods). In particular, he is bullish on a form of explanatory inference that involves *elimination of alternatives* (called, predictably, *eliminative inference*). This is the kind of thing that Sherlock Holmes was doing when he identified the criminal by systematically ruling out all other possible suspects: if nobody else could possibly have done it, then the one left must be the guilty party. Eliminative Inference resembles Strong Inference, and Godfrey-Smith approvingly cites Platt's program in this context. On the other hand, as we saw in Chapter 2, Platt's program depends on falsification to eliminate alternative hypotheses, and Godfrey-Smith is no fan of falsification. If eliminative inference is more palatable than Conjectures and Refutations, then we need to know why and we need to know how it works. In the end, Godfrey-Smith guesses that "we may have to get used to the idea of a mixed or pluralistic theory of evidence."

In summary, if holism were the insurmountable barrier to theory refutation that these philosophers think that it is, how could scientists ever reject theories? The philosopher Sandra Harding poses precisely this question in her collection of essays on the Duhem-Quine thesis.[28] Yet almost all philosophers acknowledge that scientists do reject theories; hence, for them, the Duhem-Quine Thesis is a big problem.

3.B.9 How Do Scientists Resolve the Conundrum of the Holism?

Scientists, on the other hand, deal with the challenges posed by Duhem-Quine every day; surmounting it is baked into our bones, even if we've never heard of it. The threefold solution adopted by science is to (1) do control experiments, (2) be aware of your assumptions, and (3) use a variety of tests for each hypothesis. Science, in other words, draws on its usual battery of checks and balances—comparing control groups that are as alike as possible to the experimental group, making key assumptions explicit and testing them systematically, using several dissimilar techniques and output measures, replicating experiments in different laboratories, etc. Though none of the fixes alone is perfect, in the aggregate they ordinarily work well (moreover, as we'll see in Chapter 8, the results of aggregate testing of a hypothesis is more secure than we often recognize). As a result of these strategies scientists routinely, and successfully, test and reject hypotheses despite Duhem-Quine.

3.B.10 Are Negative Data Worthless?

Critics scoff at the Conjectures and Refutations program because it generates "negative data" which, evidently, they deem to be virtually worthless. But one of the beauties of hypothesis-based research is that it teaches that negative results can be extremely valuable; they are what you get when you successfully test and reject a hypothesis or otherwise rule out a research dead end. When a friend of Thomas Edison's was commiserating with him over his apparent lack of results in finding the right filament material for his newly invented electric light bulb, Edison responded: "Results! Why, man, I have gotten a lot of results! I know several thousand things that won't work."[29] If you know which ideas are not right, then you can get on with trying to find the right ones.

Do journals publish negative data? While there are exceptions, as a rule they don't, which does hinder science. The *Reproducibility Crisis* (Chapter 7) has provoked discussion about removing the stigma associated with negative data, including ways of making such data more respectable and widely available.[30] These efforts are highly commendable, even if a listing of "things that won't work" may only be of limited use.

Still, it is undeniable that a lack of respect for negative data is currently a drag on scientific progress. I suggest that the phrase "negative data" itself is part of the problem. For one thing, there is the occasional connotation that negative data are uninformative data. In various contexts, "negative data" can refer to several kinds of outcome: the results of a rigorous, well-designed, and carefully conducted experiment that falsifies a hypothesis; a failure to replicate a previous finding; or a thoroughly inconclusive experimental outcome (e.g., the experiment was poorly designed, measurements were invalid, etc.). The value of information provided by these three classes is obviously not the same, yet the term "negative data" is applied to all of them. Unfortunately, the term is so customary that it is unlikely to go away any time soon, so one step forward might be to define it more specifically: Perhaps "negative data" could be reserved for results that are genuinely informative (i.e., they falsify a hypothesis or demonstrate a failure to replicate a previous finding). Test results that are inconclusive because the experimenters didn't execute the experiments well or because of confounding effects, etc. are uninformative or, at best, *weakly informative*; they probably shouldn't be regarded as "data" at all. You might refer to them for clues about what not to do or how the experimental design could be improved. But it may be too much to hope that recasting the problem of "negative data" in less ambiguous terms will make it disappear, a topic we'll take up in Chapter 11 when we consider the kinds of biases that affect scientists' behavior.

Another, and I think better, strategy for doing away with at least part of the problem of negative data is the one that follows from the central topic of this book: namely, to make the process of hypothesis testing explicit and overt. State hypotheses and predictions explicitly; relate results directly to hypotheses and conclusions; and abolish the notion that when we rigorously test and reject a hypothesis, we are generating "negative data" of any kind. The steps would help make the point that in testing and rejecting a solid hypothesis, you are making a *positive* contribution to our fund of knowledge. If the perceptions and attitudes of reviewers, journal editors, and members of the scientific community could be coaxed to shift in this direction, many of the difficulties associated with negative data will in fact go away.

3.B.11 Not All Science Is Hypothesis-Based

Is the Conjectures and Refutations program somehow invalidated if scientists engage in science that does not involve explicit hypothesis testing (we can't avoid the deep implicit ones)? It's true that science does take different forms; indeed, there are kinds of sciences that don't require the hypothesis or falsification testing, and we will discuss examples in Chapters 4 and 10. Popper did not discuss non–hypothesis-based science much, and perhaps he can be faulted for the omission; however, the fact that not all science is based on hypothesis testing doesn't negate the value of Conjectures and Refutations.

3.C Further Criticisms of Popper and Platt (Optional)

Up to this point, I've addressed the specific philosophical objections to Conjectures and Refutations that I think are the most germane to scientific thinking. A few additional concerns that I'll consider have a philosophy of science slant that not every lab scientist will find necessary, and so, depending on your interests, you may want to skip ahead to Chapter 4.

3.C.1 Popper Versus Platt?

As I noted, the philosopher, Peter Godfrey-Smith favors Platt's program over Popper's. Although Godfrey-Smith has objections to falsification, he suggests that, if we can't really rule out alternative hypotheses, *a la* Sherlock Holmes, maybe we can at least identify most of them as being very unlikely. Or perhaps we can find "partial support" for some of the remainder. As we noted earlier, partial support is not really what science is after—the Truth is. Nevertheless, we can ask how the search for partial support might be carried out.

One difficulty for John Platt's eliminative approach to hypothesis testing is that, unlike the relatively small group of suspects that Sherlock Holmes typically had to process, scientists have an unknown number of alternative hypotheses to sort through. Platt himself does not say much about the universe of alternatives—he alludes to "all" of them but doesn't worry about how many there might be. His program is open-ended and easily accommodates new ones as they crop up.

Theoretically speaking, things are not so simple. In principle, scientists have to confront an infinite number of alternative hypotheses that could account for a phenomenon. Godfrey-Smith agrees, but expresses the hope that "maybe there are ways" of getting down to a manageable number of them. You might wonder, if we are going to place our bets on hope, why not hope that scientists can find manageable ways of falsifying hypotheses? Or acknowledge that, to a large extent (Section 3.B.6), that's what they already do? In short, his stance—pro-Platt, anti-Popper—seems arbitrary; he rejects falsification without explaining how to eliminate hypotheses.

3.C.2 Does Taking Rational Practical Action Demand Inductivist Justification?

When philosophers criticize Conjectures and Refutations they do it on the basis of rational argument, but, as we'll see in Chapter 11, it is not always obvious how we are to understand "rational." In Chapter 2 we considered two standards for rationality: one which is consistent with the narrow demands of probability and logic, and one that is better adapted to the needs of our ancient hominid ancestors. For Popper, the essence of scientific rationality is embodied by Conjectures and Refutations. If you believe that science cannot achieve absolutely unchallengeable results, then the most rational state of mind is the "readiness to accept criticism," and the soundest hypotheses are those that have been subjected to the severest criticisms and tests and therefore are, to the very best of your knowledge, true. Acting rationally means to be guided by true statements.

The philosopher Wesley Salmon[31] thinks it is irrational to base practical actions on this reasoning because it omits what he believes is a mandatory link between a corroborated hypothesis and a predictable outcome: *induction*. He says that you have to make inductive predictions from theories in order to act rationally. Unless a hypothesis has received, in addition to extensive corroborating evidence, the blessing of inductive confirmation, you can't rationally choose it. He concludes that induction is inescapable and therefore that Popper's thinking contains hidden inductive elements. Does it?

Popper not only dismisses the thought that "induction" can add anything, but he *denies that predictions are the basis for taking action*. The denial that predictions can constitute a basis for action will strike some people as counterintuitive, and I'll try to make sense of it. In essence, Popper wants to ground practical actions on what we *know* to be true; *not* what we predict to be true.

Popper is a realist, which, you'll recall, is a technical term for someone who believes in the existence of an external world (external to our own minds) that is "regular," meaning that it is governed by physical laws even if we don't know what they are.[32] We cannot prove that realism is true: Popper accepts it for "metaphysical[33] reasons, but he's not alone. Every scientist accepts realism implicitly or explicitly because, well, we have nothing to go on if we don't: if nature doesn't exist, you can't study it. Popper also understands "True" to have the same sort of status as "Regular," and both apply to space and time beyond the here and now. Starting from these metaphysical elements, he intends Conjectures and Refutations to be a *deductive* program.

But because realism assumes that nature is regular (i.e., not utterly chaotic), Salmon believes that it incorporates a "version of a principle of the uniformity of nature"; ergo, the basis for valid induction is suddenly back in play. Therefore, since Popper accepts realism, he must also accept the tenets of induction and, *ipso facto*, according to Salmon, Popper is either a closet inductivist or an irrational man. Salmon believes that he has shown that "pure deductivism could not do justice to the problem of rational prediction in the contexts of practical decision-making."[34] "Pure deductivism," of course, was never at issue since Popper explicitly accepted realism on nondeductive grounds.

As an alternative to Salmon's critique of Popper, we could argue that Salmon has simply defined science as inherently inductivist because it depends on realism, and, he believes, realism *implies* inductivism. In this case, his argument would be a tautology which, you'll recall, tells us about language and logic but not about the world.

This is the point in the debate where the vast majority of scientists in the audience would quietly drift away back to their labs and do something useful, but as these issues arise in the corroboration-confirmation dilemma (Chapter 2.G.5) that is directly relevant to scientific thinking, we'll stick with it a bit more. There are two big questions still left to address: Does the past success of a hypothesis somehow strengthen it? What is it about the past success of a hypothesis that let's us base practical actions (e.g., technology) on it, if we don't believe in the power of "inductive reasoning?"

3.C.2.a The Truth of a Corroborated Hypothesis

Let's look again at the textbook issue of the bridge designs—you've got to choose between a well-tested one and an untested one. Popper says that the validity of a well-tested hypothesis is not strengthened when it passes experimental tests, yet he still thinks you should choose the best-corroborated hypothesis. For this, philosophers accuse him of irrationality. But Popper argues that, because you know more about the best-tested design, there is no policy more rational than choosing it. Salmon counters that, without accepting the proposition that the design has been strengthened by *confirmation*, there is no policy more rational than the untested design, and thus there is kind of "tie" between the policies—a stand-off and no rational way to choose between them.

To break the tie, we need to look at Popper's hypothesis-vetting program more closely. When you propose a hypothesis, you are implicitly proposing that it is *true*. If you have tested and not refuted it, you have found absolutely no reason to change your mind. It's still true. If you are forced to make a choice between the best-corroborated design or an entirely untested one, you have only their past records to go on. You must choose based on what you know and *not* on what you "predict." You'll build the bridge according to what you know about. Predictions and reliability are concepts that go beyond what you know now, to the future which is (wait for it) uncertain, so you can't know about it. Popper is not an inductivist of any kind.

Popper's argument is admittedly very hard to swallow at first, in part because induction comes so naturally to us (see Chapter 11) and because we immediately imagine the social or political fallout that would occur if a bridge builder publicly admitted that he was not 100% certain the bridge would work perfectly. And philosophical critics say that Popper's argument is unacceptable: that without additional inductive assurances that "it will work," they themselves could not make a choice. We protest, however, that meaningful assurances of any kind would have to be based on actual evidence from past experience with the design. If an assurance were pure opinion entirely divorced from genuine evidence), then it shouldn't enter into the decision process.

3.C.2.b Hypotheses as a Basis for Practical Action

A major problem is that, as humans, our choices are generally more or less colored by emotion, and we have trouble separating the emotional and cognitive parts of our decisions. For instance, another philosophical critic who chides

Popper for not stating that hypotheses are strengthened by experience claims that Popper gives us no reason to believe in the well-corroborated law of gravity. If you were told about an untested hypothesis that says you'll be OK if you jump off the Leaning Tower of Pisa, he says, we wouldn't know what to do without inductive reasoning, Actually, even preverbal toddlers who have never experienced a fall are not tempted to test the law: *ecological rationality* kicks in to save them. Our healthy fear of heights is embedded in our genome. Confusingly, so is *our need to justify our behavior*, so when we instinctively decide against testing the hypothesis of gravity with our own bodies, we create a reason—induction!—for not doing so.

Even a coldly analytical scientist wouldn't be tempted to jump and wouldn't have to rely on inductive reasoning in making her decision. Why? As a scientist, she would understand the law of gravity has been postulated to be a true statement (within the terrestrial realm of falling bodies), and, as it has been well-corroborated by past experience, she has no reason to think that it isn't true. Since the law of gravity applied to falling human bodies predicts a bad outcome if she were to jump, she'd be unlikely to do so. Enumerative induction plays no part in her reasoning.

Philosophers such as Salmon think the difficulty is in finding a link between the search for valid hypotheses—what I've been calling "basic science"—and technology, "applied science." He is correct in thinking that you behave differently in the two realms, but he mischaracterizes the practical problem. In the applied science case of bridge building that we've been considering, you face a forced choice—you must build a bridge and have only two alternatives. If you don't choose one design, you must choose the other one. Hence, calling it a "tie" between them, as Salmon does, is not an option. A philosophical analysis that cannot decide between tested and untested designs is irrelevant to the tangible needs of applied science.

3.C.2.c How Do Popper's Methods Mesh with "Levels" of Scientific Explanation?

It may have struck you that the concept of "levels" of scientific analysis (Chapter 1) pose a potentially serious challenge for Karl Popper's philosophy. How can we square the concept of "Truth" in science with the possibility that questions have distinct answers at several levels of inquiry? Is there a whole battery of truths, one for each level, and, if so, where does that leave the overarching search for Truth and the falsification method?

Let's start with falsification: Can a hypothesis be falsified at one level and not at another? Obviously, in a practical sense it can: we've noted the case of Newton's law of gravity. At the deepest levels of understanding, the law has been falsified by the revelations of modern physics beginning with Einstein. But Newton's law still works well for countless purposes: if you want to launch a satellite into orbit around earth, Newton's law is what you need. It works, so it must be true[35] at that level. Has it been falsified by modern physics or not?

Once again, the answer depends on whether you're doing basic (Popper's word is "pure") science or applied science. When the Nobel Prize-winning physicist Werner Heisenberg remarked that "we do not say [Classical] mechanics is false," but rather that "Classical mechanics . . . is everywhere exactly 'right' where its concepts can be applied."[36] At first, Popper bristles. Among other sins, he feels that Heisenberg's reasoning would be a disaster for pure science because you could always turn to "ad hoc for rescuing a physical theory that was in danger of being falsified"; for example, you could decide that the theory was not actually false, just inapplicable to the case in question. Nevertheless, at the end of his argument Popper, I think grudgingly, allows Heisenberg's reasoning, which is "like that of applied science" when it comes to the "success of applications." Even though he still doesn't like Heisenberg's approach, as long as it is confined to the domain of "application," away from the search for Truth, he can live with it.

In a way, the dispute about levels of science is linked to the inadequacy of pure "prediction" to act as a test of a theory. A bad theory might happen to make a good prediction, and so success in predictions doesn't guarantee that a theory is correct. Meanwhile, in the real world of action, good predictions can undeniably be useful wherever they come from, even rejected hypotheses. Our theory of the slipperiness of ice may be ever so flawed, but as it predicts that Aunt Minnie will be safer if we put sand down on the ice, the hypothesis is useful. When it's time to take action, Popper allows a standard that appeals to the utility of a corroborated hypothesis—"try it and see"—in place of severe testing, and his pragmatic stance connects his thinking to levels of science.

3.C.2.d Does the Observation of Black Swans Prove Anything?

Popper's famous, if overly simplistic, argument against induction is that, although observing any number of white swans cannot prove the truth of the statement, "all swans are white," you can disprove (i.e., falsify) that statement by observing one black swan (repeatedly with proper controls, etc.). Deborah Mayo, who feels that Popper "came up short" in his logical analyses, argues that that we can extract more information from observations of black swans and analogous falsifying tests of hypotheses than the mere fact that our hypothesis was incorrect.

Assume, she says, that you're testing the hypothesis, H_1, "all swans are white," and an alternative hypothesis, H_2, "some swans are not white." Mayo asserts that observing a black swan not only falsifies H_1 but "proves" H_2,[37] deftly turning Popper's reasoning inside out: if we can't *prove* a hypothesis by obtaining confirming evidence for it, can we at least prove a contrary hypothesis by disconfirming the first one? If her argument were universally applicable, it would knock the foundations right out from under Popper's philosophy! His main premise would be wrong—science could prove some hypotheses after all, we'd have more than corroborated hypotheses, etc.,—and the rationale for Conjectures and Refutations would collapse like a house of cards.

However, before giving up on Popper, we need to examine Mayo's argument. Superficially, it seems valid: the existence of a black swan would certainly mean that some (in logic, "some" means "at least one") swans are not white. The question is, can this conclusion benefit scientists generally? I am afraid that it can't. The statement about swans as it stands is a description (a "there is" statement in Popper's terms) not a hypothesis, and Mayo's statistical argument depends on its being a description. However, reasoning about descriptions doesn't generalize to scientific hypotheses, which are explanations. If you interpret the statement about swans as a hypothesis, then her argument is no longer valid. We'll examine both points.

Let's assume that the implicit assumption about classifying swans that both Popper and Mayo make is valid; namely, you can classify swans unambiguously as "white" or "nonwhite" and that these two categories are mutually exclusive and exhaustive: each swan can be put into only one color bin, and all swans can go into one of the two bins. Mayo's argument is sound: if you can falsify H_1 ("all swans are white") for one swan, you simultaneously prove H_2 ("not all swans are white"). So far so good, but what does this argument hold for science?

Scientific hypotheses are explanations for phenomena, and we never know if any two genuine scientific hypotheses are mutually exclusive and exhaustive as possible explanations for a phenomenon. Hence, if you're a fallibilist, then even falsifying one hypothesis does not entitle you to conclude that you've "proved" another one; there will always be other possible explanations and unforeseen complexities.

Let's look at the swan problem as a scientific hypothesis. We first have to ask what swans are. After all, if the large black bird is not a swan, all bets are off; finding one tells you nothing about swans. If you guessed that all things called swans are members of the same species of bird, you'd be wrong; there are six or seven[38] species in the genus *Cygnus*. A familiar white swan would be the trumpeter swan, *Cygnus buccinator*, while the Australian black swan is *Cygnus atratus*. So white swans and black swans are not classified as members of the same species! We reevaluate the hypothesis: Was it about swans as a species or swans as a genus? (Actually, white and black swans are classed as members of different "subgenuses," but never mind.)

You can't begin to answer these questions without knowing precisely what "species" and "genus" mean. At this point, you might be surprised to learn that these terms aren't unambiguously defined despite decades of attempts to do so. A variety of properties ranging from reproductive compatibility to behavioral traits, ecological niches, and genetic make-up have all failed to yield a consistent, uniform picture of what a species is. And the notion of "genus" is fuzzier still; there is an active debate as to whether a genus is a natural category at all. (The discussions of the biological complexities involved by, for example, Ernst Mayr[39] and Peter Godfrey-Smith,[40] are fascinating and well worth reading.) Nevertheless, without going further, we can see that falsifying, "All swans are white" is not trivial, and it does not entitle you to conclude that you've proved anything.

3.D Coda

This chapter addressed objections to the program of Conjectures and Refutations, espoused in slightly different ways by Popper and Platt. Several objections dealt with questions about falsification—chiefly, about its significance and conclusiveness. We reviewed the distinct functions that falsification has in the methods and contents of science. Popper uses the concept of corroboration to refer to tested-and-not-falsified hypotheses, and we examined the differences between corroboration and confirmation (or verification) in the contexts of basic science and applied science. Corroboration is important primarily in applied science as it provides the basis for action that distinguishes Popper's philosophy from other philosophies of science. Despite the accusations of his critics, Popper is not an inductivist, and his philosophy does not assume or rely on induction. We saw that scientists deal pragmatically with the logical complexities of the Duhem-Quine Thesis that posits that the intricacies of holism, the interlocking nature of scientific hypothesis in a network, precludes the testing of any one hypothesis in isolation. Science manages to progress by doing control experiments and taking other measures to isolate and test individual hypotheses. Finally, we considered "negative data" as related to hypothesis testing and falsification, and we evaluated a few technical concerns that philosophers have raised about Conjectures and Refutations. The chapter concludes that, despite occasional shortcomings, the hypothesis-testing programs of Popper and Platt remain viable.

The material in Chapters 1–3 defines the modern scientific hypothesis that forms the foundation for the rest of the book. The next chapter picks up a theme that appeared at the end of this chapter: that is, that there are different "kinds" of science and that a major distinction involves the extent to which the different kinds depend on the scientific hypothesis.

Notes

1. Carol Cleland, "Historical Science, Experimental Science, and the Scientific Method," *Geological Society of America* 29:987–990, 2001.
2. Peter Godfrey-Smith, *Theory and Reality: An Introduction to the Philosophy of Science* (Chicago: University of Chicago Press; 2003), pp. 57–74.
3. Samir Okasha, *Philosophy of Science: A Very Short Introduction* (Oxford: Oxford University Press; 2002).
4. Massimo Pigliucci, *Nonsense on Stilts: How to Tell Science from Bunk* (Chicago: University of Chicago Press; 2010).
5. Godfrey-Smith, *Theory*, pp. 57–58.
6. Pigliucci, *Nonsense*, p. 302.
7. See, e.g., Godfrey-Smith, *Theory*, chapters 4 and 10.
8. Okasha, *Philosophy*, p. 23.
9. David J. Glass, *Experimental Design for Biologists*, 2nd Ed. (New York: Cold Spring Harbor: Cold Spring Harbor Press; 2014).
10. Thomas Kuhn, *The Structure of Scientific Revolutions*, 2nd ed. (Chicago: University of Chicago Press; 1970).
11. Godfrey-Smith, *Theory*, pp. 57–62: "puzzling."
12. "Mob psychology," Imre Lakatos, quoted in Godfrey-Smith, *Theory*, p. 103.
13. See Godfrey-Smith, *Theory*, pp. 150–154, for discussion of Quine's views and the issue of uniquely philosophical answers to questions of scientific belief.
14. Scientific consensus on human activity being the main cause of climate change; see http://iopscience.iop.org/article/10.1088/1748-9326/11/4/048002; Naomi Oreskes and Erik Conway, *The Merchants of Doubt: How a Handful of Scientists Obscured the Truth on Issues from Tobacco Smoke to Global Warming* (New York: Bloomsbury Press; 2011), epilogue.
15. Kuhn, *The Structure of Scientific Revolutions*
16. Max Planck: "one funeral at a time." Visit https://en.wikiquote.org/wiki/Max_Planck.
17. David A. Miller, *Critical Rationalism: A Restatement and Defense* (Chicago: Open Court; 1994), p. 7; see also Karl Popper, *The Logic of Scientific Discovery* (New York: Routledge; 2002), p. 17.
18. Deborah G. Mayo, *Statistical Inference as Severe Testing* (Cambridge: Cambridge University Press; 2018). Mayo believes that Popper errs by not specifying exactly what he means by "severe" testing of hypotheses, which leads him to an untenable "demarcation problem" because it focuses his attention on whether or not a theory, rather than the methods of evaluating it, are "unscientific." Her argument gives rise to a straw man about the likelihood of cluttering of science with "nutty theories."
19. Irene Lacal and Rossella Ventura, "Epigenetic Inheritance: Concepts, Mechanisms and Perspectives," *Frontiers in Molecular Neuroscience*, 11:article 292, 2018.
20. K. Laland, T. Uller, M. Feldman, K. Sterelny, G. B. Müller, A. Moczek, et al., "Does Evolutionary Theory Need a Rethink? *Nature* 514:161–164, 2014.
21. Miller, *Critical*, p. 7.
22. Ibid., pp. 38–45.

23. Malachi H. Hacohen, *Karl Popper: The Formative Years, 1902*–1945 (New York: Cambridge University Press; 2000).
24. Bryan Magee, *Philosophy of the Real World: An Introduction to Karl Popper* (La Salle, IL: Open Court; 1985), pp. 1–5.
25. Godfrey-Smith, *Theory*, pp. 60–61.
26. Robert M. Sapolsky, *Behave: The Biology of Humans at Our Best and Worst* (New York: Penguin Press; 2017).
27. Adolph Grübaum, "The Duhemian Argument," reprinted in *Can Theories Be Refuted: Essays on the Duhem-Quine Thesis*, Sandra Harding (Ed.) (Boston: D. Reidel; 1976).
28. Sandra Harding (Ed.), *Can Theories Be Refuted?: Essays on the Duhem-Quine Thesis* (Boston: D. Reidel; 1976).
29. Thomas Edison, quoted in Frank Lewis Dyer and Thomas Commerford Martin, *Edison: His Life and Inventions* (New York: Harper & Brothers; 1910), volume 2 of 2, chapter 24: "Edison's Method in Inventing," pp. 615–616. The anecdote is due to a long-time associate of Edison's named Walter S. Mallory. Cited in http://quoteinvestigator.com/2012/07/31/edison-lot-results/.
30. Oswald Steward and Ruth Balice-Gordon, "Rigor or Mortis: Best Practices for Preclinical Research in Neuroscience," *Neuron* 84:572–581, 2014. S. C. Landis, S. G. Amara, K. Asadullah, C. P. Austin, R. Blumenstein, E. W. Bradley, et al., "A Call for Transparent Reporting to Optimize the Predictive Value of Preclinical Research," *Nature* 490:187–191, 2012.
31. Wesley C. Salmon, "Rational Prediction," *The British Journal for the Philosophy of Science* 32:115–125, 1981.
32. In some contexts, "realism" implies more than just the existence of a regular external world. Then, a *scientific realist* is someone who believes that the unobservable entities that scientists talk about—e.g., electrons, quarks—truly exist, whereas *scientific anti-realists* use concepts of unobservable entities as "convenient fictions" that enable theories to make accurate predictions. Anti-realists argue that we shouldn't worry too much about whether the unobservables actually exist or not. See Note 3, Okasha, *Philosophy*, pp. 58–76.
33. Karl Popper, *The Logic of Scientific Discovery* (London: Routledge Classics; 1959/2002), p. 250.
34. Salmon, "Rational Prediction," p. 125.
35. Philosophers and physicists tend to see the problem that I'm calling "levels" of science in terms of the concept of *limiting cases*; see, e.g., https://plato.stanford.edu/entries/physics-interrelate/#PhysSensRedu. Roughly speaking, when one theory, T_2, subsumes another, T_1, and T_2 explains everything that T_1 explains and goes beyond it to account for phenomena that T_1 cannot explain, then T_1 is said to be a limiting case of T_2. So when the theory of quantum mechanics is applied to ordinary earthly masses and speeds, it predicts the same phenomena that Newton's mechanics predicts, so Newton's theory is a limiting case of quantum mechanics. Or, for a less rigorous biological example, take the *Neuron Doctrine* of Ramón y Cajal, which says that neurons are discrete, independent elements. Indeed, if you look at neurons

through a conventional light microscope, then each one does seem to be an isolated individual. However, if you use an electron microscope or modern, computation-intensive, super-resolution optical methods you see that neurons are highly interconnected at the molecular level. I think it is fair to conclude that the Neuron Doctrine is a limiting case of the modern understanding (not yet a theory, I'm afraid) of neurons. Thus far the Neuron Doctrine provides a reliable framework for addressing problems in neuroscience, but already new higher resolution observations show that its simple framework will need to be replaced by a more powerful hypothesis.

36. Karl Popper, *Conjectures and Refutations* (New York: Routledge; 2002), p. 152.
37. Mayo, *Statistical Inference*, Excursion III, Tour II note 3.
38. What is a swan? https://en.wikipedia.org/wiki/Swan. The indication that there are "six or seven" species of swan nicely makes the point here; evidently there is uncertainty about how many swan species there are.
39. Ernst Mayr, *Towards a New Philosophy of Biology: Observations of an Evolutionist* (Cambridge, MA: Harvard University Press; 1988).
40. Peter Godfrey-Smith, *Philosophy of Biology* (Princeton, NJ: Princeton University Press; 2011). Godfrey-Smith (*Philosophy, p.* 107) also emphasizes that, although he is uncertain as to whether the concept of "species" represents a "real unit in the natural world," "species talk can be useful in biology."

4
Kinds of Science

4.A Introduction

There are two major hurdles to understanding the place of the hypothesis in science: one is that "hypothesis" has many meanings, and the second is that "science" has many meanings. I've reviewed the hypothesis and settled on the conceptual framework advanced by Karl Popper, but I haven't said much about science itself. Is it one enterprise or many? And, if there is more than one "kind" of science, how are the different kinds related to the hypothesis? Popper largely skirted the matter, which has gained prominence recently as huge areas of science have opened up that don't depend on explicit hypothesis testing. This issue frequently is hidden in discussions of science and, because it is unacknowledged, it can cause problems in communication. Hypothesis testing coexists peacefully with other modes of conducting science, but you wouldn't always know it from the way people sometimes talk about them. Before getting into that part of the debate, we can begin with the matter of whether science is one unified activity or a group of them.

4.A The Unity of Science

The vision of all scientific knowledge as forming a grand, coherent whole is ancient. The biologist E. O. Wilson dreams of a glorious unification—his word is *consilience*[1]—of all knowledge in the sciences and the humanities. Wilson called it the "Ionian Enchantment," after Thales of Miletus, a Greek philosopher from Ionia, who lived about 2,500 years ago and was perhaps the first to engage in scientific thinking along these lines. Thales thought that we could understand the world in empirical, rather than mythological terms; he proposed, for example, that water was the building block of all matter. While true consilience is an amazing image, it seems overly ambitious—it is hard to believe that the great diversity of human knowledge will come together as a set of related scientific concepts.[2]

4.A.1 Science and Nonscience: The Demarcation Problem

If there are limits to the kinds of knowledge that science can accommodate, how should we distinguish what is in its bailiwick from what is outside? Drawing the line between science and nonscience is the famous *demarcation problem*[3] (Chapter 2). Karl Popper named the problem and argued that the distinction between science and nonscience should be made according to the criterion of *falsifiability*. A scientific statement can, in principle, be empirically tested in such a way that one conceivable test outcome would show that the statement is false. In other words, scientific statements are open to the *possibility of disproof*. We accept that, since all knowledge is uncertain, falsifiability itself is also provisional, but we do the best we can. By saying that hypotheses are testable in principle, we mean that we might not be able to test them directly. "Water is formed by chemical reactions on the dark side of the moon" is a scientific statement; we can imagine doable tests that would falsify it. The philosophical notion of *solipsism* (i.e., that there is no objective, physical reality outside of my mind or, maybe, outside of your mind) is an example of a nonfalsifiable idea.

Nonfalsifiable statements, though not scientific, might stimulate scientific inquiries, or they might be meaningful in another domain of human experience; they are not necessarily "meaningless," as the Logical Positivist philosophers said they were. Popper thought such statements were nonscientific, not nonsensical. He accepted that certain metaphysical concepts are inescapable, even in philosophy and science; we've said that we accept the principle of realism, that the external world does exist, although we can't prove or falsify it.

Even with the standard of falsification as a guide, there are "gray areas."[4] We accept that "science is characterized by a fuzzy borderline."[5] In some instances, the testability of hypotheses is highly problematical, such as "the Astonishing Hypothesis"[6] that human consciousness is created by the brain or the "Super Strings"[7] that modern theoretical physics suggests are the most elemental building blocks of the universe. We encounter fuzziness when we try to tell science, such as evolution, apart from "almost-science" such as the search for extraterrestrial intelligence (SETI).[8] And fuzziness is present when we try to separate kinds of science from each other.

Fuzziness is unacceptable to certain philosophers who reject falsifiability as a standard for demarcation and insist on rigid distinctions between categories. Yet fuzziness at the borders does not mean there are no borders. It is good to distinguish between day and night, even though we can't pinpoint when the change-over happens. Similarly, if you rely on falsifiability as a standard for telling whether you're talking about science or something else, you'll usually be on safe ground. In any case, this does not answer the question of whether, within the wide boundaries established by falsifiability, science is a unified intellectual realm or a disorganized jumble of unrelated fields. Are we talking about *science* or *sciences*?

Why should it matter? Scientists go about their jobs in a variety of ways and have a variety of goals. If there were a fundamental pluralism of sciences, we could be faced with incompatible systems of thought, of ways of comprehending the world. A science unified by a set of shared premises, values, goals, and general operating principles would make for a more harmonious view of nature and a more straightforward task of understanding it than if science were disunified. The *unity of science* question has consumed much intellectual energy, and philosophers have identified many possible unities of science[9] without arriving at a definitive answer. We'll start by considering two alternative ways of thinking about the unity of science.

4.A.2 An Untidy Account?

An extreme interpretation of unity is that science aims to wrap up everything in one "tidy account."[10] Probably no one who knows anything about science would say that it is "tidy"—much of it is messy business even on a good day—but is it one account? Visions as grand as Wilson's are, in one way or another, forms of *reductionism*: the idea that we can understand a complex subject in terms of its most basic constituents. "Quarks" are subatomic particles that are realities for both physicists and psychologists, but physicists can *use* the concept of quarks to achieve their goals of scientific understanding and psychologists can't, in spite of the fact that people are just complicated arrangements of atoms and subatomic particles. The most "greedy"[11] or "uncompromising"[12] forms of reductionism can't solve the unity of science problem because not all facts are equally useful to all sciences. Some complexity is *irreducible*: we cannot comprehend why a computer gives the correct answer to a math problem based on our knowledge of the computer's electronics alone because math is a product of the mind and culture, and physics cannot explain them.

One interpretation is that science is unified by its major objectives (explanation and understanding of nature) and values (empirical testing, reproducibility, etc.) and that different branches of science have much in common from this perspective. This is sometimes called a *container* model of unity[13]: all forms of inquiry that share these objectives and values are within the container.

Another way of viewing the relationships among different sciences is *the integrated causal model.*

4.A.2 The Integrated Causal Model

Evolutionary psychologists Leda Cosmides and John Tooby[14] were concerned about the disarray and lack of progress that plagued the social sciences and that finally broke out in the "Science Wars."[15,16] The Science Wars represented a perceived split between culture and science that divided academia during the 1990s, and, as part of a plan to restore order, Cosmides and Tooby proposed that science is unified via a framework that they called *conceptual integration*—a consistency of principles throughout science. Their Integrated Causal Model holds that one branch of science cannot be based on principles that are flatly contradicted or forbidden by well-established knowledge developed in another branch. The natural sciences are already conceptually integrated: for example, although biology is separate from chemistry and physics, biology is compatible with chemistry and physics. Yet biology cannot be predicted from or entirely explained by chemistry and physics; conceptual integration is not a front for a reductionist argument. According to the Integrated Causal Model, psychology and the social sciences should be compatible with each other and with the natural sciences; psychology should not, for example, create hypotheses of human behavior based on cellular mechanisms of learning that neuroscience has ruled out. Sociologists should not erect theories of group interactions that depend on interpretations of individual psychology that psychologists can show are not true. Within the Integrated Causal Model, each branch of science retains its identity, message, and mission because, at each level of organization of the world, complexities *emerge* that cannot be explained in simpler terms. We cannot explain traffic jams by studying the behavior of individual drivers; the jams emerge from countless subtle and collective interactions among individuals that occur spontaneously and unpredictably.

The exchange should go both ways because science and culture are produced by the human mind. Evolutionary psychology approaches the study of the mind in the context of its evolutionary history. The central hypothesis of evolutionary psychology is that our present mental capabilities owe more to adaptations acquired during the 2 million years that we spent as Pleistocene Age hunter-gatherers than to changes that have occurred during the few thousand years since the dawn of civilization. From this vantage point, psychological investigations can lead to insights into such socially meaningful phenomena as male jealousy,[17] the perceptual organization of color,[18] and human preferences for certain landscapes.[19]

Since the mind is shaped by its built-in information processing mechanisms, cognitive science could conceivably discover principles that limit the intellectual constructs of physicists or help us understand cognitive phenomena, such as biases, that confront science. Anthropologists could investigate how earth scientists arrive at consensus decisions about global warming. Or it might be that problems such as human consciousness remain forever beyond our grasp because our ancestors' survival never depended on the type of thought processes that would be required to comprehend consciousness.

The Integrated Causal Model readily accommodates the cross-pollination and hybridization that creates new scientific fields—"socio-biology," "psycho-genetics," "geo-chemistry"—and makes fixed classifications of individual sciences obsolete. It sees science as being cohesive, rather a splintering collection of unrelated fields. Instead of policing or trying to legislate exactly what makes up one or the other kind of science—is it more "socio" or more "biology"?—the Integrated Causal Model accommodates all of them and is compatible with the venerable picture of scientific pluralism as branches on a tree. Branches, though separate organizational elements, are joined in the tree trunk. The names of the branches—"chemistry," "neuroscience," "psychology," etc.—are convenient labels, not signs of deep incompatibility.

The container and connectionist models of scientific unity are not mutually exclusive. Both recognize that relationships exist among different branches of scientific fields and that all fields of science are part of one giant endeavor. Another tie that binds fields of science together is the *methodological unity* that Karl Popper proposed.

4.A.4 Methodological Unity

We've seen that Popper's Conjectures and Refutations is a program of *trial and error*; it is a formal way of learning from our mistakes. Proposing, testing, falsifying, and inventing better hypotheses is how science grows and tends to its knowledge base. For Popper, science is held together by this *methodological unity*,[20] which is founded on falsification. At first glance, falsification may seem to be an element of narrowly defining characteristics of a container model, but it is broadly applicable to all of science.

To appreciate the scope of Popper's methodological unity, we need to keep in mind that falsification plays two separate roles in science: it is the key to the overarching concept of demarcation, which is how we distinguish science from non-science; and it is the centerpiece of Conjectures and Refutations, which is how we conduct certain kinds of science on a daily basis. A statement in a particular field is scientific if it can be falsified (demarcation), even if the scientists in that field are not always engaged in testing hypotheses (conduct). Imagine a biologist out counting monarch butterflies in the American Southwest. Everything about her field of study, entomology, passes the falsification test for demarcation, but she might be occupied in taking a butterfly census, gathering data, not testing a hypothesis. Thus what she's doing at the moment is not within Popper's definition of methodological unity of science as it relates to her conduct. Popper does not emphasize the distinction between the two roles of falsification, but it is important when we come to distinguishing among different branches of science.

Popper took it for granted that science would constantly discover new phenomena to investigate. He "readily admits that only observation can give us knowledge concerning facts, but this knowledge does not establish the truth of any scientific statement." Plain observations are "there-is" statements about the world (e.g., a physical object exists in such-and-such a place at such-and-such a time). Once we have observations, then we can invent and test hypotheses to explain them.

In summary, although Popper proposed that falsification is the essence of the methodological unity of science, he recognized that there are clearly scientific activities that do not adhere to the process of hypothesis testing at all times. Being mindful of the two roles of falsification makes sense of the differences between kinds of science while simultaneously seeing them as part of the unification of science.

4.B Kinds of Science: Subject Matter

Thinkers have long been divided into "lumpers" and "splitters," people who find similarities and group things together and people who focus on distinctions and divide things into categories.[21] Lumping sciences together by identifying commonalities or relationships, as we've been doing, only gets us so far; we'll have to do some splitting if we are to understand the roles of the hypothesis in science. Naming branches of science helps divide university faculty members into departments and fill up college course catalogs. It does not tell us how to distinguish among *kinds* of science or to choose between "science" and "sciences."

In the next section, I'll review a few of the conventional ways in which science has been split up and argue that they are not the primary divisions that we need to make; I'll get to more fundamental ones in Section 4.C.

4.B.1 Hard Versus Soft

Hard and soft sciences (e.g., physics and chemistry versus psychology and sociology, respectively) differ in the kinds of objects they investigate, the degrees of quantitative rigor they demand, and the predictive power of theories they produce. At the same time their ideals and goals of achieving rational understanding of the world are largely the same. Scientists in hard and soft fields rely on the

Scientific Method, formulate and test hypotheses, strive for objectivity and universality in their theories, and are open to correction and rejection. The hard–soft split, though common and sometimes convenient, is not fundamental.

4.B.2 Historical Versus Ahistorical

Can you do a properly controlled experiment where you have great influence over the experimental conditions? If so, you're doing *ahistorical* science. Ahistorical sciences include all the laboratory sciences plus a few more (e.g., biology and botany). If you have to deal with the universe as it is and cannot manipulate the variables you'd most like to manipulate, then you are probably doing *historical* science (e.g., evolutionary biology, astronomy, geology). Astronomers can look at galaxies, count, measure, and analyze them, but can't move them around or change their size or composition; evolutionary biologists and geologists face similar limitations, so these are all historical sciences.

Still, the similarities of background information, values, and methods of evolutionary biologists and geologists are greater than their differences. The famous study of the extinction of the dinosaurs is a good example. The father–son team of scientists, Walter and Luis Alvarez, hypothesized (the Alvarez Hypothesis) that a large asteroid struck the earth about 65 million years ago[22] and its impact triggered cataclysmic environmental events, including an earth-enveloping dust cloud that blocked the sunlight for several years, as well as fires, acid rain, and increased vulcanism, all of which combined to cause worldwide extinctions. This is clearly historical science.

Despite the antiquity of the event that the Alvarez Hypothesis tried to explain, the hypothesis was consistent with known geological data, and it made predictions that could be tested by future observations. In other words, the mere fact that historical sciences study past events doesn't mean they can't test hypotheses. For instance, the Alvarez Hypothesis accounted for the thin layer of iridium that is found everywhere in the world—iridium is an element in much higher concentrations in asteroids than is typical on earth—and for the presence of tectites—small, rounded, glassy rocks—associated with the iridium. The geological age of these phenomena was roughly in agreement with the onset date of the known extinctions. Most dramatically, the hypothesis predicted that there should be a huge asteroid impact crater somewhere on earth. Eventually a crater having the predicted age and size—110 miles across, 12 miles deep—was found near Chicxulub in the Yucatan peninsula of Mexico. The reasoning and procedures in historical science and ahistorical laboratory science are so much alike that it makes little sense to classify them as different kinds of science.

4.B.3 Natural Versus Social

For some authors there is an unbridgeable gap between natural and social sciences. Indeed, an extreme point of view is that social sciences are not, and should not to aspire to be, scientific in the way that natural sciences are. Under this interpretation, the subject matter of social sciences—society or culture—should be interpreted as a "text" analogous with the study of literature rather than rigorously interrogated for empirical scientific truths.[23] Many social scientists would not go so far; nevertheless, they would argue that interpretations in their fields must be conducted with qualitative methods that are necessarily less objective than the quantitative methods of the natural sciences. And indeed, it does seem reasonable to conclude that, if there is no way of making or desire to make social studies scientifically objective, then these studies truly have little or nothing in common with natural sciences. But then, what would be the purpose of calling them "sciences" in the first place? Why not group them with other careful, scholarly inquiries such as art, music, or literature that are not considered sciences?

On the other hand, many social scientists do embrace the ideals of modern science, including hypothesis testing and the standard of, at least, "weak objectivity."[24] And, increasingly, social scientists are adopting Big Data analyses and Bayesian methods to propose and test hypotheses (example in 4.E.2). On what grounds could we deny that these social scientists are engaged in real science?

In summary, although intuitively appealing, traditional divisions among kinds of science won't provide much guidance in following the debates about science that this book is concerned with. Instead, I suggest that an alternative set of distinctions will be more helpful.

4.C Kinds of Science: Objectives and Methods

You can distinguish among kinds of science not only by their subject matter or methods, but also by their ultimate objectives. I mentioned (Chapter 2.G.) that many philosophers have difficulty in understanding why scientists hold Popper's and John Platt's programs in such high regard, given that scientists generally pay almost no attention to any philosophical teachings. You may recall Godfrey-Smith's comment about scientists liking Popper because he holds up a heroic mirror for them to admire themselves in—violin-playing cowboys!—but he doesn't explore the motivation for scientists to invent and test theories in the first place. In a sense, scientists always have to *do* something; their job is to figure out nature. And, says the philosopher Bryan Magee,[25] Popper's philosophy is "a philosophy of action." Popper describes how scientists can get their job done. This perspective suggests that we can distinguish among kinds of science according to the kinds of action that scientists take.

4.C.1 Basic Science Versus Applied Science

Many discussions of modern science either take it for granted that we're talking about *basic research science*, or they blur—even reject (Chapter 10.A)—the distinction between basic and *applied science*. The express purpose of basic science is to discover Truths about nature, solely for the sake of understanding it. Naturally, basic scientists and society, especially research funders, hope that *basic* science discoveries will translate into solutions for societally significant problems, but concrete payoffs are not its immediate goal.

Applied science is directed toward concrete payoffs; it seeks to solve specific problems, often in the service of technology development paid for by commercial enterprises or governments. Despite the fact that applied science typically relies on or extends the findings of basic science, it is not primarily dedicated to advancing knowledge about the world per se. Applied science questions may or may not be answered by testing hypotheses. Clinical science ("translational research" refers to findings meant to go "from bench to bedside") is applied science that aims to find cures for diseases via the study of human subjects, for example. Consider the study of HIV/AIDS. While both basic and applied scientists are hoping to contribute to a cure, the basic researcher might be trying to identify novel biochemical mechanisms that viruses use to replicate (and which could be new therapeutic targets), whereas the applied researcher might be trying to make a drug to block viral replication as it is currently understood. An applied scientist may be happy with a drug that is 95% effective in blocking viral replication because that could represent a major improvement in treatment. A basic scientist, however, might be encouraged by such a finding and yet not satisfied until she figured out everything about the viral replication process. Basic and applied sciences differ in their goals and the kinds of problems they try to solve.

Of course, it may be difficult or impossible to say precisely when basic turns into applied science. Scientists working on practical problems do make groundbreaking basic discoveries, as Arno Penzias and Robert Wilson did when their efforts to modify a radio telescope accidentally led them to discover Cosmic Microwave Background radiation[26] thereby gaining evidence for the Big Bang (and eventually winning a Nobel Prize to boot). Conversely, we support basic cancer research because we hope that it will lead to a cure for cancer. To a large extent, though, it makes sense to consider basic and applied science to be different kinds of science.

4.C.2 Confirmatory Versus Exploratory

Unfortunately, the ill-defined "confirmatory–exploratory" terminology[27] has come into vogue as another way of distinguishing among different kinds of science. The underlying impulse is not bad: it can be useful to differentiate between large-scale, rigorous studies that are intended to produce definitive evidence for or against an important prediction, and small-scale studies, where you're just trying to get a foothold in a new area of research. As it is used, the confirmatory–exploratory distinction has a number of drawbacks.

You might do a "confirmatory" study if you had a lot of data and wanted to test a prediction of a well-developed and -corroborated hypothesis. An ideal confirmatory study would have large sample sizes, be tightly controlled, *double-blinded, preregistered*, and have impeccable statistical design. So far so good. However, some confirmatory studies are also called "hypothesis-testing" studies, while others do not test an explicit hypotheses. A drug company might do a confirmatory study to test a prediction, such as, "drug X will reduce the frequency of heart attacks," if it hopes to take X to the marketplace and treat patients, not because it cares about the hypothesis underlying X's cellular mechanism of action per se. Its main objective is to establish whether or not X works so it can either continue to its program on X or cut its losses and move on to a new project. Calling them "hypothesis-testing" creates a misunderstanding that is compounded by the ambiguity of the term "confirm" (Chapter 1L.). A confirmatory study could quite unexpectedly reveal that drug X is worthless or even harmful.

A neutral and descriptive name (e.g., "decision-making" studies) would call attention to the defining characteristics of confirmatory studies, which is to decide something *once and for all*.

"Exploratory" on the other hand, has been applied to preliminary pilot-level studies that are sometimes also called "hypothesis-generating" or even "pre-hypothesis" studies, and, in this context, "exploratory" seems apropos. Unfortunately, "exploratory" is sometimes applied to small-scale rigorously conducted hypothesis-testing studies, which is not at all helpful.

You can probably already guess that the biggest problem with the "confirmatory–exploratory" terminology is that it suggests a dichotomy that entirely omits the vast middle ground between primitive exploratory studies and well-advanced, dedicated decision-making investigations. In trying to shoehorn all of science into "confirmatory–exploratory," the terminology misses a huge swath of experimental science, which is focused on the hypothesis-based work that gives rise to most of our currently accepted scientific knowledge. At a minimum, I suggest that "decision-making," "hypothesis-testing," and "exploratory." categories would capture many significant distinctions among kinds of science. Nevertheless, attempting to divide up science like this requires even more categories, and several have come into common use.

4.C.3 Scale: Big Science/Small Science, Big Data/Little Data

If you search for information about the status of the hypothesis in today's biomedical sciences, you'll eventually encounter pundits who proclaim that science no longer "needs" the hypothesis and that the successes of *Discovery Science, Systems Biology*, or *Big Data* prove it. In the next sections, I'll review definitions and point out how they fit into the discussion about hypothesis-based and non–hypothesis-based science.

Big Science differs from *Small* Science[28] by its scale: Big Science projects can have industrial-sized magnitude, multisite collaboration, thousands of researchers, and budgets and administrative structures to match. Big Science takes advantage of automated analytic devices to collect massive amounts of data and relies heavily on computer-based algorithms for analysis. It strives for "throughput," the ability to obtain and process information quickly and efficiently. But, beyond its scale, Big Science is characterized by its large fund of accepted general principles and methods, as well as by the "invisible colleges of researchers"[29] who interact mainly with each other. Big Science is mature; there is wide agreement within a Big Science community about the major scientific problems that it must solve.

Small Science[30] labs tend to be self-contained and have bare-bones administrative support and budgets. Small Science is typified by its multiplicity of research topics and experimental methods. Scientists in these labs disagree on the fundamental questions that face them, and within their community there is a lot of flux in research directions, with many questions being actively pursued at once. As a whole, Small Science is a mile wide and an inch deep.

If you were to change perspectives and look only at the sizes of the datasets (a "dataset" is a collection of data about a specific topic) that scientists deal with, rather than the size of the laboratories or extent of their collaborations, you might see science as divided into Big Data and Little Data investigations. If you were to plot the size of the dataset that a given laboratory works with against the numbers of laboratories working with that dataset, you'd get a relationship that resembles a falling exponential function (Figure 4.1). There are only a few labs working with huge datasets (on the left of the x-axis) and increasingly many labs working with smaller and smaller ones as you move to the right of the axis. This plot, said to illustrate the "long tail"[31] of science, gives a global impression of how the entire scientific enterprise sorts itself out on these dimensions. How does the information in the long tail line up with the other classifications? Big Data and

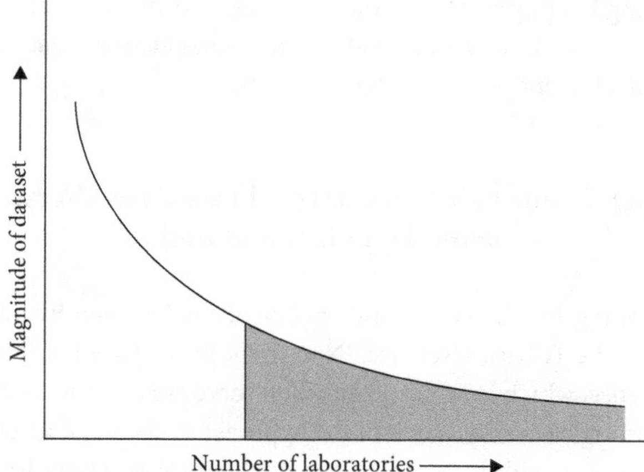

Figure 4.1 The "long-tail" of science. First named by Chris Anderson (quoted by C. L. Borgman, see Note 31) the graph shows that as the number of laboratories grows, the size of the datasets that each laboratory uses and generates decreases. Only a relatively few large labs are dealing with the extremely large datasets that characterize Big Data. Small Science is carried out in the long tail of the distribution and generates the large majority of scientific findings today.

Little Data are only "awkwardly analogous"[32] to Big Science and Small Science. While all Big Science projects make use of Big Data, some Small Science labs do as well. Big Data strategies are now open to laboratories of all sizes since powerful, inexpensive computers and access to massive datasets have become widely available.

"Little Data" is a logically necessary though rarely used term for the raw material and final output of Small Science. Little Data are generated by the materials, methods of data collection, and analyses that most Small Science projects employ; small datasets with relatively few subjects and variables, and people, rather than machines, to do much of the work. Small Science using Little Data methods is not easily "scalable," meaning that you can't readily increase the rate of progress from the start of data collection to the generation of conclusions. For instance, for the past 50 years or so the premier way to measure the electrical properties of neurons in the brain has been for a single investigator to insert a fine glass probe with a submicroscopic tip into a single neuron in an in vitro slice of brain tissue. The method yields extremely high-resolution information; it is also a pitifully slow method, and so far no one has come up with a better one. Single-cell recording like this represents a typical Small Science/Little Data scalability bottleneck. Despite such drawbacks, Little Data projects make up the long tail of the distribution and represent the principal sources of research information in many fields, including biology and biomedical science, neuroscience, and psychology. And, even today, Little Data and Small Science generate the great majority, perhaps 85%,[33] of all scientific knowledge.

4.D Kinds of Science: Hypothesis-Based Versus Non–Hypothesis-Based

A critical dividing line between kinds of science is between hypothesis-based and non–hypothesis-based science. Non–hypothesis-based sciences include *Discovery Science*, which is undertaken when there are too few data for focused hypothesis testing, as well as some of the approaches classified as Big Data that, ultimately, dispense with hypotheses, in part because there are too many data and they are too complex. However, Big Data is also used in conventional tests of hypotheses, which I'll discuss in this chapter, and I'll postpone consideration of the "Big Data Mindset" until Chapter 15.

Whereas Discovery Science and hypothesis-based science are mutually exclusive, they both may be associated with Big and Small Science and with Big and Little Data. The tree diagram in Figure 4.2 is one way of seeing their relationships.

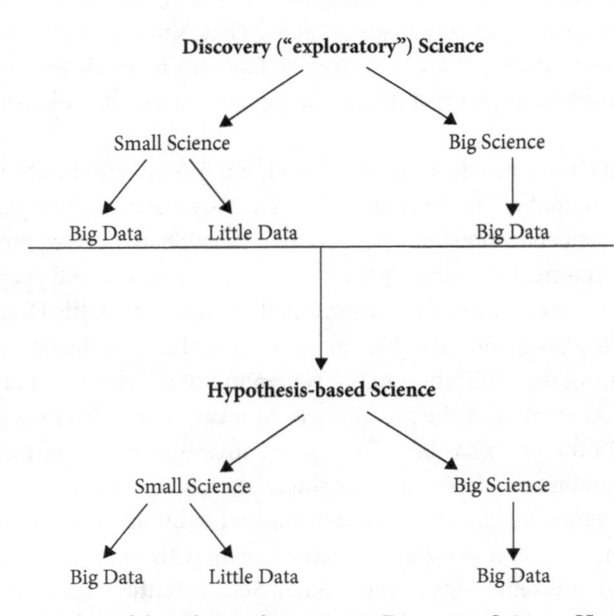

Figure 4.2 Diagram of the relationships among Discovery Science, Hypothesis-based science, Big and Small Science, and Big and Little Data.

4.D.1 Discovery Science

Discovery Science opens up new areas of inquiry; it identifies and catalogs the core elements of an area of research and characterizes their interactions. In principle, Discovery Science ranges in scale from small exploratory studies to gigantic Big Science projects: it seeks to classify and catalog, not to explain. You need to have some discovery before you can do hypothesis testing. I capitalize Discovery Science merely to set it apart; I don't know any self-identified "Discovery Scientists" and, on the contrary, believe that most scientists shift fluidly back and forth between Discovery and hypothesis-testing modes. Examples will follow.

4.D.1.a Discovery Science Versus Hypothesis-Based Science

Almost no one today seriously suggests that hypothesis-based science does not need basic information to work on (see Chapter 10.B for a counterexample). Even the classical Greek philosophers who wanted to understand planetary motion did not merely sit and contemplate. Their Discovery Science inquiries consisted of looking up at the night sky and keeping records.

While a purist might insist that Discovery Science and hypothesis-based science do not overlap, these dissimilar modes of conducting science are related in two ways. First is the one we've been talking about. Discovery Science is the feeder mechanism: it provides the data that hypotheses are designed to explain. Second, Discovery Science necessarily assumes the truth of certain (deep implicit) hypotheses before it can discover anything.

In biology, you can often identify formal Discovery Science projects by the "ome" or "omics" in their names[34] (e.g., "genomics," "proteomics," "metabolomics," and "connectomics"). Discovery scientists typically start by inventorying the constituent elements of a system, identifying, and organizing them. Like European explorers in Western history during the Age of Discovery who set sail with only the vaguest notions of what was before them, Discovery scientists cannot be confident about what they might find. No explicit hypotheses are necessary or even possible at this stage.

4.D.1.b Discovery Science/Big Science

The archetype of a Discovery Science project on the Big Science scale was the Human Genome Project,[35] a multibillion dollar, multinational, multiyear collaboration among thousands of scientists. The human genome consists of about 3 billion molecules called *base pairs* that encode the genes and bits of DNA that regulate the genes. Strings of genes make up chromosomes, and scientists originally used biochemical methods to investigate how they go together in sequence on the chromosomes, one molecule at a time. The Human Genome Project adopted the then-new tactic of breaking down ("shot-gunning") the long genomic DNA strings into small pieces, deducing the gene sequences of the pieces, and conceptually reassembling them into the original strings. The Human Genome Project procedure was like manufacturing a machine by assembling small parts into modules and putting the modules together to make the final product, whereas previous methods resembled traditional manufacturing methods of building the machine from scratch, one bolt at a time. The Project required enormous effort and quantities of computing power. It combined DNA samples from multiple individuals because its goal was to determine the sequence of an average human genome, not that of any particular person.

4.D.1.c Discovery Science/Small Science

Small Science also engages in Discovery Science, as the example of an investigation of RNA viruses inhabiting spiders found in western Washington state[36] showed. The investigators in a small lab were motivated by a general concern with the spread of viral disease, which can infect vertebrates, including humans, as well as spiders. Spiders in general are "undersequenced," and nobody knew much about the viruses that the target spiders harbored, except that their genomes were made of single strands of RNA, not double strands of DNA, and had about 10,000 bases instead of the 3 billion base-pairs in humans. The investigators obtained rough viral genome sequences, which enabled them to identify six new species of *Picornavirus* constituting a putative new virus clade; apart from providing a window into "the greater invertebrate virosphere," the work had no immediate application.

Like the Human Genome Project, the spider virus project investigated a scientifically significant problem and obtained information that could later lead to hypotheses for testing, though, at this stage, nothing was tested and there was nothing to falsify. The final important point here is that although Discovery Science projects are, strictly speaking, not within the scope of the methodological unity described by Popper, they are clearly integral components of science. They complement rather than supplant his vision.

And yet the data provided by Discovery Science do more than provide the foundation for the hypothesis; they constitute a virtual provocation to form hypotheses, as the next section shows.

4.D.2 From Discovery Science to Hypothesis

We cannot resist generating hypotheses once we have some data; it is an innate urge that is part of our built-in survival skill set (Chapter 12). We continually try to understand the world around us. Once scientists have discovered a new phenomenon, they immediately get to work trying to explain it. Discovery Science motivates the formation of hypotheses. A perfect example is the study of the human gut microbiome, which began with a general survey and classification of the varieties of bacteria inhabiting our intestines.

4.D.2.a The Human Microbiome from Discovery to Hypothesis

To understand the mammalian gut biome, investigators are sequencing the genomes of thousands of species of microbes.[37] Initially, the microbiome project was pure Discovery Science. Scientists in the Human Microbiome Project isolated microbes, sequenced their DNA, deduced and identified their genes; in all, there were more than 10,000 species, and particular kinds were associated with particular physiological functions. For instance, bacteria can digest a polysaccharide, a kind of starch-like carbohydrate molecule found in plants that we humans can't digest by ourselves. We need bacterial help to be able to use the starch for food.

Importantly, the Human Microbiome investigators found that some bacteria were more effective than others at various tasks. Although Discovery had been their initial motivation, the investigators immediately guessed that variation in the number or types of bacteria could affect human health. One specific hypothesis was that obesity could be caused by the gut microbiome, and it predicted that the microbiomes of obese and lean people would differ systematically. This prediction was soon confirmed. But the hypothesis also predicted that the microbial differences were causal, not merely correlational; it wasn't just that specific populations of bacteria go along with obesity or leanness but that the bacteria *help cause* these body types. To test this prediction, the investigators turned to animal models because they could manipulate the animals' microbiomes and they couldn't do that with people. There are three strains of genetically programmed mice that were useful: one was obese, one was lean, and one had no gut bacteria at all.

First, the hypothesis predicted that mice that completely lacked bacteria would be at a severe metabolic deficit because they couldn't efficiently extract energy from food, and indeed, these mice were thin and frail. Second, the hypothesis predicted that obese and lean microbiomes would differ, and that was also true. Third, to test the prediction that the gut microbiomes determined the body types, the investigators transferred microbes from obese or lean mice into the mice with no gut bacteria. The results again were consistent with the causal hypothesis: the mice that received bacteria from obese mice gained weight, but if they received bacteria from lean mice, they didn't.[38]

Finally, to test the prediction that the microbiome influenced human obesity, investigators turned to patients who were suffering from an infection of *Clostridium difficile* (*C. diff.*). The gut microbiomes of these patients are seriously disrupted but can be successfully treated with a technique—fecal microbiota transplant[39]—that transfers intestinal microbiota from healthy individuals to patients. Amazingly, as in the mice, obese transplant donors pass on a tendency to obesity to human recipients.[40]

The story of the human gut microbiome research and its application to disease reveals a seamless transition between pure Discovery Science investigations and conventional hypothesis-testing experiments. It demonstrates that, while the methods and goals of Discovery Science and hypothesis-testing science are quite distinct, these modes are complementary, not opposed to each other. There is one last link between Discovery Science and hypothesis-based science that I want to bring out before leaving the subject.

4.D.2.b Discovery Science and the Implicit Hypothesis

I've quoted Ramón y Cajal's comment that "No one searches without a plan,"[41] and it's very instructive; a plan often depends on informal assumptions. Say you're wandering around looking for your apartment keys; even if you have "no idea" where you could have left them, you probably wouldn't look inside the box of cereal that you had for breakfast. You'd probably assume that, though perhaps forgetful, you're not yet completely out of it, and there is no way that you'd have put the keys there. Discovery Science is informed by substantive, deep implicit hypotheses of this kind. In their selection of methods and read-outs, Discovery Scientists reveal what they believe to be the most important features of nature to study.

In the early days of molecular biology researchers spent a lot of effort counting genes because they thought that genes alone determined a species' biology. It was an article of faith (i.e., a deep implicit hypothesis) that highly sophisticated animals would have many more genes than simple ones; it was an ostensibly reasonable hypothesis that made a straightforward prediction that turned out to be false.

Initial calculations suggested that, given their quantity of DNA and an assumed average size of a *coding gene* (which carries the instructions for—"codes for"—how to produce proteins), humans would have 50,000–140,000 genes.[42] In fact, it now appears that each of us has about 20,000 genes[43]; that is, roughly the same number as a mouse. Surprisingly, the pool-dwelling water flea, *Daphnia pulex*, has about 31,000 genes,[44] which, to put the situation in perspective, means that a tiny crustacean, about 1 mm in length, that expends most of its energy avoiding predators during the day and eating phytoplankton at night, has about 50% more genes than Albert Einstein did. The hypothesis that the number of genes alone determines an organism's biology is false, and the falsification immediately stimulated other investigators to wonder what was going on with the quantity of our DNA that was supposed to be making up all those other genes?

One hypothesis is that we have a large amount of *non-coding* DNA; up to 98% of total DNA in humans might not be part of identified genes. It could be "junk DNA," a by-product of spontaneous molecular duplication with no biological function,[45] or it could be part of the system of *gene regulation*. In any case, the gulf between Einstein and a mouse or a water flea may pertain to the amounts and functions of non-coding DNA, rather than to the genes themselves. The episode of gene counting is a prime example of how the implicit testing of deep hypotheses underlying a pure Discovery Science spawns new discoveries and new hypotheses to explain them.

4.D.2.c You're Not Necessarily Doing Discovery Science Just Because You Don't Have a Hypothesis

The prominence and success of Discovery Science projects like the Human Genome Project have had some unfortunate consequences. My experience in reviewing research papers and grant applications suggests that the label "Discovery Science" is occasionally pasted onto poorly organized and incompletely thought out hypothesis-based science. It is unclear whether the authors of these productions wanted to associate their work with a popular trend, whether they believed that they were genuinely doing Discovery Science or, maybe, whether they even knew what Discovery Science was. Whatever the explanation, the result was typically an unfocused publication or application that rambled from one topic to another. Science has plenty of room for trying out hunches to see if anything turns up or doing pilot, exploratory experiments. I suspect that reserving the designation "Discovery Science" for thorough, focused investigations bent on pure discovery with the aim of producing a novel database would go long way toward clarifying scientific communications and policies.

4.E Big Data: What Is It?

While the origins of the term "Big Data" are murky, it gained academic respectability in 2003,[46] although in 2001 Laney[47] had already defined the "bigness" of data by the famous "3 Vs": volume, velocity, and variety. The alliteration provoked a stream of imitators, and the number of V's describing Big Data grew to 4, 7, and 10 within a few years, all the way to the current record: "The 42 V's of Big Data and Data Science"[48] a list that includes the plausible "visualization" and "veracity" as well as the cynical "vagueness" and "varmint" (the number of software bugs increases with program size), apparently indicating that enough is enough. The point is that Big Data is not solely about dealing with large quantities of data.

Big Data is now a "phenomenon and a discipline." Big Data has transformed science in both\ concrete and abstract ways. In this chapter, I want to take a quick look at how Big Data is influencing conventional science practice; in Chapter 15, we'll examine its disruptive, abstract influences on science and the philosophy of science.

4.E.1 Hypothesis-Based Big Science/Big Data

Although the Human Genome Project typifies Discovery Science in the context of Big Science and Big Data, Big Science is not confined to the realm of Discovery Science. The search for the Higgs boson carried out at the Large Hadron Collider is a good example of Big Science/Big Data testing of a quantitative hypothesis: the Standard Model of physics. Another recent example is the test of the prediction of the existence of gravitational waves predicted by Einstein's General Theory of Relativity,[49] but these examples get a lot of popular coverage so they are probably well-known to many readers. The employment of Big Data in Small Science projects is probably not as familiar.

4.E.2 Hypothesis-Based Small Science/Big Data

Small Science also uses Big Data to test hypotheses. Psychologist Morteza Dehghani and colleagues[50] were interested in the social "glue" that holds communities together and hypothesized that shared moral values, especially "moral purity," which is related to feelings of disgust or uncleanliness, caused people to form social networks. That is, their hypothesis predicted that the closer two people were in their views on moral purity, the closer their social bonds. To test predictions of the hypothesis, the authors collected and analyzed 700,000 Twitter tweets to determine the degree of homophilily ("self-liking") among groups of people. They investigated tweets occurring around the time of the government shutdown brought about by Congress in 2013 because of political disagreements over healthcare ("Obamacare") funding. The public debate took place along party lines, and, indeed, Deghani's group, with the help of a "community-detection algorithm," identified clusters of tweets representing the two major US political parties. The investigators then used copious computer processing of the data to determine the social distances among individuals and the "moral content" of their tweets (identified by key words). People who were closely connected socially shared moral purity values but did not necessarily share other moral concerns (e.g., on fairness, loyalty, or authority). The authors also found that moral bonds were tighter than political ones and that, within political parties, people also sorted themselves into groups of shared value systems. Finally, the authors tested other predictions of their hypothesis in behavioral "laboratory studies" (actually online surveys conducted via Mechanical Turk[51]) with a few hundred people. In the laboratory studies, the authors directly asked the participants about their views of moral similarity and differences along with their other personal traits and compared them. The laboratory findings corroborated the conclusions of the Big Data analysis.

This example shows that Big Data can find a place in a Small Science setting. It also demonstrates that, whereas Big Data greatly expanded the universe of this social science study, it did not alter the study's conceptual framework: the authors used Big Data to test a conventional hypothesis in a straightforward way.

4.E.3 Systems/Computational Biology

Complex systems have numerous component elements that interact in nearly limitless ways. You might know a great deal about an individual element—for example, a gene, a protein, an atmospheric aerosol particle—in isolation and still know nothing about how that element will behave when it is part of a huge population of elements, let alone how the population as a whole will behave. The "bland" *reductionist* assumption that all physical phenomena depend on simple physical entities doesn't mean that a *constructionist* program of building up to the more complex systems from simple ones will work. Novel behaviors emerge when the elements interact. "More is different," says the physicist P. W. Anderson.[52] Anderson, a committed reductionist, firmly rejects constructionism and uses a chemical example: both ammonia and sugars are made up of similar atoms; however, a detailed atomic-level comprehension of the atomic motions within the ammonia molecule does not help you understand the atomic motions in the larger sugar molecules.

Tackling the large assemblies of elements head-on is often referred to as a "systems" approach to emphasize that the goal is to understand the system as a whole. *Systems biology* (aka *computational biology*) tries to understand how its parts interact by developing computer models that reproduce the emergent properties of the system.

Does the interest in the system as a whole preclude a role for hypotheses? Is its success proof that science does not need hypotheses to advance? On the contrary, many systems biologists consider hypotheses to be integral parts of their field. For instance, the biochemists Daniel Beard and Martin Kushmerick[53] state that the high-level unifying generalizations, the hypotheses, and Platt's program of Strong Inference (Chapter 2) provide the logical and scientific structure that are necessary for understanding the formidable system complexities of biology.

However, Beard and Kushmerick disagreed with Platt's downplaying the importance of mathematical models in scientific reasoning. Platt felt that compelling logical models were more powerful; he wanted to have a chain of nonmathematical reasoning that would decide in favor of one hypothesis by eliminating others. For Platt, a mathematical model was simply a convenient way of summarizing your conclusions. The systems biologists argue that you can't get a meaningful grasp of an entire system without using powerful mathematical tools; that "the simple qualitative framework of an earlier age of molecular biology" is no longer up to the task. Nowadays, they say, reasoning is not based on the crude box-and-arrow diagrams of 50 years ago, but on sophisticated computer models, and "the fundamental key is to recognize that a computational model, however simple or complex is a *hypothesis* [emphasis added]." Hypotheses cast as quantitative models in their view are precise and explicit, as they must be so you can test and potentially falsify them. Falsification is important because "successful disproof is the key to progress." Systems biology lets us solve enormously complex problems without abandoning the fundamental logic-based procedures of Popper and Platt.

In these past sections, I've reviewed kinds of science that have been proposed as alternatives to hypothesis-based science. Before leaving the topic, I want to consider one final kind of science. So far, we've been reviewing forms of *modern science*. One overlooked class of knowledge does not fit into this category because it was not generated by modern societies. It is *Indigenous science*.

4.F Modern Versus Indigenous Science

For countless generations, Indigenous peoples have studied nature and developed systems of knowledge that are independent of Western science. "If we understand 'indigenous' to relate to people who have a long-standing and complex relationship with a local area and 'science' to mean a systematic approach to acquiring knowledge of the natural world, then Indigenous science is the process by which Indigenous people build their empirical knowledge of their natural environment."[54] Naturally, as a traditional body of knowledge, Indigenous science does not use the formal thinking tools, recording methods, or instruments that characterize modern science.

Indigenous science seeks "valid and efficacious" knowledge that is reliable and predictive. An example comes from the Polynesian mariners who navigated great distances over open ocean guided by the stars and patterns of the ocean swells. Today the unique experiences of Indigenous peoples are sought after by international bodies such as the International Panel on Climate Change and others to assess and document the extent and severity of climate change on remote environments[55] where instrumental records are scarce or absent. For instance, in the Northern Hemisphere, many Indigenous groups have collected information regarding changes in native animal number, habitat change, invasion of exotic plant and animal species, wild fires, etc. Indigenous science produces practical knowledge that is aligned with applied rather than basic science.

Often Indigenous knowledge provides the jumping off point, the raw materials, for modern science. In other cases, Indigenous knowledge may be directly useful to modern society: at least that is the hope for vetiver grass, a hardy, deep-rooted, tall (up to 8 feet) grass from Asia. Vetiver grass has been used for centuries to prevent soil erosion[56] and is being evaluated for use in shoring up levees in New Orleans (although its advantages are uncertain).

Times are changing, however, as is illustrated by the story of the Chinese pharmaceutical chemist, Tu Youyou.[57] Seeking to find a cure for malaria, Tu and her co-workers investigated thousands of traditional Chinese herbal treatments before homing in on an extract from *Artemisia annua* ("sweet wormwood," *qinghao*). When her initial preparations were only sporadically effective, she got a hint from an older text (circa 340 C.E.) to do the extractions under cold temperatures, rather than boiling them, as was the modern practice. Her team isolated a compound, artemisinin, that killed the most deadly malaria-causing parasite, *Plasmodium falciparum*, first in animals, then in people (she insisted on being the first human test subject). Tu Youyou and her team later developed variants of artemisinin that permitted the chemical synthesis of the molecules and, eventually, drugs which are credited with saving millions of lives. In 2011, Tu was recognized with the Lasker-Debakey Clinical Medical Research Award, and, in 2015, she became the first Chinese scientist to win the Nobel Prize in Physiology or Medicine.

Does its frequent inclusion of supernatural beliefs and religious or magic rituals preclude Indigenous knowledge from being considered science? If the knowledge is attainable and reliable when accessed by a nonbeliever, then the ritual itself is not an essential component. The psychoactive drug cannabis has been used in cultural rituals throughout the world for thousands of years. Among other things, cannabis preparations are effective pain relievers and appetite stimulants.[58] Modern science now tries to understand and take advantage of its physiological effects. The ritual remains outside of science while the knowledge remains within Indigenous science.

Alternatively, rituals can affect the body in the way an effective placebo can, via the nervous system. In a controlled clinical study, researchers found that hospitalized patients who participated in months of group drumming practice had decreased depression and anxiety.[59] The drumming experience also affected biochemical measures of immune system function. Knowledge may be significant and reliable despite being couched in supernatural explanatory terms.

A major distinction between modern and Indigenous science is that Indigenous science tends to regard knowledge as sacrosanct and resistant to change, whereas openness to change is a hallmark of modern science. Indigenous science prizes belief and adherence to custom, while doubt and skepticism are the driving forces for modern science. The concepts of hypothesis testing and falsification are evidently not accorded prominent roles in Indigenous science, which may represent its most important distinction from modern science.

4.G Coda

Modern science is not a simple branch of knowledge; it is unified by common objectives and values, as well as by concepts, but science is not monolithic. This chapter reviewed several ways of classifying kinds of science. The principles of hypothesis testing and falsification formed a dominant theme in the classification. Falsification played two roles in sorting science into different bins. Falsification distinguishes science from nonscience and, within the confines of bona fide science, it's pragmatic role distinguished hypothesis-testing–based science from non–hypothesis-testing-based science. Distinctions between basic and applied science are also fundamental. Discovery Science does not employ hypothesis testing, however it leads smoothly into hypothesis testing and is, itself, based on a network of associated, implicit hypotheses. The Big and Small Science modes do not map neatly onto the Big and Little Data dichotomy, and each can be linked to Discovery Science and hypothesis-based science. Indigenous science is best seen as a form of applied science practiced by native peoples.

There is much more about the hypothesis that we need to cover. Generally, we haven't said why the hypothesis is beneficial, why hypothetical thinking comes so naturally to us, how it affects the Reproducibility Crisis, and related topics. Before we can move on to these broad topics, there is one more major kind of hypothesis that we need to think about: the *statistical hypothesis*. What is it and how is it related to the scientific hypothesis? What kind of statistics are we talking about? The concept of statistical hypothesis is so large and complex that it gets two chapters, which come next.

Notes

1. Edward O. Wilson, *Consilience: The Unity of Knowledge* (New York: Vintage Books; 1998).
2. H. Allen Orr, "The Big Picture," see http://bostonreview.net/archives/BR23.5/Orr.html, 1998.

3. Karl Popper, "Survey of Some Fundamental Problems," in *Logic of Scientific Discovery* (New York: Routledge; 2002), pp. 3–26. This essay is a brief, exceptionally clear and readable introduction to Popper's thinking.
4. Peter Godfrey-Smith, *Theory and Reality* (Chicago: University of Chicago Press; 2003).
5. Massimo Pigliucci, *Nonsense on Stilts.* (Chicago: University of Chicago Press; 2010).
6. Francis Crick, *The Astonishing Hypothesis: The Scientific Search for the Soul* (New York: Touchstone; 1994); also, Daniel Dennett, *Consciousness Explained* (New York: Back Bay Books; 1991); Christof Koch, *Confessions of a Romantic Reductionist* (Cambridge, MA: MIT Press; 2012);
7. Brian Greene, *The Elegant Universe: Superstrings, Hidden Dimensions, and the Quest for the Ultimate Theory* (New York: Vintage Books; 2003).
8. Pigliucci, *Nonsense*, pp. 33–40. Although I disagree with Pigliucci's conclusions regarding the demarcation problem, his chapter on "Almost Science" is a nice overview of the issues confronting modern physics and other scientific projects in the "transitional zone" between science and pseudoscience.
9. Jordi Cat, "The Unity of Science," in *The Stanford Encyclopedia of Philosophy* (Winter 2014 ed.), Edward N. Zalta (ed.), http://plato.stanford.edu/archives/win2014/entries/scientific-unity/.
10. The notion of science as aiming to give one "tidy account" has been much debated in the context of the pluralism of science and in the relationship of science to feminist and multicultural concerns. Scientific *monism* is opposed to pluralism, and the concepts encompass issues such as *demarcation* and *reductionism* among others. Cf. Sandra Harding, *Objectivity and Diversity: Another Logic of Scientific Research* (Chicago: University of Chicago Press; 2015); see, e.g., chapter 5, "Pluralism, Multiplicity, and the Disunity of Science," for an overview of some of these matters.
11. Daniel Dennett, *Darwin's Dangerous Idea: Evolution and the Meanings of Life* (New York: Simon & Schuster; 1995).
12. Steven Weinberg, "Against Philosophy," in *Dreams of a Final Theory: The Scientist's Search for the Ultimate Laws of Nature* (New York: Vintage Books; 1994), pp. 166–190. Weinberg makes essentially the same distinctions between kinds of reductionism as Dennett does; Weinberg adds that the findings of particle physicists are simply not useful to less fundamental science. This essay of Weinberg responds to a notorious tract by the philosopher Paul Feyerabend, *Against Method*, that argued that the principle for understanding "science" was "anything goes."
13. John Tooby and Leda Cosmides, "The Psychological Foundations of Culture," in Jerome Barkow, Leda Cosmides, and John Tooby (Eds.), *The Adapted Mind: Evolutionary Psychology and the Generation of Culture* (New York: Oxford University Press; 1992).
14. Tooby, Cosmides, *Adapted Mind*, chapter 1.
15. Steven L. Goldman, *Science Wars: What Scientists Know and How They Know It* (Chantilly, VA: The Teaching Company; 2006).
16. Paul R. Gross and Norman Levitt, *Higher Superstition: The Academic Left and Its Quarrels with Science* (Baltimore: Johns Hopkins University Press; 1994).

17. Margo Wilson and Martin Daly, "The Man Who Mistook His Wife for a Chattel," in Barkow, Cosmides, and Tooby (Eds.), *Adapted Mind*, chapter 7.
18. Roger Shepard, "The Perceptual Organization of Color: An Adaptation to Regularities in the Terrestrial World," in Barkow, Cosmides, and Tooby (Eds.), *Adapted Mind*, chapter 13,
19. Gordon F. Orians and Judith Heerwagen, "Evolved Responses to Landscapes," in Barkow, Cosmides, and Tooby (Eds.), *Adapted Mind*, chapter 15.
20. Popper, "The Problem of the Empirical Basis," *Logic*. Chapter 5 is the source for the discussion in this section.
21. Glenn Branch, "Whence Lumpers and Splitters?," https://ncse.com/blog/2014/11/whence-lumpers-splitters-0016004
22. Alvarez hypothesis; see Paul R. Renne, Alan L. Deino, Frederik J. Hilgen, Klaudia F. Kuiper, Darren F. Mark, William S. Mitchell, et al., "Time Scales of Critical Events Around the Cretaceous-Paleogene Boundary," *Science* 339:684–687, 2013; https://en.wikipedia.org/wiki/Alvarez_hypothesis.
23. Clifford Geertz, "Local Knowledge: Further Essays in Interpretive Anthropology," https://monoskop.org/File:GGeertz_Clifford_Local_Knowledge_Further_Essays_in_Interpretive_Anthropology_1983.pdf.
24. Sandra Harding, *Objectivity and Diversity: Another Logic of Scientific Research* (Chicago: University of Chicago Press; 2015).
25. Bryan Magee, *Philosophy and the Real World: An Introduction to Karl Popper* (London: Open Court; 1985).
26. Robert W. Wilson, "The Cosmic Microwave Background Radiation," Nobel Prize lecture, http://www.nobelprize.org/nobel_prizes/physics/laureates/1978/wilson-lecture.pdf, 1978. Penzias and Wilson intended to measure fundamental cosmological properties; however, their discovery of the CMBR resulted from their attempts to improve their telescope's performance.
27. J. Kimmelman, J. S. Mogil, and U. Dirnagl, "Distinguishing Between Exploratory and Confirmatory Preclinical Research Will Improve Translation," *PLoS Biology* 12:e1001863, 2014. See also S. C. Landis, S. G. Amara, K. Asadullah, C. P. Austin, R. Blumenstein, E. W. Bradley, et al., "A Call for Transparent Reporting to Optimize the Predictive Value of Preclinical Research," *Nature* 490:187–191, 2012.
28. Christine L. Borgman, *Big Data, Little Data, No Data: Scholarship in the Networked World* (Cambridge, MA: MIT Press; 2015)
29. "Big Science–Little Science" distinction made by Derek de Solla Price, quoted in Borgmann, *Big Data*, pp. 5–7.
30. Size is, of course, relative. Years ago, some faculty colleagues and I heard that a US Senator from Maryland had expressed interest in visiting a "small laboratory" at our university. To the question of how small a lab the Senator had in mind, the answer was, "Oh, nothing with a budget over $3 million dollars a year." We had a good laugh at that. Our labs were subsisting on annual budgets of roughly one-tenth of $3 million. The thought of anything with a budget in the Senator's range could be considered small exposed the gulf separating our ordinary lives from the Olympian remove of Washington, DC.

31. In a typical plot, the magnitude of the dataset, y = f(x), is plotted against the numbers of laboratories (x), increasing to the right on the x-axis. Working with a dataset that large; $f(x) = x^{-2}$; i.e., the size of the dataset decreases as the negative square of the numbers of laboratories.
32. Borgman, *Big Data*, p. 5.
33. Ibid., p. 8. See also Adam R. Ferguson, Jessica L. Nielson, Melissa H. Cragin, Anita E. Bandrowski, and Maryann E. Martone, "Big Data from Small Data: Data-Sharing in the 'Long Tail' of Neuroscience," *Nature Neuroscience* 17:1442–1447, 2014.
34. "Omics" like "omes," are English neologisms loosely related to a Greek language stem. The "ome" suffix has come to refer to the totality of a subject matter, and "omics," is the study of an "ome." Evidently they are not recent coinages, however. Reportedly, "biome" appeared in print in 1916; see https://en.wiktionary.org/wiki/Appendix:Suffixes_-ome_and_-omics.
35. International Consortium Completes Human Genome Project, "All Goals Achieved; New Vision for Genome Research Unveiled," http://www.genome.gov/11006929
36. Ryan C. Shean, Negar Makhsous, Rodney L. Crawford, Keith R. Jerome, and Alexander L. Greninger, "Draft Genome Sequences of Six Novel Picorna-Like Viruses from Washington State Spiders," *Genome Announcement* 5: e01705-16, 2017.
37. S. R. Gill, M. Pop, R. T. Deboy, P. B. Eckburg, P. J. Turnbaugh, B. S. Samuel, et al., "Metagenomic Analysis of the Human Distal Gut Microbiome," *Science* 312:1355–1359, 2006; P. J. Turnbaugh, R. E. Ley, M. Hamady, C. M. Fraser-Liggett, R. Knight, and J. I. Gordon, "The Human Microbiome Project," *Nature*.449:804–810, 2007; Human microbiome project at the NIH; see https://hmpdacc.org/hmp/
38. P. J. Turnbaugh, R. E. Ley, M. A. Mahowald, V. Magrini, E. R. Mardis, and J. I. Gordon, "An Obesity-Associated Gut Microbiome with Increased Capacity for Energy Harvest," *Nature* 444:1027–1031, 2006.
39. https://en.wikipedia.org/wiki/Fecal_microbiota_transplant
40. https://www.scientificamerican.com/article/fecal-transplants-may-up-risk-of-obesity-onset/
41. Santiago Ramón y Cajal, *Advice for a Young Investigator*, translated by Neeley Swanson and Larry W. Swanson (Cambridge, MA: MIT Press; 1999).
42. Initial guesses as to the number of human genes ranged from 50,000–140,000 to as high as 2 million in the mid-1960s (https://www.genome.gov/human-genome-project/What)As an intriguing twist, Gill et al., "Metagenomic Analysis" (Note 37), point out that the gut microbiome has about 100× more genes than the standard human genome, meaning that each of us is essentially a "human supra-organism" with a vastly larger genome than we realize.
43. I. Ezkurdia, D. Juan, J. M. Rodriguez, A. Frankish, M. Diekhans, J. Harrow et al., "Multiple Evidence Strands Suggest that There May Be as Few as 19,000 Human Protein-Coding Genes," *Human Molecular Genetics* 23:5866–5878, 2014.
44. The animal with the most genes? The water flea! See https://www.wired.com/2011/02/water-flea-genome; https://www.nsf.gov/news/news_summ.jsp?cntn_id=118530
45. A. Cipriano and M. Ballarino, "The Ever-Evolving Concept of the Gene: The Use of RNA/Protein Experimental Techniques to Understand Genome Functions," *Frontiers in Molecular Bioscience* 5:20, 2018.

46. Francis X. Diebold, "A Personal Perspective on the Origin(s) and Development of "Big Data": The Phenomenon, the Term, and the Discipline," Second Version. Penn Institute for Economic Research Working Paper No. 13-003, http://ssrn.com/abstract=2202843.
47. Doug Laney, https://blogs.gartner.com/doug-laney/deja-vvvue-others-claiming-gartners-volume-velocity-variety-construct-for-big-data/, 2001.
48. 42 V's of big data: https://www.elderresearch.com/company/blog/42-v-of-big-data.
49. https://www.space.com/38471-gravitational-waves-neutron-star-crashes-discovery-explained.html.
50. M. Dehghani, K. Johnson, J. Hoover, E. Sagi, J. Garten, N. J. Parmar et al., "Purity Homophily in Social Networks," *Journal of Experimental Psychology, General* 145:366–375, 2016.
51. Mechanical Turk is an Amazon website where businesses can go to contract with workers who are paid to do a variety of online tasks; visit https://www.mturk.com/worker/help.
52. P. W. Anderson, "More Is Different," *Science* 177:393–396, 1972.
53. D. A. Beard and M. J. Kushmerick, "Strong Inference for Systems Biology," *PLoS Computational Biology* 5:e1000459.
54. http://livingknowledge.anu.edu.au/html/educators/02_questions.htm. The term "Indigenous Science" was coined by Dr. Apela Colorado (www.wisn.org); see also https://www.culturalsurvival.org/publications/cultural-survival-quarterly/indigenous-science; http://blogs.nwic.edu/briansblog/files/2011/02/Discovering-Indigenous-TEK-Implications-for-Science.pdf.
55. https://www.forbes.com/sites/davidbressan/2017/07/05/indigenous-knowledge-helps-scientists-to-assess-climate-change/#5721d4e25527, Indig science and global warming; http://www.nature.com/nclimate/journal/v6/n4/full/nclimate2954.html?WT.feed_name=subjects_climate-change-adaptation&foxtrotcallback=true; https://academic.oup.com/bioscience/article/61/6/477/225035/Linking-Indigenous-and-Scientific-Knowledge-of.
56. https://en.wikipedia.org/wiki/Chrysopogon_zizanioides. Description of vetiver grass (*Chrysopogon zizanioide*) and many of its uses; http://www.vetiver.org/ICV3-Proceedings/SA_stemborer.pdf. Van den Berg et al. found that female stem borer moths preferred to lay eggs in vetiver grass than in corn (maize), but the survival of their larvae was much lower in this grass than an alternative or in the corn. The authors suggest that vetiver grass can act as a protectant for crops such as corn and perhaps rice. Vetiver is also reputedly repellent to termites; see B. C. Zhu, G. Henderson, F. Chen, H. Fei, and R. A. Laine, "Evaluation of Vetiver Oil and Seven Insect-Active Essential Oils Against the Formosan Subterranean Termite," *Journal of Chemical Ecology* 27:1617–1625, 2001.; http://lacoastpost.com/blog/?p=21639 (and correspondence at end). The blog cites evidence that vetiver may not have advantages over native grasses.
57. https://www.nobelprize.org/nobel_prizes/medicine/laureates/2015/tu-lecture.html; https://en.wikipedia.org/wiki/Tu_Youyou. Tu's successes were rooted in Indigenous science, and yet they relied heavily on modern science techniques, including the

Scientific Method, careful record-keeping, controlled studies, animal models, clinical trials, and synthetic drug development. Owing to traditional cultural values and the atmosphere created by Mao Zedong's "cultural revolution," Tu Youyou's name did not appear on the first papers describing artemisinin. For most of her life she lived in relative obscurity until two Western scientists eventually identified Tu as the person most responsible for discovering artemisinin. She had been known as the Professor of Three "Withouts"—without a graduate degree, without research experience abroad, and without membership in a prestigious scientific society.

58. David Casarett, *Stoned: A Doctor's Case for Medical Marijuana* (New York: Penguin Books; 2015). A layman's summary by a skeptical physician of current information about effects and potential uses of medical marijuana; liberally laced with anecdotal, often witty, personal inquiries into the world of medical marijuana.
59. D. Fancourt, R. Perkins, S. Ascenso, L. A. Carvalho, A. Steptoe, and A. Williamon, "Effects of Group Drumming Interventions on Anxiety, Depression, Social Resilience and Inflammatory Immune Response among Mental Health Service Users," *PLoS One* 14;11(3):e0151136, 2016.

5
Statistics and Statistical Hypotheses

5.A Introduction

At this point you might be wondering about another prominent kind of hypothesis that I seem to be ignoring: the statistical one. This is not because scientific and statistical hypotheses are more or less interchangeable but because they are so different that the statistical hypothesis merits its own coverage.

I have three interrelated goals in this chapter: to distinguish between statistical and scientific hypotheses, to show how the principle of falsification coexists with probability and statistics, and, finally, to review statistical ideas that will come up again later in the book. I will focus on general concepts that few scientists were exposed to in their introductory statistics class, not on the computational details of the methods that we learned, which, as I'll show, are a mash-up of contradictory philosophies that have divergent implications for science.

Why do we have to get into statistics at all if we're interested in the scientific hypothesis? First, I've argued (Chapter 2) that basic science is not ultimately interested in "probable truth." To appreciate why it isn't, we need to understand the various meanings of "probability," a subject that recurs in several chapters and is within the purview of statistics. Second, if you think about the pervasiveness of uncertainty, you'll see there could be a problem in the scientific process involving falsification of scientific hypotheses. Thus far, I've implied that test outcomes are neatly cut and dried, with tests either confirming or falsifying a prediction. Science is almost never so straightforward, however. Variability resulting from sampling error, instrument resolution, or random noise, constantly rears its ugly head. Change is constant, and no two things are the same (in the macro world; all electrons may be identical). We need statistics to help us cope with variability. Third, several fundamental statistical matters arise in the context of the Reproducibility Crisis (Chapter 7), and we'll need to be familiar with them in order to evaluate the claims of crisis. Let's start with the comparison of scientific and statistical hypotheses.

5.B Scientific Versus Statistical Hypotheses

The scientific hypothesis tries to explain a natural phenomenon. The statistical hypothesis is *nonexplanatory*; it is part of a mathematical, inferential, decision-making procedure that determines what your experimental results mean for your clever scientific hypothesis (i.e., whether the results could have occurred by random chance alone or whether they are consistent with the hypothesis).

The term "statistical hypothesis" is unfortunate because scientific and statistical hypotheses are so unalike. Why is this a problem? After all, words usually have multiple meanings, and this usually doesn't create difficulties. Typically, however, context helps to sort things out so that, for instance, the record industry executive, mob boss, boxer, and baseball player can all refer to a "hit" unambiguously. Scientists test statistical hypotheses and scientific hypotheses in contexts that overlap and make it hard to tell them apart.

Statistical and scientific hypothesis share a couple of features: both are definite statements about the world that can be tested, and neither can be proved to be true, only false (or left in limbo). On the other hand, statistical and scientific hypotheses diverge in purpose, conceptual content, and in the ways in which they are tested.

How are scientific and statistical hypotheses related? Hierarchically. The scientific hypothesis is a conjectural explanation of a natural phenomenon that strives for generality, even universality. We assess its truth indirectly by experimentally testing the logical predictions it makes, and, if they're false, we reject it. Scientific hypotheses predict how tests of statistical hypotheses will turn out.

Statistical hypotheses are tied to particular circumstances and datasets, not to universal truths. Hence, statistical hypotheses are subordinate to scientific hypotheses; in fact, statistical hypotheses are akin to scientific predictions. We design an experiment to test a prediction, adopt a statistical hypothesis, collect data, run an appropriate statistical test, and interpret the results. Once its practical job of helping us evaluate the scientific hypothesis is over, the statistical hypothesis goes back into the analysis toolbox. Here's a quick example.

5.B.1 Statistical Versus Scientific Hypothesis: Example

For this example, I'll assume that we're in the null hypothesis significance testing mode (NHST) because it is in wide use and most of us are acquainted with it enough to get the gist of the argument. Soon I will step back and reexamine NHST and its shortcomings from a perspective you won't get from statistics class. (If a refresher of NHST would be helpful, please consult any introductory statistics textbook.)

Suppose that you've heard about the obesity epidemic that is sweeping the country[1] and are worried about the effects of excess body weight, "metabolic syndrome," on people's health. When you learn that there is a roughly parallel rise in obesity and in doughnut consumption,[2] the scientific hypothesis occurs to you that "Obesity is caused in part by excessive doughnut consumption."

To test your hypothesis, you recruit doughnut-seeking males (DSMs) from lists of "top consumers" at doughnut shops. ("Top consumers" amass bonus points—redeemable for more doughnuts—as rewards for their purchase records. I'm making this up.) For comparison, you recruit doughnut-avoiding males (DAMs) from "doughnut-eaters anonymous," a support group for recovering doughnut addicts. One prediction that follows from your hypothesis is that "DSMs will weigh more than DAMs."

To test your scientific hypothesis, you develop a statistical hypothesis, a *null hypothesis*, $H_{0,}$ that the "DSMs weigh about the same as DAMs." Basically, you reformulate your *scientific prediction as a statistical hypothesis*. There is nothing explanatory about it. It concerns one possible consequence of doughnut eating and addresses the question: Are doughnut-eating habits associated with differences in male body weight or not?

We'll assume that your DAMs come from the general population of American males over the age of 20 and therefore have an average body weight of 195.7 ± 0.9[3] lbs (mean ± standard deviation). If your DSMs weighed 275 ± 10 lbs and only approximately 5% of DAMs weigh that much or more, the chance of getting a group of such heavy DSMs by random chance alone would be small; $p < 0.05$. In line with the usual convention for biology, you'd reject H_0.

In rejecting H_0, you are saying that your statistical hypothesis of no significant difference between DSMs and DAMs was wrong. According to your test and the convention for significance testing that you adopted, the groups do differ in weight. And that's all you get from the statistical hypothesis; it tells you nothing specific about your scientific hypothesis. The information from your statistical test does allow you to conclude that the prediction of your scientific hypothesis was correct and that, therefore, your data were consistent with your scientific hypothesis.

Incidentally, in putting forward this problem, I assumed that eating too many doughnuts would lead to weight gain (i.e., DSMs would be heavier than DAMs). It made sense given what I suspect about the nutritional properties of doughnuts. Therefore, I considered a *one-tailed* significance test, sometimes called a *directional test*, that essentially bets that any deviation from H_0 can only be either greater than or less than the mean assumed by H_0. In this case, we planned to reject H_0 only if DSMs weighed more than DAMs. However, it is possible that compulsive doughnut consumption would trigger biochemical or neurological reactions (e.g., loss of vital nutrients, chronic nausea and diarrhea) that would have the opposite effect: DSMs could weigh less than DAMs. A one-tailed test of significance directed toward detecting only heavy DSMs would miss abnormally skinny DSMs. If you want to keep an open mind as to which way the chips will fall, you can do a *two-tailed* test that is sensitive to significant deviations in either the plus or minus direction with respect to H_0. Basic scientists generally favor two-tailed tests, so they don't miss something unexpected.

Note that we test the statistical hypothesis *directly*. We test the scientific hypothesis only *indirectly*, by testing the predictions that it makes. This is not hair-splitting; it is one of the chief distinctions between them, as it illustrates the linkage between the statistical hypothesis and the scientific prediction. Because it is long-established custom, I will refer to "statistical hypotheses" throughout the chapter; still, we should keep this distinctions between the two kinds of hypothesis in mind. There are at least three further distinctions between them that we need to know about.

5.B.2 Essential Empirical Content

Statistical and scientific hypotheses also differ in their *empirical content*. Their empirical content refers to the things in the world that they refer to. The empirical content of a scientific hypothesis is essential because it determines how we understand and evaluate it; empirical content is not the same as *numerical content*. This point is probably obvious but is worth emphasizing. Consider the scientific hypothesis, "Obesity is caused in part by excessive doughnut consumption." If you were to switch "excess" to "occasional," "consumption" to "avoidance," or "doughnut" to "hot dog," you'd have an entirely new hypothesis. Apart from substituting synonyms, changing any word in a scientific hypothesis alters what the hypothesis asserts, what predictions it implies, how we should test it, and, finally, how we decide whether to reject or accept it. The pivotal role of word meaning is what makes empirical content *essential* for a scientific hypothesis.

The *statistical* hypothesis, as part of a practical mathematical testing process, has no *essential* empirical content. The distinction between statistical and scientific hypotheses is like the distinction between numbers and the things you count with numbers. Once we have data and are ready to test a statistical hypothesis, its empirical elements—the terms that tell us what it is about—become irrelevant. You can strip away the word meanings and still do the testing.

In the example, you can answer questions such as, "do their body-weights (of DSMs and DAMs) differ significantly or not?" without knowing anything about weights. You can freely alter the external referents of the words without affecting either what tests would be appropriate for making the statistical comparison or for the results themselves. If you took the same numbers and declared that they represented the counts of green scales on the bodies of space aliens from Planet Zeflon as compared with those from Planet Yorlou, your statistical test would dutifully spit out a significance level for the difference. You could state with probabilistic certainty whether or not Zeflonians are scalier than the Yorlouese. Of course, while the concepts expressed in the statistical hypothesis are inessential for testing it, or for interpreting the results, they do determine how seriously we take the outcome.

The saying "garbage-in, garbage-out" originated in the early days of the computer revolution to remind users that powerful computers give silly answers when loaded up with silly data or silly programs. But the warning also applies to ordinary statistical testing. If the initial assumptions or the numbers are faulty, or the wrong test is chosen, statistical testing will generate a wrong answer as readily as a right one. And this, too, is because the statistical hypothesis itself is devoid of essential empirical content.

5.B.2 Testing Scientific and Statistical Hypotheses

A corollary to the preceding argument is that a statistical hypothesis can *only* be tested mathematically. No independent, nonmathematical criteria can be applied to evaluate its truth value, unlike a scientific hypothesis, which has an irreducible empirical foundation and logical consequences that open it up to testing in a myriad of ways. To test the doughnut hypothesis, we could assign groups of people randomly to include or avoid doughnuts in their diets, construct longitudinal dietary histories, compare the metabolic status of doughnut-seekers and doughnut-avoiders, perform controlled laboratory experiments to measure biochemical responses to doughnut consumption, and so on.

5.B.3 Relationship to the External World

In a sense, because of its fundamentally mathematical nature, the statistical hypothesis exists in an abstract world apart from the external world of the scientific hypothesis. You might object that the scientific hypothesis often refers to unobservable underlying mechanisms. Aren't these abstractions, too? They are. The distinction becomes clearer when we break down the process of testing a statistical hypothesis. As we'll review in detail later, when testing our statistical hypothesis, we compared the measured body weight data of DSMs and DAMs to imaginary, theoretical populations of body weights to look for possible untoward effects of habitually overindulging in doughnuts. We asked if both experimental sample groups were likely to have come from this same imaginary population. And, regardless of the outcome of testing the statistical hypothesis, we wouldn't learn anything about the external world reasons for any differences between the experimental groups.

In contrast, the true test of a scientific hypothesis is how well it performs in the world and what it tells us about actual physical phenomena or events, not about mathematical abstractions. The scientific hypothesis may refer to unobservable entities or processes—electrons, force fields, "chemical imbalances in our brains," etc. Nevertheless, as scientific realists, we believe that the unobservables in a scientific hypothesis do exist, and the test bed, the court of last resort, for a scientific hypothesis is empirical testing.[4]

If testing the statistical hypothesis revealed that DSMs were heavier than DAMs, the result would be consistent with our hypothesis. We might, therefore, be inspired to go beyond the data and infer something about the world. Since abnormally heavy body weights are associated with numerous health problems, maybe excessive doughnut-eating isn't good for you. Nonetheless, this conclusion would not follow from testing the statistical hypothesis alone. Instead, we would have to interpret the test results in a separate cognitive process that connects back to the scientific hypothesis. This is one example of how judgment enters into statistical testing—it is not a mechanical procedure that we can casually offload to a computer.

5.C Why Statistics Matters

When you measure something, the measurement includes the true value that you're interested in, plus unknown variable factors that are usually combined and called *error* or *noise*. Remember trying to measure the pH of a solution in high school chemistry lab? You had to worry about whether the pH electrode was clean, what the room temperature was, whether you'd calibrated the meter correctly, whether your shaky hands had pipetted the correct amounts into the test tube, etc. Variation in any of these factors could throw off the measurements from one trial to the next or explain why your lab partner's results did not agree with yours.

Scientists never get to study the entire population of the objects or events that they're interested in; they can only measure samples from the population and hope that their samples fairly represent the whole. Some fields have to cope with much wilder noise than others. In addition, the living things that biologists study vary intrinsically. Even identical twins are not exactly alike. Experimental physicists can exert great control over their experimental conditions and reduce noise to small values, which lets them make precise and accurate measurements. Regardless of the circumstances, noise is an unavoidable fact of scientific life.

Fortunately, much of the noise is not utterly unpredictable. It follows patterns, and we can often deal with it tolerably well with probability theory and statistics, the branch of mathematics that deals with noise. We use statistical reasoning to decide what a test shows and what its results mean. But that's getting ahead of the story. Let's start with one of the primal ideas in statistics: probability. Probability is a fundamental concept although it is not an easy one to grasp.

5.C.1 Frequentists and Bayesians

There are two major schools of statistics, the *frequentists* and the *Bayesians*, and they differ both practically and philosophically. What you learned in science classes in school was almost certainly frequentist statistics: measures of central tendency (mean, median, mode), distributions of data, variance and standard deviation and how to compute these parameters, how to do t-tests, and the NHST procedures that I've mentioned. We are not going to worry about the procedural details, although they are important and are not always understood or properly used by scientists.[5] Instead, we are going to consider the statistical concepts in the context of bigger ideas.

Gerd Gigerenzer[6] makes the distinction between chance as "risk" or chance as "uncertainty." (Be aware that "uncertainty" here has a narrower sense than it does elsewhere in the book.) *Risk* is what we face when we have a fixed set of alternative outcomes and can calculate their probabilities. Gambling in casinos is the classic case of engaging risk: you do not know exactly what will happen on any given turn of the wheel or roll of the dice, but, in the long run, you are guaranteed to lose money at a rate determined by the odds inherent in the particular game and, usually, also by the local government, which decides in advance how much it wants to be paid by the casino. After the government gets its take, the rest is left for the casino owners' profits and players' "winnings." In West Virginia, for instance, odds of winning at the slot machines are adjusted so that enthusiasts can be assured of parting with, on average, about $20 of every $100 worth of quarters that they pour into the machines.[7] The $80 is returned to them in the form of "jackpots" that the state, of course, then taxes. To calculate risk, we use the frequentist interpretation of probability.

In contrast, in many daily circumstances, there are no standard odds, no "risk" as defined above, and we are in a state of "uncertainty." Consider this classic example: "There is a 30% chance of rain tomorrow." What does that even mean? Or, supposing you knew what "chance of rain" meant, how would you go about determining what that uncertain chance actually was? It seems that we have to think about "probability" in other ways when dealing with risk and uncertainty. There are statistical ways of interpreting and reckoning with uncertain statements, and they exist mainly within the branch of mathematics called *Bayesian statistics* that we'll get to in Chapter 6. For now, we'll stay with the frequentist interpretation and see where it leads.

5.C.2 Probability: A Frequentist Viewpoint

For frequentists, probability is a way of characterizing the real world. Events recur with a frequency which, in the long run, is their *probability of occurrence*. Properties of the world have definite values that, because of variability, are generally unknown. Under the frequentist interpretation, you can often calculate the exact value of a probability. If you know all of the elements of a group—the numbers of spots on the sides of a pair of dice, the percentage of females in a population, the incomes of American workers—you can calculate precise probabilities involving the elements. The chance of rolling a seven with a pair of dice (i.e., a total of seven spots showing on their upper surfaces when the dice come to a halt) is equal to the number of ways that the sum of the spots on the two dice can add up to seven—there are six ways with two ordinary dice—divided by the total number of possible outcomes of rolling them—36 possibilities—so the chance of getting a seven with two dice is 6/36 (1/6). In this case, the sources of variability include countless unknown factors: the detailed contours of each die, the velocity and angle of the toss, the collisions the dice undergo, what they're made of and what they land on, etc. Or, the chance of randomly picking an American worker with a very high income is equal to the total number of workers in the very high income category divided by the total number of American workers.

It is important to stress, though, that you can only calculate precise probabilities in the long run (i.e., essentially, the infinitely long run); there is no guarantee that on any given trial or group of trials you'll get an expected outcome. Rolling two dice six times does not mean that a seven will show up on one of those rolls, and assuming that a seven is "due" just because you haven't rolled one in a while, is called *the gambler's fallacy*. In the limit, however, sevens should show up on one-sixth of the total throws.

While frequentists aspire to be entirely objective in their calculations, they do take some things for granted, and subjective judgment enters into their procedures. Frequentists assume that variability (in measurement, sampling, noise) is distributed according to a *bell-shaped curve*, meaning they follow a mathematical *Gaussian distributions* or, as it is commonly known, a *normal distribution*. Statisticians know all about the mathematical properties of a normal distribution, and, if values are distributed normally, statisticians can carry out tests that are based on the distribution. This is especially handy for biologists since many biological parameters do follow a normal distribution. If you assume a normal distribution of human heights, for example, you can infer that the average professional basketball player in the National Basketball Association, who is 6′7″, is taller than 99.95% of Americans.

It is true that your assumption of normality will often rest on sketchy information. Fortunately, frequentist statistics is backed up by the *Central Limit Theorem of Statistics*,[8] which allows you to tap into the benefits of the normal distribution even if the underlying data that you are interested in aren't truly normal.

As an example of how the Central Limit Theorem simplifies things, assume that many people in the small town of Averageville, USA, are of Dutch descent, and, since the Dutch are a very tall people, the distribution of heights in the town is skewed toward taller than average (i.e., it is not a normal distribution). The Central Limit Theorem says that if you take many sample groups of Averagevillians, calculate the mean of each group, and plot all of the means, the plot will tend to follow a normal distribution. The Central Limit Theorem is a pillar upholding many frequentist procedures, although we tend to take its applicability for granted if we're aware of it at all.

In summary, frequentists assume that a real-world parameter has a single, fixed-though-unknown value. They think about probability because variability obscures the true values that they want to find. To draw their inferences, frequentists compare actual observations with observations predicted by their scientific hypotheses. If this sounds familiar, it is because the frequentist school of thinking was the default mode in your education. The frequentist viewpoint is intuitively appealing and powerful in many contexts, such as when we want to how fast the casinos will drain our wallets or testing the predictions of a scientific hypothesis. As Karl Popper is our touchstone for this topic, we should step back and look at the subject of probability in hypothesis testing as it first appeared to him.

5.D Popper and Probability

Recall that, because scientific facts can't be proved to be true, philosophers who felt that it was important to justify scientific truths had to give up on the idea of conclusive verifiability and be satisfied with "verifiability in principle," or probable verifiability (Chapter 2) replacing the ideal of Truth with a watered-down version. Many people found this to be reasonable; if we can't really be sure of anything, they thought that settling for probable truth would be an acceptable compromise. Popper found the thought of compromise completely unacceptable; it violated the principle that science aimed to discover Truth, and he took his principles seriously. Whatever "probable truth" might be, it definitely was not Truth, yet he could not arbitrarily dismiss it. Indeed, probability itself posed a serious challenge to his program of falsification.

5.D.1 Probable Truth and Popper

Popper began by analyzing the notion of the *probable truth* of a hypothesis from a frequentist perspective in the longest chapter in *The Logic of Scientific Discovery*.[9] For example, he asked how you could define "probable" truth. Could the probable truth of a hypothesis—its *probability*—be the number of correct predictions a hypothesis makes divided by the total number of predictions it makes? At first glance, this might make sense: a "probably true" hypothesis should certainly make many correct predictions. The problem is that the frequentist probability, a ratio, will be strongly determined by the denominator, which in this case is the total number of predictions that a hypothesis makes. Every hypothesis potentially makes an infinite number of predictions,[10] and we can only ever test a finite number of them. If you divide any finite number by an infinite number the result is zero, and, hence, the probable truth of a hypothesis assessed in this way is zero. Not a very helpful insight.

Taking another approach, Popper asks, "How, using the frequentist definition, would we interpret a hypothesis that has a probability of 0.5?" Would it mean that half of its predictions are true and half are false? This, too, is nonsensical. Or consider, if we don't know the Truth in advance, how would we know if we're halfway to it? After exploring a number of logical possibilities, Popper concludes that "probable truth" is meaningless within a frequentist framework.

If probable truth meant anything, we'd have to understand it subjectively, but subjectivity has no place in Popper's worldview. For one thing, subjectivity would imply that truth would "depend on the skill and training of an experimenter," rather than on "objectively reproducible and testable results." The thought demolishes any hope of rigorously defining the probable truth of a hypothesis and, for Popper, also firmly closes the case against inductivism, which considered that repeated observations made a conclusion more likely to be true. But Popper's failure to define probable truth was the beginning, not the end, of his real struggles with probability.

5.D.2 The Problem of Probability and Hypothesis Testing

The bedrock notion of Conjectures and Refutations is that, to be scientifically meaningful, a statement must be falsifiable in principle. Yet, even in principle, a probable statement cannot be absolutely falsified. Consider, when the local high school student asks you to buy a raffle ticket to support the school's chess team, you mentally write it off as a donation; knowing that 1,000 tickets will be sold and the chances are 999 to 1 against winning, you assume you have "no chance."

Nevertheless, "against all odds" you win! Amazing! Still, winning was always a possibility, and, in fact, if the chance of occurrence of any physical process is not literally zero, it may happen. This means that, no matter how improbable your experimental results are, they could not, strictly speaking, falsify your hypothesis. The hope that falsification would provide a sound foundation for scientific reasoning was faltering. Probability, the stick that Popper had used to fend off his nemesis, came back to beat him.

5.D.3 Popper, Probability, and Falsification

One of the seminal events in the evolution of Popper's thinking was the test of Einstein's groundbreaking General Theory of Relativity that was carried out by the astronomer Arthur Eddington (see Chapter 2.I.2.). The theory, you recall, predicts that light travels through space along curved paths. Eddington's observation that starlight did, in fact, curve as it traveled through space, thus falsifying Newton's theory, greatly impressed Popper. Physics became his ideal of how science should be carried out. There was only one problem. Physics, especially the branch of physics called *quantum mechanics*, is probabilistic to its core (at least in its traditional interpretation, see Chapter 10.C, for a nonprobabilistic interpretation). An electron is never *in* any one place; it only has a *probability* of being there. How could Popper reconcile the great strides that physicists were making with the fact that statements having a non-zero probability of being true can't be falsified?

Popper's solution was pragmatic. What physicists actually did was to consider some explanations as so improbable that they could safely, if tentatively, be ruled out. Physicists have, Popper said, adopted a convention, a *methodological rule*, that went something like this: we should not explain things as being the way they are because of a series of unlikely, unaccountable accidents. Instead, we should agree to disregard explanations that are sufficiently improbable that we can't (for the time being) accept that they are true.

Suppose that you're driving down a highway one night, and you notice that all of the hundreds of cars coming toward you have only their left headlights lit. While it is physically possible that all of the right-side headlights randomly happened to burn out, such a thing would be so extremely improbable that you should assume that there was a simpler causal explanation. Maybe the line of left-lighted cars was a staged event of some kind, or maybe teenaged boys had smashed all of the right-hand lights of cars parked in a lot because that's the sort of thing that teenaged boys would do. If you concluded that all of the right headlights just happened to have burnt out, you would be violating the methodological rule not to accept extraordinarily improbable explanations for phenomena. Physicists established rules for how unlikely a result could be before they invented a theory to explain it.

Popper fastens on the notion of a methodological rule and proposes that scientists in any field could agree to adopt their own convention for making decisions where probability came into play. The crux of the idea is that, if their experimental results were so unlikely that they fell outside the probability limit of what an existing theory could explain, then the results would falsify the theory. These ideas are close in spirit to the statistical testing that scientists today do. For example, the "methodological rule" for considering a result in biology to be significant is commonly accepted as $p < 0.05$. Yet Popper was working in the 1930s, before the establishment of modern statistical methods, and he doesn't refer to proper statistical testing.[11]

You may find Popper's proposal unremarkable, even obvious, but it has subtle and far-reaching implications that must be understood by anyone wishing to know how science works and how statistics influences our knowledge, so I'll take a minute to emphasize it here and come back to it later as well.

What we're saying is that Popper's methodological rule, as it has been extended and refined by statistical analysis, is a *tool* for making decisions; as a tool it is vitally important, though it is still only a tool. A statistical test can reveal that a particular result is so extremely unlikely in light of a given hypothesis that, according to our judgment and the conventions adopted in our branch of science, we can regard it as effectively impossible. Thus, if such a result occurs, we can consider it as evidence that falsifies the hypothesis. Statisticians and philosophers still actively debate the meaning and validity of the rules, as well as the distinct but related matter of what conclusions and opinions are "warranted" by statistical test results.[12] These elevated conversations notwithstanding, scientists today consider a statistical significance level (nowadays, a *p-value*) to be a rational device for evaluating the predictions made by a hypothesis.

We keep in mind, though, that getting a result with a low p-value, say $p < 0.001$, only means that that result would only rarely occur by chance alone; it does *not* mean that the result is special or, more importantly, that it is true. A significance level is a handy indicator, not an end in itself; it is like the oil level indicated by the dipstick in your car's engine, not a merit badge or an achievement. It gives you information that, provided you're willing and able to interpret it, may be useful to your scientific, decision-making process. We'll return to the concept of p-value later in this chapter and again in Chapter 7 on the Reproducibility Crisis.

In summary, despite the fact that non-zero probability statements in themselves are literally unfalsifiable, when combined with a methodological rule for interpreting them, they are compatible with falsification tests of predictions. (Remember that statistical tests are testing scientific predictions directly, and scientific hypotheses only indirectly.) Right statistical reasoning is entirely consistent with the rest of Popper's thinking: uncertainty is the ultimate reality, and, as long as the ultimate goal is the search for Truth and the results permanently subject to criticism and revision, we can learn from our mistakes. Statistical testing fits right in.

The question then becomes, "what kind of statistics should we use?" The standard answer for biologists, psychologists, and others has been the frequentist-based, NHST program, which we'll turn to next. While I'll start with a few basic concepts, even readers with solid practical experience in statistical methods may find something new here. We scientists need to know a little about the complex history of what we've come to think of as "statistics" in order to appreciate how we've been misled and, in many cases, gone wrong. First, what is NHST all about?

5.E Frequentist Statistics: The Rocky Road to NHST

Scientists are taught the NHST strategy as a natural and intuitively obvious statistical framework; they are not exposed the messiness at its core.[13] Although we learn it as a coherent subject, the NHST program is an "inconsistent hybrid" system[14–16] that was kludged together from two rival schools of thought. The founders of these schools, Sir Ronald A. Fisher, and Jerzy Neyman and Egon Pearson, disagreed sharply about fundamental principles. Their unresolved dispute continues to cause trouble for our comprehension of scientific procedures today.

5.E.1 Fisher and Null Hypothesis Testing

Fisher (1890–1962), who invented the concept of testing the *null* hypothesis, was a geneticist who was initially interested in answering questions such as, "does this manure really help the potatoes grow?" He was at a well-established agricultural station and had extensive records to work with. At the time, there was no systematic way of figuring out if a particular fertilizer mixture made a difference to plant growth. Fisher's insight was to put forward and test a null hypothesis, H_0 (i.e., the statistical hypothesis that there is nothing beneficial about a given mix). Basically, H_0 said that the same amount and quality of potatoes would grow in both treated or untreated fields. The *null* hypothesis, the one that may be *nullified* by your experiment, is technically distinct from a *nil* hypothesis (i.e., one that says that there is *zero difference* between your sample and the population you are testing it against). In fact, the *null* is not necessarily a *nil* (you could test for a definite, non-zero difference between the two groups); however, it often is, and we won't distinguish between them.

If you were following Fisher's procedure, you would start with the assumption that some biological parameter of the potatoes, say their weight, is distributed according to a normal distribution in an ideal, imaginary population of untreated potatoes. You estimate the properties of this ideal population by taking a sample of untreated potatoes and compare it with a treated sample. You randomly sample both groups to be sure that potentially influential variables would be scattered around and not bias the outcome. Your H_0 would be that your experimental sample of manure-treated potatoes could have come from the untreated population. If the treated potatoes weighed about the same as the untreated ones, you would not reject H_0 because you'd have no reason to think that the manure did anything. In contrast, if you'd expect to encounter potatoes as heavy as your sample's in an untreated population only very rarely, you'd suspect that the fertilizer did something and you'd reject H_0.

Fisher invented the *significance test* to answer the question, "should we reject H_0 or not?" In his scheme there is no explicit alternative hypothesis, only H_0, and what we reject is the *conclusion that H_0 can explain our data*. The ideal population of values assumed to exist under H_0 wouldn't change whether we reject H_0 or not. Despite there being no explicit alternative hypothesis, the implicit alternative is "*not H_0*." If we reject H_0 we might decide to do a full-scale investigation. Testing H_0 for Fisher is, effectively, a pilot study, a screening procedure that you do with small samples, just to see if there might be anything worth investigating, not to answer meaty research questions.

How do you know whether or not a result is unusual enough to justify rejecting H_0? In the doughnut-eating example, we asked whether DSMs and DAMs differed in their body weight. H_0 was that they wouldn't differ. Fisher said we should pick a level of improbability that seems "significant"; he called the significance level "sig," but nobody else does anymore: it is the "p-value." Since the DAMs' weights are distributed according to a normal distribution, and we know a lot about the normal distribution, we can choose a range of values away from the mean as our significance level. Fisher believed that values above or below 95% of the population would be interesting for many purposes.

For Fisher, there was nothing sacred about a significance level; how to interpret the results of a significance test was a judgment call, not something that followed a fixed rule. Even though encountering a group like your sample of heavy DSMs by random chance alone would be unusual, it would not be impossible. Under Fisher's theory, rejecting H_0 simply means that you should consider the possibility that doughnut-seeking affects adult male body weight.

Each branch of science could adopt a *convention* for a significance level to reject H_0, but strict adherence to conventions should be avoided. Indeed, Fisher felt that we should report exact p-values (e.g., $p = 0.0432$) so that each reader could judge how important the results were. The exact probability is not the same as your significance level—if you base your decision to accept or reject H_0 on $p \leq 0.05$, then $p \leq 0.05$ is your significance level, not $p = 0.0432$. The smaller the p-value, the more unlikely the result would be and therefore the more significant the result. Fisher encouraged *post facto* analyses of the data because he considered data interpretation to be so important. For him, this would not introduce unhealthy bias because significance testing was only a preliminary exploratory step.

5.E.2 Neyman-Pearson Decision-Making

Neyman and Pearson thought that null hypothesis testing was meaningless; it was far less valuable to know that a statistical hypothesis was wrong than to know if it was right. Their program is about decision-making. In their scheme, there are always at least two hypotheses—usually H_1 and H_2—and you assume that one of them is correct before you do any testing. Statistics helps you decide which one it is. You test just one hypothesis, the one that you think is most likely to be right, and, given that one of the two has to be right, after the test you know by elimination which. If you reject H_1, you accept H_2.

Let's look at the excessive doughnut-eating problem from the Neyman-Pearson perspective. Now your H_1 might be that "the weights of DSMs and DAMs have the same mean weight, 195.7 ± 0.9 lbs" and H_2 could be that "on average, DSMs weigh 30 lbs more than DAMs (225 ± 0.9 lbs)." These hypotheses presuppose that the body weights of males fall into the distributions of doughnut-seekers or of doughnut-avoiders, which characterize the two populations. In testing H_1 versus H_2, you are deciding whether your subjects came from the same or separate populations. Under Neyman-Pearson rules, you know the difference in the means of the two underlying distributions, which is called the *effect size*[17] (we'll take up effect size in Section 5.G). If your statistical test shows the groups differ, then, unlike the case in Fisher's program, you also know by how much they differ on average.

5.E.2.a Decision-making according to Neyman-Pearson

Significance testing for the Neyman-Pearson program determines which hypothesis is likely to be correct. When you know which hypothesis to reject, you can then act as if the other is true. You do not have to believe it is necessarily true in every aspect or that it constitutes the whole story, but it is reasonable to act as if it were true. If you find that doughnut-eating is associated with heavier body weights, then you should take this into account when making health policy recommendations, even if you suspect that other factors might also be at work. Unlike Fisher, who saw significance testing as a preliminary tool, Neyman and Pearson made it a key factor in the conduct of research and decision-making. As such, the small sample sizes that sufficed for Fisher are unacceptable in their program; Neyman-Pearson theory demands large experimental samples to assure greater confidence in the results.

5.E.2.b Statistical Errors

Neyman and Pearson also originated the idea of analyzing errors and their consequences. Statistical error, like sin, comes in at least two forms. There are *errors of commission*, commonly known as *α*, *alpha*, or *Type I errors*, and *errors of omission*, *β*, *beta*, or *Type II errors*). The α level determines the chance of *wrongly rejecting* your main hypothesis, H_1. If you reject H_1 when you shouldn't, you make an error of commission; you mistakenly conclude that the variable that you were investigating (e.g., seeking doughnuts) is associated with increased body weight when it isn't. The α level is superficially similar to Fisher's significance level or p-value but differs in crucial ways. Neyman-Pearson theory assumes that the experimenter will be doing multiple measurements and tests, and, therefore, α is the likelihood of committing the error of wrongly rejecting H_1 *in the long run*.

The converse error, the one you make if you *don't reject H_1 when you should*, is a β error: you conclude that doughnut-seeking is not associated with heavier body weights when it actually is. A major feature in Neyman-Pearson experimental design is that these error levels must be decided on before doing the experiment because you're going to base meaningful, practical decisions on the outcome of the testing.

5.E.2.c Statistical Power

Neyman and Pearson thought that we should know if a test could correctly identify and accept a true H_2. The probability of correctly accepting H_2 is called *statistical power*, or just *power*, and is defined as $1 - \beta$. This makes sense: the probability of incorrectly rejecting H_2, which is β, plus the probability of correctly accepting H_2, which is power, must equal 1. You've got to do one thing or the other—accept H_2 or reject it—so, $\beta + power = 1$, and, rearranging things, $power = 1 - \beta$.

To assess statistical significance, you have to take into account how big a difference you're trying to detect and the chance of error you're willing to tolerate. And how much you can tolerate depends on what the error will cost you. The problem is that you can't avoid error altogether, and reducing the chance of making one kind of statistical error automatically increases the chance of making the other. Although you can't escape statistical error, you do care which one is more likely because they come with varying costs. Fortunately, you are not totally helpless; you can influence how much each error will affect the results.

You could think of the situation like this: imagine you are sitting across from a colleague at one of those tiny café tables. Two hypotheses can account for control of the table top: H_1—"your space," and H_2—"his space." To claim the area you believe you are entitled to under H_1, you stake it out by subtly arranging your utensils, water glass, coffee cup, etc., along a border between you and him. Your colleague, acting under H_2, does the same until you both reach an equilibrium. The border is like your significance level; you accept that space beyond the border as not being within H_1. On the other hand, it is a small table; if he were not there, it would all be yours. With him there, your spatial claims overlap. Not staking out enough territory is like committing an α error: if, for example, you set your boundary so that you are 95% certain of claiming of your actual territory, there is a 5% chance that you're giving up control over area that is rightfully yours. In effect, you risk erroneously rejecting H_1. To reduce the chance of this mistake, you might expand your claim, nudging your utensils, water glass, etc., slightly forward, to encompass what you are 99% certain should be within your domain. This would decrease the possibility of understating the true extent of your H_1, and yet stretching your boundary could encroach on territory that is legitimately part of his H_2, thus potentially provoking his annoyance and retaliatory moves. Wrongfully claiming area for H_1 is like committing a β error by erroneously failing to accept H_2. Obviously, the harder you try to avoid making an α error, the greater your chance of making a β error. The balance you strike will depend on a cost-benefit analysis of the entire situation, including, in this case, the importance of remaining on speaking terms with your table mate.

Let's see how this works experimentally. At the $\alpha \leq 0.05$ level, you have a 1 in 20 chance of being wrong in concluding that abnormal doughnut consumption has no downsides. If this risk is unacceptably high—maybe denouncing doughnut-eating as unhealthy would threaten your state's doughnut tax revenues, decrease doughnut industry employment, or unfairly stigmatize DSMs—and you want more assurance before drawing it, you can reduce the significance level of your test to $\alpha \leq 0.01$; then your probability of incorrectly rejecting H_1 drops to 1 in 100. With this more stringent level you would expect to encounter fewer heavy DSMs by random chance alone. If your sample of DSMs weighed in with the heaviest 1% of the population, you could be pretty confident that they did not reflect normal variability. But your low α level means that you could miss detecting a genuine problem with doughnut-eating, a β error. If DSMs do gain weight, yet not enough to put them into the top 1%, then you'd incorrectly accept H_1 and infer that eating lots of doughnuts doesn't affect body weight. If β error goes up, statistical power goes down because power = 1 – β. So, decreasing α error also decreases statistical power and, with it, your ability to identify a true H_2.

5.E.2.d Errors, Power, and Convention

To be sure, there is still no such thing as a free lunch. While increased statistical power by itself would be a good thing, when power goes up, generally the chance of "finding" something that isn't there also goes up. When power drops, the chance of missing a real effect goes up. The best way to increase power while minimizing error is to increase the sample size, *n*. This, too, involves a balancing act, however. Increasing *n* is comparatively easy for physicists hunting the Higgs boson at the Large Hadron Collider who collect and analyze enormous masses of data automatically. It can be more difficult and costly for small-scale biology studies to increase sample sizes enough to meet optimal power requirements. In Chapter 7, we'll see that critics of neuroscience research[18] blame tests having with low statistical power for many of the woes of the Reproducibility Crisis.

Parenthetically, we note that Benjamin and colleagues[19] propose decreasing the p-value in, for example, biological sciences to $p < 0.005$ for "new discoveries" (not for "confirmatory or contradictory" effects). Chief among the anticipated advantages of this change would be decrease in the *false positive rate* (i.e., the number of erroneous claims of discovery due to too low a significance level). Lakens et al.[20] respond that the drawbacks of decreasing the significance level across the board outweigh its advantages. Drastically reducing the p-value (α level) will, among other things, increase the *false negative rate* (i.e., the number of genuine effects missed because of a too-conservative level). Rather than a one-size-fits-all approach, Lakens et al. favor flexibility, exhorting scientists to focus instead on rigorous, well-thought-out experimental design that encourages them to "justify your alpha" rather than see it as a goal in itself.

However this debate is resolved, it is clear that the level of α error that a given branch of science generally adopts is not purely arbitrary; there are strategic considerations. Much basic biological and social research tolerates a 1/20 ($p < 0.05$) chance of α error, whereas particle physicists won't announce the discovery of a new subatomic particle unless their chance of making an α error in identification is approximately 1/3,500,000 ("5 sigma"; i.e., $p < 0.0000003$). Biological scientists don't want to risk missing a new finding by being overly cautious at first, and the cost of an α error is often relatively low in their laboratory experiments. They may not want to miss possible harmful effects of, for example, doughnut consumption on human health, and consider the potential costs of erroneously telling people not to eat so many doughnuts to be acceptable.

On the other hand, particle physicists can't take the chance of erroneously accepting the existence of a phenomenon that could wreak havoc with their beautiful quantitative theories. As the Higgs boson is a cornerstone of the Standard Model of physics, physicists could not afford to be overeager in thinking they'd found it.[21,22] The cost of an α error for physicists is high, and they choose an extremely low p-value.

Box 5.1 Two Statistical Standards for Quality Control

Assume that a precision rotator gear is a key component in the small ultralight aircraft called "gyrocopters" that can be used by political protestors to land on the US Capitol lawn.[23] And assume that defective gears can fail without warning, thus imperiling pilots' lives and that, like every manufacturing process, gear-making is imperfect. Now, requiring acceptable gears to be perfect would be economically infeasible so, to stay in business, gear manufacturers balance the costs of defective products against the costs of achieving perfection. In this example, H_1 says that a given gear is good (i.e., that its properties are well within the variability of the population of acceptable gears); H_2 says that the gear is bad. The Quality Control Department in the gear manufacturing plant, encouraged by the attorneys in the Product Liability Department, may dictate a high α error rate for H_1, say $\alpha \leq 0.2$ (two-tailed), and reject 1 gear in 5. In other words, to be on the safe side, they'll be conservative and reject some gears that would probably be all right in order to catch a high percentage of gears that are genuinely no good.

H_2 is set to identify definitely bad gears. Here, the Quality Control folks might call for a low β error rate, say ≤ 0.05, which translates to a power $(1 - \beta)$ of the test of $1 - 0.05 = 0.95$ (i.e., quite high given that a power of 0.8 is considered reasonable).[23] This high power would mean that there is an excellent chance of accepting H_2—deciding that the gear is bad—when the gear really is bad. The properties of normal and defective gears still overlap, but having two standards for keeping bad gears out of the aircraft would build in a greater margin of safety thus saving lives, not to mention money for the gyrocopter company owners. And maybe they'll throw in a free parachute with each purchase, just to be on the safe side.

Note that the Neyman-Pearson approach, with its two independent hypotheses, H_1 and H_2, allows you to have separate α and β levels for each one and this, in turn, lets you design especially powerful tests. An example is given in Box 5.1.

5.E.2.f Summary: Fisher Versus Neyman-Pearson

Fisher:

- Null hypothesis testing for exploratory studies on small experimental samples
- Significance testing only on H_0, which you can only reject
- No alternative hypothesis

- Calculate and report exact probabilities so readers can determine the significance of results; lower p-values may signal more important outcomes
- Judgments based on experimental results; significance levels are only guidelines
- *Post hoc* analyses of the data are OK (studies are preliminary)
- No provision for α error, β error, or effect size

Neyman-Pearson:

- Decision-making program; significance testing determines which hypothesis to accept
- No null hypothesis; both hypotheses are substantive; one is assumed to be correct
- Testing reveals magnitude of the difference between hypotheses, the effect size
- Requires large samples and assumes repeated sampling
- Calculates the magnitudes of α, β errors, and statistical power of tests
- Details of analyses and α, β, and significance levels are set in advance of testing
- No post hoc analysis allowed

5.F The NHST Program: A Little of Both and a Lot of Shortcomings

By the middle of the twentieth century, many scientists realized that they needed a systematic, objective method of evaluating their experiments, but the conceptual foundations for Fisher's and for Neyman and Pearson's programs were barely compatible. Rather than choose between them, textbook authors and researchers gradually settled on a merger of the two, which became NHST, even though NHST had critics from the start (one wanted to rename it "Statistical *H*ypothesis *I*nference *T*esting"[24]; emphasis added).

From Fisher, we get null hypothesis testing, but we apply it for purposes that it was never intended for, and we do it in a rote, unthinking way that Fisher abhorred. We learn NHST as a "ritual,"[25] as a series of steps to be memorized and slavishly followed, not as a flexible tool for fostering informative insights. We presume that its dictates are meant to be obeyed, not judiciously considered. For these reasons, many of us lack a sophisticated appreciation of the strengths and limitations of statistical testing. Worse than that, because we learn to carry out NHST as an empty ritual, we are subject to a range of "delusions" that can contribute to the Reproducibility Crisis, as we'll see.

Many statistics textbooks directed at scientists essentially define H_0 as a *nil* hypothesis, which was not Fisher's intent; as noted earlier, we can specify a non-zero difference between groups as H_0. We use null hypothesis testing to make substantive "yes-no" decisions that it was not designed to make. And by misinterpreting Fisher, NHST ratifies the reliance on small samples; users of NHST seldom hesitate to draw sweeping conclusions from what he meant to be pilot studies. We ignore the Neyman-Pearson demand for large sample sizes and the need to estimate certain critical parameters. For example, we might guess that the variability in the distributions of the weights of DAMs and DSMs is the same, but since we don't know much about DSMs, our calculations may be only as accurate as the guess is right. We rarely take the concepts of statistical errors and power seriously, and, in fact, many of us don't learn about statistical power or its importance.

In any case, we cannot meaningfully calculate the probabilities of α and β errors or statistical power without having substantive alternative hypotheses or adequate sample sizes. When it comes to alternative hypotheses, we have at most "H_a," which simply states that "H_0 is wrong." When we're testing a null hypothesis against an empty alternative, we're avoiding the deeper analysis of the scientific question and the need to formulate explicit alternative scientific hypotheses. If we only test a hypothesis of the deleterious effect of doughnuts by comparing DSMs and DAMs with H_0 of "no difference" versus H_a of "difference," we may be ducking the underlying biological (or social) issues involved. Testing H_0 might be a reasonable place to start if we have no information whatsoever, but we shouldn't consider it an end unto itself.

Null hypothesis testing can easily generate nonsensical conclusions because there is no such thing as absolute equivalence between two groups. With a sufficiently fine-grained analysis and large enough sample sizes, you can almost always find statistically significant differences between groups. Paul Meehl[26] first called attention to the flawed conclusions that can result from overlooking this fact. For example, do you think girls and boys have the same IQs? Get big enough groups of kids and test the null (nil) hypothesis with the most rigorous, finest-grained analysis possible and you are mathematically guaranteed to reject H_0 because their IQs will not be absolutely identical. Even if your hypothesis is directional—"Girls have higher IQs than boys"—you have a 50% chance of being right[27] since H_0 is certain to be rejected. So much for $p \leq 0.05$!

Conflating Fisher's and Neyman-Pearson's theories makes it hard to understand critical statistical principles. Take, for example, the concept of the p-value itself; what exactly does it refer to? The p-value is the percentage of the imaginary distribution of values that you expect to observe *if H_0 is true*; convention condones rejecting H_0 if the sample values are at least as extreme as the p-value. If the mean weight of DSMs is 230 lbs, and this is significantly different from the population mean of 195.7 lbs at a p-value of ≤0.05, this means only that the probability of randomly drawing a sample group weighing 230 or more from the normal population is ≤5%. It does *not* mean that you have a 5% chance of making a wrong decision if you reject H_0 or that, if you repeated the experiment many times, you would get a significant result on 95% of them. Yet, when quizzed about these issues, majorities of psychology students and large numbers of teachers, even statistics teachers, misinterpreted the p-value.[28] It would not be surprising if bioscientists had similar misconceptions about their statistically derived scientific conclusions. As I noted earlier, the p-value is at most a useful tool, one of many, that can help evaluate the predictions of our hypotheses.

These examples hint at the corrosive influence that the rote application of NHST can have on scientific thinking abilities. Focusing narrowly on p-valued significance testing causes us to lose sight of other kinds of assessments. Indeed, you might think that if the problem is p-valued testing, we could solve it by avoiding these tests. Is that true? Would non–p-value-dependent testing strategies circumvent the problems of NHST?

5.G Alternatives to P-Valued Testing: Effect Size, Confidence Intervals, and Errors

Measures of *effect size* and *confidence intervals* are often advocated as alternatives to the p-value approach; again, the technical details are easy to find, so we'll concentrate on the larger picture. Sometimes you need to know how big the difference is between your experimental groups. For instance, if you are concerned about public health policy, it would be good to know not only that DSMs weigh more than DAMs, but how much more. Because the significance of the p-value test depends on sample sizes, a relatively small difference in weight might show up as statistically significant with very large samples, while only a much greater weight gain would be significant with small groups. If DSMs weigh 198.2 ± 7.2 lbs and DAMs weigh 196.8 ± 8.3 lbs, this small (1.4 lbs) difference is not statistically significant with groups of 30 people ($p \leq 0.05$, one-tailed test). But with groups of 300 people, it would be. The difference of 1.4 lbs might have meaningful medical implications; then again, it might not. Basing a decision on the outcome of a large group study without knowing how big an effect you have is inherently risky.

5.G.1 Effect Size and *Cohen's d*

A frequently used index of effect size, *Cohen's d*, assesses the difference between two groups according to some property they share, and it is independent of the number of subjects in the groups. In the doughnut example, the effect size would be the difference in mean weights between the two groups, normalized by their average standard deviations.[29] Normalization converts the raw numbers into the common currency of effect size, which is dimensionless (i.e., not expressed in pounds, microns, newton-meters, etc.) and allows you to compare two dissimilar treatments directly. Effect size usually ranges from zero (no effect) up to a small number, say 3, on a scale where 0.8 is considered a "large" effect.[30] An interactive online calculator[31] allows you to manipulate Cohen's d and get an intuitive feel for the relationships between the populations being compared.

Consider the apparently burning issue[32] of whether or not some European groups are taller than Americans. If true, this would represent a remarkable shift: in 1840, Americans were taller than most other peoples, including, for example, the Dutch, who at the time were shorter than Americans and are now reckoned to be the tallest people in the world today. But what about others, say, the Swedes—maybe Americans still hold the edge? On average, American women aged 20–29 are 64.2 ± 2.82 inches tall,[33] whereas Swedish women of that age are 65.5 ± 0.08 inches tall.[34] Are Swedish women actually taller? The difference in mean heights of the women is 1.3 inches, and the effect size is 0.52,[35] which would be considered a medium-sized effect.[36] The effect size suggests that the Americans may be shorter, but can we say more?

5.G.2 Confidence Intervals

After determining effect size, you would want know if it is statistically significant. You can get this information from the *confidence interval*, which is an estimate of the range of possible values that the true value of a population parameter might have. You can calculate a confidence interval for many parameters (e.g., mean, median, even effect size). If you assume that the underlying population values are normally distributed, then the confidence interval is a portion of the distribution that extends above and below your experimental sample mean. For example, the 95% confidence interval extends from 1.96 standard deviations above the sample mean to 1.96 standard deviations below it (the values reflect mathematical properties of the normal distribution; don't worry if they're not intuitively obvious). What you can be confident about is this: if you repeat this procedure many times—take a sample, calculate its mean and confidence interval—in 95% of those cases, the true mean of the population will be in the calculated interval.

The 95% confidence interval around the sample mean of the American women's heights (64.2 inches) extends from 58.7 to 69.7 inches, and the true mean height of the population of American women will fall within 95% of the confidence intervals that you calculate. While you can't say there is a 95% chance that the true mean *is within* the confidence interval, you can be *95% confident* that it is.[37]

5.G.3 Confidence Intervals for Effect Size

How do you use the confidence interval to estimate statistical significance? If the mean of one group is outside the 95% confidence interval for the mean of another, you can be 95% confident that the groups are not the same. On average non-Hispanic white American men aged 20–39 are 70.2 ± 3.95 inches tall, and, as this mean is outside the 95% confidence interval for the 20 to 29-year-old women's heights, you can be 95% confident that the men are actually taller than the women. Things are not always so straightforward.

To see how confidence intervals can help when you have effect sizes of inter-group differences, let's return to the question of the American–Swedish female height disparity. The 95% confidence interval for the American women's height is 58.7–69.7 inches, and, as the mean height of the Swedish women (65.5 inches) is well within this interval, you can't immediately say that the groups do differ. Still, the substantial effect size between them, 0.52, is suggestive; you'd be tempted to think that they do differ. To take the next step, you can determine the 95% confidence interval for *effect size*—the calculation is less obvious than for the mean[38] but not difficult—and, in this case, you'd find that it ranges from 0.41 to 0.63. This means that you can be 95% confident that the true effect size is within this interval (i.e., that there is a noteworthy difference between American and Swedish women's heights).

Note that this confidence interval does not include zero. This fact is significant because it means that you can be 95% confident that there is indeed a genuine difference between the groups. If zero fell within your confidence interval, you could not exclude the possibility that there was zero difference between them (i.e., that the two groups are really the same). This is like rejecting the null hypothesis of no difference between the groups at $p \leq 0.05$.

The grand conclusion here is that you can get the same kind of information from effect sizes and confidence intervals as you get from the typical p-value tests while steering clear of the deficiencies of NHST. And, as a bonus, you get an estimate of the actual magnitude of the effect you're interested in. Nevertheless, even with effect sizes and 95% confidence intervals, you have a 5% chance of being wrong. You could reduce this risk—analogous to an α error—by widening the confidence interval, say to 99%, which would decrease the chance of wrongly concluding that the groups differ ("α error") to 1%, but, as you expect by now, this would simultaneously increase the chance of failing to pick up a genuine difference between them ("β error"). There are no magic shields to protect you against error.

In summary, there are problems with, and possible fixes for, the frequentist testing program. Given the inertia built up over decades of using NHST, it will probably remain the default standard for a time, although there are moves afoot to diminish its prominence, if not abolish it altogether.[39] As an alternative to NHST, Bayesian methods, are making inroads into experimental sciences in addition to sociology, as we'll see in Chapter 6.

5.H Coda

This chapter distinguished statistical from scientific hypotheses on practical and philosophical grounds. The statistical hypothesis is part of a mathematical testing procedure and is tested by purely mathematical methods. As a rule, statistical hypotheses are more like the predictions made by scientific hypotheses than the scientific hypotheses themselves.

We also introduced the two distinct interpretations of probability: the objective (frequentist) one and the subjective (Bayesian) one. In this chapter we focused on frequentist statistics, principally NHST, and how these influence scientific reasoning. Despite its popularity, NHST has notable shortcomings. Because it is an awkward amalgamation of two distinct statistical schools of thought (Fisher's and Neyman and Pearson's) that is treated as though it is a coherent whole, its pitfalls are seldom taught to, or appreciated by, lab scientists. The inconsistencies of NHST contribute directly to the statistical confusion that plays a role in the Reproducibility Crisis.

In accordance with Popper's recommendation to adopt conventional standards for determining if a test prediction is falsified, probabilistic propositions can be falsified within the limits of the uncertainty of the test. We covered some frequentist alternatives to the usual p-valued tests that avoid some of the drawbacks of NHST, although none of them is entirely free from drawbacks.

We will continue the introduction of statistical ideas in next chapter, which covers the basics of Bayesian statistics, and we'll draw on the fundamental statistical concepts (largely frequentist) in later chapters, where they play supporting roles in the examination of the Reproducibility Crisis (Chapter 7) and help illuminate the advantages of the scientific hypothesis (Chapter 8).

Notes

1. Obesity epidemic in the United States; see https://journals.lww.com/nutritiontodayonline/Abstract/2003/03000/The_Supersizing_of_America__Portion_Size_and_the.4.aspx.

   "Metabolic syndrome is not a disease in itself. Instead, it's a group of risk factors—high blood pressure, high blood sugar, unhealthy cholesterol levels, and abdominal fat"; see https://www.webmd.com/heart/metabolic-syndrome/metabolic-syndrome-what-is-.
2. Doughnut consumption in United States: https://www.statista.com/statistics/283198/us-households-consumption-of-donuts--doughnuts-trend/
3. Table of American adult (≥20 years old) male body weights: http://www.cdc.gov/nchs/data/series/sr_11/sr11_252.pdf
4. Theoretical physics may extend beyond the bounds of empiricism. See David Deutsch's discussion in Chapter 10.
5. K. S. Button, J. P. Ioannidis, C. Mokrysz, B. A. Nosek, J. Flint, E. S. Robinson, and M. R. Munafò, "Power Failure: Why Small Sample Size Undermines the Reliability of Neuroscience," *Nature Reviews Neuroscience* 14:365–376, 2013; R. Nuzzo, "Scientific Method: Statistical Errors," *Nature* 506:150–152, 2014; P. Bacchetti, "Small Sample Size Is Not the Real Problem," *Nature Reviews Neuroscience* 14:485, 2013; J. C. Ashton, "Experimental Power Comes from Powerful Theories: The Real Problem in Null Hypothesis Testing," *Nature Reviews Neuroscience* 14:585, 2013; and C. Hoppe, "A Test Is Not a Test," *Nature Reviews Neuroscience* 14:877, 2013.
6. Gerd Gigerenzer's theories relating to statistics, heuristics and biases, risk and uncertainty, and related matters were the major resource for the history of statistics discussed in this chapter. For convenience, I will reference the collections instead of the original articles, and, where themes in the collections overlap, I will cite one that seems particularly apt without intending to imply that that source is the only, or original, reference. See G. Gigerenzer, *Heuristics: The Foundations of Adaptive Behavior* (New York: Oxford University Press; 2011); G. Gigerenzer *Simply Rational: Decision Making in the Real World* (New York: Oxford University Press; 2015).
7. Report in the Baltimore Sun (http://www.baltimoresun.com/business/bs-bz-slots-payouts-20150826-story.html) about the deliberations of state regulators that determine the percentage payout from state-sanctioned slot machines.
8. Central Limit Theorem (CLT); see https://en.wikipedia.org/wiki/Central_limit_theorem. "In probability theory, the central limit theorem establishes that, in most situations, when independent random variables are added, their properly normalized sum tends toward a normal distribution . . . even if the original variables themselves are not normally distributed." The CLT allows us to refer the results of statistical tests to the normal distribution when the underlying distribution is unknown or known to be non-normal.
9. Karl Popper, *The Logic of Scientific Discovery* (New York: Routledge; 2002).
10. If it seems counterintuitive that a hypothesis entails an infinite number of predictions, consider this: your hypothesis predicts that protein Y is a receptor for drug A. You

test 10 concentrations of A and find that their binding to Y follows a sigmoidal curve that is consistent with your hypothesis. However, your hypothesis predicts that an infinitely graded series of concentrations of A would fall along the line of the curve. Each concentration is a distinct prediction, and there are an infinite number of points on the line.

11. Moreover, Popper was Austrian by birth and lived in Europe. He spoke and wrote in German until, because he was a Jew, he fled Europe to escape the Nazis and emigrated to New Zealand in 1937. While in New Zealand he resolved never to write in German again, although he continued to follow closely the work of colleagues who did. Indeed, his masterwork, *Logic der Forschung*, was written in German in the 1930s and not published in English until 1959. Fisher and the other founders of statistics were publishing in English, and it is possible that Popper was simply unaware of their work. Biographical information from the definitive text: Malachi H. Hacohen, *Karl Popper: The Formative Years 1902–1945* (New York: Cambridge University Press; 2000). The philosopher of science Deborah Mayo (see Note 12) reports corresponding with Popper and learning that he "regretted not studying statistics," and she takes him to task for the omission.
12. Deborah G. Mayo, *Statistical Inference as Severe Testing: How to Get Beyond the Statistics Wars* (New York: Cambridge University Press; 2018).
13. Gerd Gigerenzer, Zeno Swijtink, Theodore Porter, Lorraine Daston, John Beatty, and Lorenz Krüger, *The Empire of Chance* (Cambridge: Cambridge University Press; 1989) p. 107. The authors surveyed 30 statistics textbooks and found that none discussed the issues or the authors. In a check of six standard statistics texts, I corroborated their finding in five of them (I can't be sure that our lists didn't overlap); only one book that I reviewed alluded to the divided origins of statistics at all.
14. J. D. Perezgonzalez, "Fisher, Neyman-Pearson or NHST? A Tutorial for Teaching Data Testing," *Frontiers in Psychology* 6:223, 2015. http://journal.frontiersin.org/article/10.3389/fpsyg.2015.00223/full.
15. J. D. Perezgonzalez, "P-Values as Percentiles. Commentary on: 'Null Hypothesis Significance Tests. A Mix-Up of Two Different Theories: The Basis for Widespread Confusion and Numerous Misinterpretations,'" *Frontiers in Psychology* 6:341, 2015. http://journal.frontiersin.org/article/10.3389/fpsyg.2015.00341/full.
16. C. Lamdin, "Significance Tests as Sorcery: Science Is Empirical, Significance Tests Are Not." http://tap.sagepub.com/content/22/1/67; J. Cohen, "The Earth Is Round (p < .05)," *American Psychologist*, 49:997–1003, 1994.
17. Robert Coe, "It's the Effect Size, Stupid: What Effect Size Is and Why It Is Important. " Paper presented at the British Educational Association annual conference, Exeter, September 12–14, 2002.
18. See Note 5. We'll re-visit criticisms of statistical power in Chapter 7.
19. Daniel J. Benjamin, James O. Berger, Magnus Johannesson, Brian A. Nosek, E.-J. Wagenmakers, Richard Berk, et al., "Redefine Statistical Significance," *Nature Human Behavior* 2:6–10, 2018.
20. Daniel Lakens, Federico G. Adolfi, Casper J. Albers, Farid Anvari, Matthew A. J. Apps, Shlomo E. Argamon, et al., "Justify Your Alpha." *Nature Human Behavior* 2:168–171, 2018.

21. Lisa Randall, *Higgs Discovery: The Power of Empty Space* (New York: HarperCollins; 2 012).
22. Not all physicists were elated by the confirmation of the Higgs' appearance; see http://www.newyorker.com/news/news-desk/i-think-we-have-it-is-the-higgs-boson-a-disappointment; http://www.theatlantic.com/technology/archive/2012/07/why-the-higgs-boson-discovery-is-disappointing-according-to-the-smartest-man-in-the-world/259468/. Because the Higgs was a definite prediction of the Standard Model, its confirmation was expected; yet for that reason it didn't provide new insights into the shortcomings of that model. Physicists were hoping to observe an unpredicted particle that could lead them to exciting new physics.
23. According to newspaper reports, retired mailman Doug Hughes was on his way to deliver letters to members of Congress to protest the existence of corruption in the US government. His gyrocopter landed safely on the West Lawn of the Capitol, and he was eventually sentenced to 4 months in prison (http://www.usatoday.com/story/news/2016/04/21/gyrocopter-pilot-sentenced-4-months-prison/83271738/; https://www.washingtonpost.com/news/local/wp/2015/04/15/a-gyrocopter-just-landed-on-the-capitol-lawn/).
24. Jacob Cohen, "The Earth Is Round."
25. Gerd Gigerenzer, "Statistical Rituals: The Replication Delusion and How We Got There," *Advances in Methods and Practices in Psychological Science* 1:198–218, 2018.
26. Paul Meehl, "Theory Testing in Psychology and Physics: A Methodological Paradox," *Philosophy of Science 34:103–115*, 1967. Available at: http://www.jstor.org/stable/186099.
27. Niels G. Waller, "The Fallacy of the Null Hypothesis in Soft Psychology: Commentary," *Applied and Preventive Psychology* 11:83–86, 2004.
28. Gerd Gigerenzer, *Rationality for Mortals: How People Cope with Uncertainty* (New York, Oxford University Press; 2008), p. 161.
29. To calculate the mean variance: $[\{(n_1\text{-}1)(sd_1)^2 + (n_2\text{-}1)(sd_2)^2]/(n_1+n_2\text{-}2)\}]^{-1/2}$, where n_1 and n_2 are the numbers in each group, and sd_1 and sd_2 are the standard deviations for the measurements.
30. Coe, "It's the Effect Size, Stupid."
31. Online calculator for Cohen's d: http://rpsychologist.com/d3/cohend/.
32. http://www.newyorker.com/magazine/2004/04/05/the-height-gap; http://www.newyorker.com/magazine/2004/04/05/the-short-americanhttp://atlanticreview.org/archives/661-Europeans-are-taller-than-Americans.html; http://www.theatlantic.com/health/archive/2014/05/how-we-get-tall/361881/.

    The study of human heights is important because national height is taken as a proxy for all-around well-being; systematic variations in average heights among nations, or within one nation over, say, hundreds of years, offers evidence of significant height-influencing societal forces at work. Although no one interpretation of this phenomenon has emerged, simple averaging of heights across a heterogenous mix of distinct groups can't explain it, because the national trends hold true within even narrowly defined groups within a country.
33. http://www.cdc.gov/nchs/data/series/sr_11/sr11_252.pdf.

34. https://ourworldindata.org/human-height/; http://www.disabled-world.com/artman/publish/height-chart.shtml. I could only find the mean of the measured heights of a group of Swedish women aged 20–29 years; I made up the group size and standard error of the mean for this example.
35. To calculate the effect size, divide the difference in heights (65.5″ – 64.2″ = 1.3”) by the mean variance, as in Note 28. Effect size for women's heights is 1.3/2.51 = 0.52.
36. Coe, "It's the Effect Size, Stupid."
37. For any given interval that has been calculated, the population mean is definitely either in it or not: this is a certainty, not a probability. You can describe your own degree of confidence as to whether the mean is in or out in terms of probability. Statisticians insist on this subtle distinction; common usage is not always as strict. Examples of misinterpretations of confidence intervals can be found at https://en.wikipedia.org/wiki/Confidence_interval. As we'll see in Chapter 6, the Bayesian approach of determining probability distributions, rather than single values, for parameters such as the mean can directly tell us whether or not the confidence interval captures the true mean. See https://scholar.google.com/citations?view_op=view_citation&hl=en&user=ln2kZXIXtMcC&citation_for_view=ln2kZXIXtMcC:u5HHmVD_uO8C.
38. Coe, "It's the Effect Size," gives, in other notation, the calculation for the 95% confidence interval around the effect size (ES) as

$$\mathrm{ES} \pm 1.96\left(\mathrm{sd}_{\mathrm{ES}}\right), \text{ where}$$

$$\mathrm{sd}_{\mathrm{ES}} = \left[\left(n_1 + n_2\right) / \left(n_1 n_2\right) + \left(\mathrm{ES}\right)^2 / 2\left(n_1 + n_2\right)\right]^{-1/2}$$

39. Gigerenzer, "Statistical Rituals."

6

Bayesian Basics and the Scientific Hypothesis

6.A Introduction

The frequentist statistics in the previous chapter dealt with countable, identifiable items and probably seemed reasonably familiar even if you don't do statistical calculations on a daily basis. Statistics is not always so straightforward, though. Getting a good grip on what "a 30% chance of rain tomorrow" means, for instance, is trickier than deducing your chances of rolling a seven with two dice. For one thing, the chance of rain is a "single-event probability."[1] You can't sample from a large set of identical tomorrows, so calculating the exact odds of rain is out of the question; obviously, the frequency interpretation of probability does not apply. Moreover, just what the "%" refers to—the *reference class*—is often unclear: Are we talking about the percentage of time that it will rain tomorrow, the area of the local region that will get rained on, or something else?

The statistical methods that can handle issues like "the chance of rain" are part of a large and complex class known today as *Bayesian statistics* after its inventor, the English clergyman Thomas Bayes. Despite Bayes' discovery of the primary rule, or theorem, of his method, his work only appeared posthumously after a friend put Bayes' notes together and published them. This branch of statistics was independently discovered and elaborated by the French mathematician, Pierre-Simon Laplace, whose contribution was notable enough that some writers feel that we should be talking about "Laplacian" instead of "Bayesian" statistics. Nevertheless, a name change is unlikely at this point.

My main goal here is to consider how we might put Bayesian reasoning together with conventional hypothesis testing scientific methods. I'll go into the fundamentals of Bayesian statistics only as much as necessary to reach the goal. Because it is usually omitted from the science curriculum, this chapter may at first look dauntingly dense and equation-filled, but you'll soon see that what looks like several equations is really just one, shape-shifting its way through the chapter; if you stick with it, your intuition about what's going on will improve. Readers with a solid background in the basics of Bayesian methods may want to skip ahead to Section 6.D, where we'll zoom out to consider how to integrate these methods with hypothesis-based thinking in ways that are rarely discussed, even by Bayesians.

6.B Probability from the Bayesian Perspective

The Bayesian notion of probability is poles apart from the frequentist one. It is *subjective*, not *objective*. An extreme statement of the Bayesian point of view[2] is that "Probability does not exist." That is, probability is a purely abstract construct that has nothing to do with events that are external to us. (Recall that frequentists believe that probability characterizes actual occurrences of physical events.) A somewhat less radical position is that "even if physical probability exists,"[3] we can know about it only subjectively. We are only sure of our own uncertainty about the world, and we express our degree of ignorance in terms of probability. When a Bayesian analyst calculates a probability, she incorporates a measure of subjective belief, in the form of a probability, into the calculation. There are many ways of incorporating beliefs, including "objective" ones,[4] but we use them because we don't have frequencies to work with.

The National Weather Service (NWS) calculates the chance of rain (the "probability of precipitation") as the *degree of confidence* that it has that rain will fall somewhere in the forecast area multiplied by the percentage of the area that will see rain.[5] The degree of confidence reflects the percentage of times the NWS was right in predicting rain on "days like tomorrow" without being too precise about what tomorrow will be like. And there are other ambiguities. While "rain" means at least 1/100 of an inch of water, the "forecast area" is not well defined. Hence, despite the formulas and numbers, weather forecasting exemplifies the *subjective* Bayesian approach to probability. It entails a degree of belief in an outcome; a belief that is usually shaped by past experience, a mathematical model, or hard evidence of the probability of occurrence of similar events. In the end, a "30% chance of rain" conveys a sense of what the weather models come up with and how your TV weather person interprets the information; basically, it means that a dry day is somewhat more likely than a damp one. Since we can't rely on precise calculations to help with decisions in cases of uncertainty, instead, we often call upon rules of thumb called "heuristics,"[6] and we will have much more to say about them and their flip sides, "biases," in Chapter 11.

More pertinent for scientists is the concept that if you combine quantitative information with appropriate assumptions supported by available data, you can make a Bayesian subjective probability calculation that predicts the chance that a hypothesis is correct. You can immediately see the appeal of the approach. Rather than being confined to assessing the truth value of hypotheses indirectly by eliminating false ones and amassing tested-and-not-falsified ones—Popper's strategy—Bayesianism offers the prospect of directly assigning (probabilistic) truth values to hypotheses. The critical prior assumptions in the Bayesian method, universally referred to as *priors*, are expressed as numerical probabilities, or *probability density functions* (confusingly abbreviated *pdfs*). Much hinges on the validity of the priors. A Bayesian analyst typically begins by finding appropriate priors to use in her calculations, so we begin by examining how priors are incorporated in a Bayesian-type problem.

6.B.1 A Bayesian Problem

Suppose you are researching the growing problem of prescription opioid addiction in the United States. While visiting your local addiction center to interview addicts, you discover that 90% of them, a probability of 0.9, use marijuana regularly. This is alarming, and you leap to the inductive generalization that regular marijuana use is a predisposing factor, a "gateway drug," to opioid addiction. You vow to spearhead a massive anti-marijuana campaign. Luckily, just before you do, a statistician friend informs you that your reasoning is flawed. The evidence from the clinic does not mean what you think it means.

In fact, you can't determine the relationship between marijuana use and opioid addiction from what you know. To decide whether or not it's time to ban pot, you have to answer the question: "What is the chance that a marijuana user will become an opioid addict?" In other words, you want to start with a group of people who are marijuana users and ask how many of them go on to become opioid addicts. This not the information that you got from the clinic visit. Instead, you learned about a group of opioid addicts who also use marijuana. Not the same thing at all, and you have to distinguish what you know from what you need to know. Besides the addicts who use marijuana, you need to know the chance (the prior probability) that a random person uses marijuana and the prior probability that a random person is an opioid addict. Once you have these data, then you can use *Bayes' Theorem*, also called *Bayes' Rule*, to put them together and figure out the chance that a marijuana user is also opioid addict, which is what you want to know.

The problem might sound complicated, but it is simple in symbols, or, if you prefer, a diagram (Figure 6.1). If we use M to stand for marijuana use and A for opioid addiction, then the probability that an opioid addict uses marijuana is $p(M/A)$. This is the probability, p, that the condition to the left of the slash (uses marijuana) is true, given that the condition to right of the slash (is an opioid addict) is true.

The expression $p(M/A)$ is called a *conditional probability* because it is the probability of one condition occurring given that the other condition already

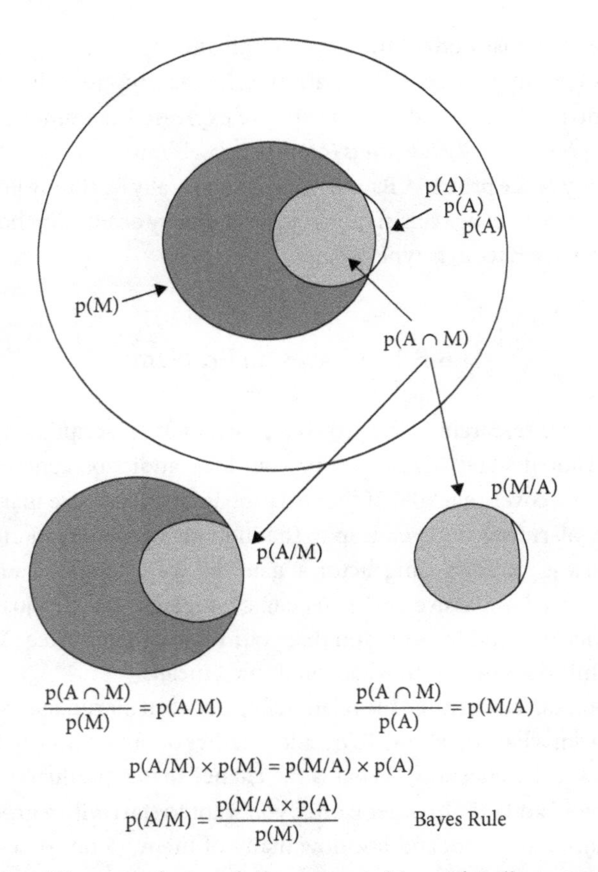

Figure 6.1 Venn Diagram of a sample Bayesian problem. This illustrates the opioid addiction–marijuana use problem described in the text. The large white circle represents the population of the United States; the dark circle, the percentage of the population that currently uses marijuana, and the small light circle the percentage that is addicted to opioids (diagram not to scale). The overlap area marked A ∩ M is the group that uses marijuana *and* is addicted to opioids. The "∩" indicates the joint probability of A and M. The left diagram below shows the percentage of marijuana users who are also opioid addicts, *p*(A/M); the right diagram shows the percentage of opioid addicts who are also marijuana users, *p*(M/A). We want to figure out *p*(A/M) given *p*(M/A), *p*(M), and *p*(A). The expressions below the diagram illustrate how to derive Bayes' Rule (or Theorem) from the relationships.

exists. This conditional probability *p*(M/A) is what you learned during the clinic visit. Its value is 0.9; however, what you're trying to find out is the probability that a marijuana user will become an opioid addict, denoted *p*(A/M). The Bayesian notation immediately shows that the quantities of interest—*p*(A/M) and

p(M/A)—are not the same. You also need p(M), the *prior probability* that a random person uses marijuana and p(A), the *prior probability* that a random person is addicted to opioids. What we want to know, p(M/A), is usually called the *posterior probability*, or the *posterior odds*, of opioid addiction given marijuana use.[7] Bayes' Theorem puts these pieces together like this:

$$p(\mathrm{A/M}) = [p(\mathrm{M/A}) \times p(\mathrm{A})] / p(\mathrm{M}) \qquad \text{Equation 6.1}$$

The equation says that the chance of someone's becoming an opioid addict, given that she already uses marijuana, is equal to the probability that a randomly selected addict uses marijuana multiplied by the prior probability that a randomly selected person in the population uses marijuana p(M), all divided by the prior probability of a random person in the population being an opioid addict, p(A). If you take a minute to go over Figure 6.1 you can see how the different concepts are related by the simple algebra underlying Bayes' Theorem.

Let's use real numbers and go through the problem. Of the estimated 249.4 million adults in the United States,[8] 2.1 million[9] (0.8%) are said to be addicted to prescription opioids (have opioid "substance use disorders"). Thus, p(A) = 0.008. According to a recent CNN poll, 1 in 8 American adults, roughly 31.2 million, "currently" use marijuana[10]; so p(M) = 0.125. The numbers of opioid users who use marijuana is surely high, though not entirely firm, so we can take your clinic data as a fair estimate (i.e., p(M/A)) is 0.9. If you plug all of this into Bayes' Theorem, you find that the probability that a marijuana user will become an opioid addict, p(A/M), is approximately 0.06; that is, only 6%, rather than the 90% you might have inferred from the clinic data! In other words, 94% of current marijuana users probably will not turn into opioid addicts. Not exactly strong support for your anti-pot campaign. And, of course, even the calculation of 6% says nothing about any genuine *causal* relationship between marijuana and opioid use. Bayesian reasoning teaches you to tread cautiously in drawing conclusions from data and to be sure that you understand what the evidence is really saying.

In general, Bayesians tend to obsess about priors because their analytic methods absolutely depend on priors. In the preceding example, p(A), p(M), and p(M/A) were priors. Clearly, we could not solve the Bayes equation and evaluate the marijuana–opioid linkage without having values for the priors, but, conveniently in this case, we had firm numerical data that gave us *objective* and *informative* priors to work with. Often there is little or no solid support for your priors, and you have to get them in other ways. Priors represent the current state of uncertainty in neutral, though quantitative, terms.

At the other extreme from objective and informative priors are *subjective* and *noninformative* (or *uninformative*) priors. Noninformative priors are not based on experimental data. A noninformative prior might be a numerical constant or other estimate based on mathematical functions. If you didn't have the clinic data, you might have made the neutral guess that 50% of addicts would use marijuana, and then your estimate of the proportion of marijuana users who become opioid addicts would drop to 3%, half of what you previously estimated.

Bayesians disagree about what is acceptable when it comes to subjective priors. The source of subjective priors ranges from plausible estimates to the quantification of expert opinion. For some Bayesians the concept that opinion alone could be a valid prior is a bridge too far; opinion, they believe, belongs to psychology. To someone raised with traditional frequentist sensibilities, formally including even expert guesswork into a quantitative equation is an odd way to proceed; on the other hand, if you recall that Bayesians equate probability with degrees of subjective uncertainty, then perhaps it is not so strange.

Sometimes people reveal their degree of confidence in a subjective probability prediction by placing a monetary bet on it. In fact, the ex-gambler, sports better, and professional election predictor Nate Silver[11] sees placing a money bet as the acid test of a prior. Because you care about the outcome—it's your money!—you have every incentive to evaluate the odds as rigorously and accurately as you can. The physicist Stephen Hawking once expressed $100 worth of confidence that the Higgs boson would not turn up in experiments done at the Large Hadron Collider and had to pay up when it did.[12] Hawking's experience highlights one of the dangers of subjective probability.

6.A.2 Bayesian Hypothesis Updating

Bayes' Theorem lets us estimate whether a given hypothesis is likely to be correct and, especially, to update and improve predictions as new information becomes available. To appreciate this feature of Bayesian reasoning, the first thing is to rewrite Bayes' Theorem in a general, scientifically relevant form. Typically, a Bayesian scientist would like to know how some data that he's gathered affect the chances that his hypothesis is true. That is, he wants to determine the hypothesis is true given the data—*p*(hypothesis/data)—and to do that he needs to know the chance that he'd get that data if the hypothesis were true (i.e., *p*(data/hypothesis)). Finally, he has to estimate the probability that the hypothesis is true in the first place, *p*(hypothesis) and the probability of observing the data, *p*(data) that he's gathered. With this information, he can use Bayes' Theorem like this:

$$p(\text{hypothesis / data}) = \left[p(\text{data / hypothesis}) \times p(\text{hypothesis})\right] / \, p(\text{data})$$

Equation 6.2

(Note that Equation 6.2 is just Equation 6.1 with new labels.)

Here's an example of how data can improve the probability that a hypothesis is true: imagine that you've made a new acquaintance, Pat, and, knowing nothing about Pat's occupation, you are curious (OK, nosy) about her salary. Knowing nothing, you can only assume that she is a typical member of the American work force and therefore that the odds of her earning less than $250,000 are 97 of 100 (97%).[13] In other words, your best guess is that she makes less $250k a year; this is your initial hypothesis, so *p*(hypothesis) = 0.97. Subsequently, you learn that Pat has just bought a new sports car with a base price of $68,000. Immediately, you reevaluate your image of Pat's resources, estimating that someone would have to earn at least $100k per year to be able to afford that car (in other words, buying the car is evidence that she makes >$100k). You want to know the probability that she can afford the car if she makes less than $250k. From what you know now, her salary is likely to be between $100k and $250k, and, given that 20% of American households make $100k or more, you estimate that about 17% (20 – 3%) earn between $100k and $250k. Thus the probability of Pat's being able to afford the sports car on less than 250k is 0.17 (which is *p*(evidence/hypothesis) in Equation 6.2 = 0.17). You can now update the posterior odds, *p*(hypothesis/evidence), of your hypothesis with the aid of Bayes' Theorem and the calculator app on your smartphone. So

$$p(\text{hypothesis/evidence}) = \left[p(\text{evidence / hypothesis}) \times p(\text{hypothesis})\right] / \, p(\text{evidence})$$

Equation 6.3

The answer is 82%, meaning that there is an 82% chance that she earns less than $250k a year. Thus, learning that she'd bought the car decreased your estimate that she is in the under-$250k group of wage earners from 97% to 82%; still high, but lower than it was. Or, turning the numbers around, the chance that Pat makes more than $250k has jumped from 3% to 18%. You could continue the process if you got more information. If you learned that Pat is an analyst-level investment banker and that these folks make a minimum of $80k, with an average annual salary in the range of $120k—350k,[14] you could further adjust your estimate. Now *p*(hypothesis) might be only 0.1 (10%), *p*(evidence/hypothesis) might go to 90% (0.9), and your estimate that she makes less than $250k would drop down to 45%. Progressively incorporating new information about Pat's circumstances would dramatically alter your hypothesis about her income.

Because you have to estimate priors, even "objective" ones, there is always an element of subjectivity in Bayesian probabilities. Even when based on previous evidence, subjective probability is a more flexible notion of probability than is frequentist probability because it depends on assumptions that are themselves uncertain. How accurate are the priors, and how applicable are they to the case at hand? Knowing her job description and its handsome salary range would still not tell you unequivocally how much money Pat makes—maybe she's an idealist who is just not that interested in making a lot of money, but a rich uncle died and left her a pile of cash and her lavish spending is entirely unrelated to her income. While Bayesian procedures give you a rational way of adjusting your confidence in your hypotheses, they also give you a way to compare hypotheses to find the stronger one.

6.A.3 Bayesian Hypothesis (Model) Selection

Frequentists test hypotheses (or "models"; we'll compare the two in Chapter 10.B) and reject improbable—effectively, falsified—ones. Classically, Bayesians don't reject hypotheses. The usual frequentist statistics compare a test result with an imaginary, idealized probability distribution to decide whether to reject their hypothesis or not. Bayesians think this makes no sense whatever.[15] How, they ask, can you possibly arrive at valid conclusions by comparing data that you didn't measure against a population that you can't observe? Bayesians prefer hypotheses that postdict actual data already in hand, and they conclude that a superior hypothesis postdicts existing data with a higher probability than an inferior one. Otherwise said, they trust a hypothesis more if it has already performed well in the past, and they select hypotheses based on past performance.

Suppose you have two hypotheses, H_1 and H_2, that can both account for existing data, D, and you want to decide which one is best. You can assess their strengths using Bayes' Theorem (Equation 6.2) to calculate the posterior probability of each H given D (i.e., [p(H/D)]). This tells you the probability that each hypothesis is likely to be correct given the data that you have. If H_1 predicts the existing data with a higher probability than H_2, then you conclude that H_1 is superior to H_2. In the usual notation, you set

$$p(H_1/D) = \left[p(D/H_1) \times p(H_1)\right]/p(D) \qquad \text{Equation 6.4}$$

against

$$p(H_2/D) = \left[p(D/H_2) \times p(H_2)\right] / p(D) \qquad \text{Equation 6.5}$$

and choose the one with the higher probability. Alternatively, you can just divide Equation 6.4 by Equation 6.5, and if the result is >1.0, then H_1 is superior; if it is <1.0, then H_2 is. Box 6.1 shows how to arrive at the same conclusion using an essentially equivalent but slightly different and commonly used parameter; the *Bayes' factor*.

Box 6.1 Bayes Factors and Odds Ratios

Notice that p(D) is the same in the denominators of both Equations 6.3 and 6.4 because you're trying to decide which hypothesis does the best job of accounting for D. Therefore, when you divide the equations, p(D) cancels out. Hence, you'll sometimes see Bayes' Theorem written with no denominator at all:

$$p(H/D) = p(D/H) \times p(H)$$ [16]

With this simplification, dividing the equations to compare H_1 and H_2 gives

$$\frac{p(H_1/D) = p(D/H_1) \times p(H_1)}{p(H_2/D) = p(D/H_2) \times p(H_2)} \qquad \text{Equation 6.7}$$

And the parts have special names: $p(H_1/D)/p(H_2/D)$ is the *posterior odds ratio;* $p(D/H_1)/p(D/H_2)$, is the *Bayes'* factor[17] and $p(H_1)/p(H_2)$, is the *prior odds ratio.*

In words,posterior odds ratio = Bayes factor × prior odds ratio.

What does this buy you? Remember, the posterior odds are what we're interested in. The posterior identifies which of the two hypotheses best predicts the data that you already have.[18] Let's start with the prior odds ratio. The ratio of the prior odds for H_1 and H_2 expresses the strengths of your convictions that each hypothesis is true. If you have no reason to favor one hypothesis over the other, then you're basically saying that the prior odds are equal, so their ratio equals 1.0, and, in that case, you can omit them since multiplying by 1.0 doesn't change anything. The posterior of the hypotheses then boils down to the Bayes' factor, which is the ratio of the probabilities that the data are predicted by H_1 and by H_2, i.e., **posterior odds ratio ≈ Bayes' factor.**

A Bayes factor greater than 3, or less than 1/3—i.e., one hypothesis predicts the data with 3 times greater probability than the other—would imply that the hypotheses genuinely differ and that the farther the Bayes' factor is from 1.0, the stronger the evidence of difference. The size of the Bayes' factor that a field accepts as significant is set by convention. We'll return to this topic when considering how to integrate Bayesian reasoning with hypothesis falsification in Section 6.E.

6.B Bayesian Objectives

In addition to comparing hypotheses (or models), you can use Bayesian methods to estimate and compare the fits of any parameter (mean, standard deviation, etc.) of a model with experimental data. In place of H and D, you'll often see the Greek letter theta (θ), standing for any parameter that you might want to estimate and *y* for the data that are being fit. Now Bayes' equation becomes

$$p(\theta / y) = \left[p(y / \theta) \times p(\theta)\right] / p(\theta)$$ [19] Equation 6.6

which, despite its forbidding look, is fundamentally our old friend, Equation 6.1 again. One important difference when you're estimating parameter values is that the posterior probability is no longer a single point value but a distribution of values, in keeping with the Bayesian view that we can know the world only probabilistically. Therefore the priors here are often candidate probability distributions (e.g., normal distributions), rather than specific, single-point values. If you wanted to determine the conditional distribution of a parameter—say the mean—given your data, for each data point you could try to find the normal distribution that has that point as its mean.

If frequentists and Bayesians wanted to determine the relationship between children's heights and their ages, frequentists would calculate the mean height for each age and then characterize the relationship by fitting a line through the means.[20] In contrast, Bayesians would find a distribution of values about the mean at each age by a probability density function—in this case, a normal distribution—and connect the dots with a linear regression model.

Formation of predictive models is a central goal of Bayesian analysis, and predictive accuracy is its highest virtue. A guiding principle of Bayesian procedures is the creation of increasingly more accurate models.

6.C The Overfitting Problem

A great risk in any approach that prizes the ability of a hypothesis to account for existing data above all else is *overfitting*. This is the error that occurs when a mathematical model, designed to fit tightly to an initial group of data points, is challenged to fit new data points. Typically, the new data fit is much worse than the original because of noise in the original data.

Imagine taking a handful of pennies, tossing them into the air, noting exactly where each one lands on the ground, and creating a mathematical model that would postdict the exact location of each one. Now imagine scooping up the pennies and tossing them into the air again. How successful do you think the model, designed to *post*dict the first pattern of penny landings, would be in *pre*dicting the second pattern? Not at all? Correct. An enormous amount of random variability would have determined the landing positions of the first bunch of pennies, and because your model was designed to fit those landings, it would be heavily influenced by the chance factors. Since random variability will also affect the pennies during the second toss, the second pattern will not duplicate the first one. The model wouldn't fit the second pattern because it had been *overfitted* to the first.

The overfitting problem gave rise to the motto that "less is more[21]"; that is, the realization that simple models that may be less accurate when applied to an initial dataset often provide greater predictive accuracy when applied to subsequent sets because simpler models are less sensitive to random noise. The desire for more effective predictive models is key to understanding part of the dispute surrounding biases and heuristics that we'll cover in Chapter 11.

In summary, Bayesian statistics offers the possibility of improving experimental predictions and is most useful under conditions of low information and little experimental control. It chiefly produces predictive models that are evaluated according to their ability to account for existing data. Traditionally, Bayesian analysts concentrate on making probability judgments about hypotheses, not rigorously testing or rejecting them. The success of Bayesian methods depends heavily on the validity of the chosen priors. With this overview in mind, we can return to the main topic and look into how Bayesian methods mesh with the practice of scientific hypothesis testing.

6.D Bayesian Reasoning and the Statistical Hypothesis

Bayes' Theorem can help you take into account the effects of new data on the *posterior odds* that your hypothesis is true without saying anything about whether it is really true. The issues of truth and certainty do not trouble Bayesians because truth and certainty are not on their radar screens. Bayesians are after probable truth, not actual truth.

Falsification is not in the Bayesian lexicon, while *probability* and *prediction*, which characterize inductivist thinking, are. For this reason, typical Bayesian procedures are incompatible with conventional falsification-based hypothesis testing. However, from a nontraditional perspective, Bayesian methods can find a place in the context of hypothesis testing.

6.D.1 Bayesianism and Induction

Are Bayesians inductivists or not? A firm answer to this simple question is surprisingly elusive. If the purpose of a Bayesian analysis is to "update our belief in the likely truth of a hypothesis in light of new knowledge," then the answer would be "no, Bayesians are not inductivists." According to the Critical Rationalist philosopher David Miller, "pure Bayesians" operate in a straightforwardly deductive way.[22] Once they have settled on which *priors* and *odds ratios* they want to use, and insert those values into their equations, their decisions "follow directly from the newly derived conditional probability distributions." That is, the conclusions follow from logical reasoning without referring to environmental regularity, generalizations of rules, etc., as an inductivist would. Hence, argues Miller, the pure Bayesian analyst evades the "problem of induction," however, this advantage comes at a cost: the Bayesian abandons the search for Truth. When typical Bayesian analysts compare hypotheses, they have no standard, apart from their probability calculation, for determining which one is objectively better. They are content with probable truth, and, for Miller, a feisty writer, a "probable truth" is no better than "rumored truth."

On the other hand, according to a noted Bayesian theorist, the Bayesian approach is fundamentally inductivist[23] in nature. This view is reinforced by Bayesians who argue that they are engaged in "inductive inference": a Bayesian scientist gathers data, makes assumptions, then builds and compares predictive models. Bayesians measure scientific progress by the successive improvements in the fits of their models to data and, vice versa, their models are improved by the evidence gathered and by the selection of priors that they rest on. The data, not manipulative experiments or challenges to existing ideas, are what drives a typical Bayesian investigation. Hence, their interpretations do aim to make probabilistically true statements about the world.

Yet if that's the case, according to Miller's analysis, if Bayesians are truly inductivists, they cannot avoid the problem of induction. There is no guarantee that the future will be like the past. Therefore, he concludes that, to the extent that Bayesianism is deductive, it is not directly related to the search for Truth. To the extent that it is inductive, it suffers the weaknesses that cripple every other inductive method. Hence, for the Critical Rationalist, pure Bayesianism is neither inductivist nor part of a quest for universal scientific Truth. It is simply irrelevant to basic science.

But, we might ask, do scientists really have to choose between Popper and Bayes? Perhaps not, as I'll try to show in the next section.

6.D.1 Bayesianism and the Hypothetico-Deductive Method

Some Bayesians, including Andrew Gelman and Cosma Shalizi[24] consider the interpretation of Bayesianism as an inductivist procedure to be "faulty." They think that, rather than incrementally increasing the strength of predictive models, Bayesians should pay more attention to active hypothesis "checking"; that is, testing and rejecting hypotheses in line with the traditional scientific *modus operandi*. For example, in studying the relationships among people's income and their voting behavior across states, Gelman and Shalizi carried out Bayesian-style hypothesis checking and correction to learn from more elaborate hypotheses what factors were most important for accurate prediction. The authors argue that "The power of Bayesian inference here is *deductive*: given the data and some model assumptions, it allows us to make lots of inferences, many of which can be checked and potentially falsified."

The prior distribution, in particular, is a testable component of a Bayesian inference process, and Bayesian hypothesis checking is like conventional frequentist-based, statistical testing of a hypothesis prediction. Gelman and Shalizi go on, "The main point where we disagree with many Bayesians is that we do not see Bayesian methods as generally useful for giving the posterior probability that a model is true, or the probability for preferring model A over model B, or whatever." Rather, they argue, that "Certain of the model's implications [should be] compared directly to the data, rather than entering into a contest with some alternative model. . . . The goal of model checking, then, is not to demonstrate the foregone conclusion of falsity as such, but rather to learn how this model fails." And a model that has failed is a model that has been falsified.

Note that Gelman and Shalizi's suggestion for interpreting Bayesian procedures deductively is quite different from David Miller's. Miller focused on how Bayesians use Bayes' Theorem to reason about the probability that a hypothesis is true; Gelman and Shalizi propose that we should use Bayesian statistics to decide whether to *accept or reject* a hypothesis. Using Bayes' Theorem to test a hypothesis would be largely free of the criticisms that plague null hypothesis significance testing mode (NHST) procedures, for example, in the context of the Reproducibility Crisis (Chapter 7).

6.D.2 A Bayesian Standard for Falsification

The potential for applying Bayesian methods to hypothesis testing evidently hasn't been exploited to any great extent, so I'd like to explore it a little more. Ordinarily, with Bayesian methods, you choose between two models by comparing their posterior odds ratios and opt for the one with the higher posterior odds of postdicting the data already at hand. However, rather than concentrating solely on the probabilities per se, as a pure Bayesian analyst would do, experimental scientists could simply agree on a quantitative standard for rejecting the weaker hypothesis. For example, following the reasoning in Box 6.1, we might agree that a Bayes factor of, say, 10, would be sufficient to declare the less probable hypothesis to be falsified (see example in Box 6.2). The critical value of the Bayes' factor would be a matter for each scientific community to set, analogously to the present $p < 0.05$ level for rejecting a hypothesis. A methodological rule like this would allow Bayesian criteria to fit into a hypothesis-testing philosophy.

6.D.3 Bayes and Explanation

While we hope that good predictive models will have explanatory force, it is true that greater predictive accuracy does not imply greater understanding. You can make good predictions without good understanding, as the Ptolemaic and Mayan astronomical systems showed. It is not that predictive power is unimportant in science; on the contrary, applied science, technology, and even basic science fields place high value on accurate predictive hypotheses. It's just that basic science is not ultimately satisfied with better predictive hypotheses unless they are primarily explanatory.

6.E Frequentists, Bayesians, and the Philosophy of Hypothesis Testing

Despite the fact that frequentist and Bayesian statistics have their own specialized niches, frequentists and Bayesians do not always coexist peacefully. Frequentists, who hold the dominant position in most areas of experimental science, tend to ignore Bayesian thinking altogether in their textbooks. Bayesians, on the other hand, often make much of the Bayesian–frequentist split and call attention to the deficiencies of frequentism. While Bayesian reasoning has not made a mark in many areas of biology, it is increasingly important in social sciences and certain kinds of Big Data analyses (Chapter 15).

Box 6.2 Falsification Using Bayesian Model Selection Methods

Here's a trivial example to show how reasoning would work. As a neuroscientist, you are investigating brain mechanisms of drug-seeking behavior in cocaine-addicted mice. Your idea is that within a group of cells, the (make-believe) nucleus Sniffens (n. Snf.), there is a dedicated neuronal circuit that regulates this behavior. There are only two sorts of neurons in n. Snf., *Locals* and *Fars*, one or both of them must mediate the behavior. The Locals outnumber Fars by 50:1. If a mouse pushes the correct lever press, a minute pinch of powdered cocaine is automatically deposited onto a tiny mirror nearby. Your H_1 says that most of the cells that are active during the behavior (B) are Locals; that is, that the posterior odds ratio, $p(B/L)/p(B/F)$ is >1.0. H_2 says that most of the cells are Fars; that is, $p(B/L)/p(B/F)$ is <1.0. Assume journal editors and grant reviewers in your specialty area now accept a posterior odds ratio of >10 (or <1/10) as meaning that the less likely of two hypotheses has been *falsified*.

When you examine each cell group in isolation, you find that 85% of the Fars are active when the animal sniffs cocaine, whereas the percentage of Locals activated is only 18%. In other words, the likelihood (see Box 6.1) of a Far's being activated by the behavior, $p(F/B)$, is 0.85, and the likelihood of a Local's being activated, $p(L/B)$, is 0.18, and the Bayes' factor, therefore, is 0.18/0.85 = 0.22. At this point, given the high percentage and low Bayes' factor, you might be tempted to suspect that in fact the Fars are the predominant group. Nevertheless, Locals outnumber Fars by 50:1. The prior probability of a random cell's being a Far, $p(F)$, is approximately 2%, and the prior probability of a random cell being a Local, $p(L)$, is approximately 98%, for a prior odds ratio of 0.98/.02 = 49. The posterior odds ratio is equal to the Bayes' factor times the prior odds ratio; that is, $0.22 \times 49 = 10.78$. Since the posterior odds ratio is >10 and favors H_1, you can conclude that most of the activated neurons are Locals. You reject H_2.

NHST remains the major statistical hypothesis testing mode in basic science. If you think that probabilities are properties of the external world, then the fixed values of frequentist probabilities form natural cutoffs for all-or-none decisions involving falsification. Of course scientific decisions, including those based on falsification, are always tentative. We use statistics to aid in making judgments, and judgments are by definition subjective. Nevertheless, frequentist methods obvious mesh nicely with the methods of Fisher and Popper.

On the other hand, the Neyman-Pearson philosophy has much in common with the typical Bayesian perspective. Both focus on practical problems and strive for good, predictive models. The Neyman-Pearson program emphasizes the importance of taking errors into account and of playing their costs of various kinds of errors off against each another in deciding what to do. Their philosophy is close in spirit to that of the Bayesian program of weighing and incrementally improving models by selecting for the better fitting ones. Moreover, since a modified Bayesian approach is compatible with hypothesis testing and rejection, scientists could create a parameter that would serve the same purpose as the p-value does now. The affinities between the Neyman-Pearson methods and Bayesian approaches suggest several potential ways of improving our scientific decision-making.

6.F Coda

This chapter reviewed the fundamental features of Bayesian statistics and methods, together with their advantages and disadvantages. Probably the most important and problematical aspect of Bayesian statistics is its absolute dependence on prior probabilities, and the chapter discussed some of the issues associated with selecting appropriate priors. We also worked through several kinds of problems to give the gist of Bayesian thinking and reviewed a few of the marked distinctions between frequentist and Bayesian philosophies of science. Given its emphasis on probabilistically secure answers, Bayesian statistics is generally more suitable for applied, rather than basic, science. Nevertheless, we also discussed Bayesian model selection procedures and I argued that a modified Bayesian approach could be incorporated into hypothesis-testing procedures.

Unless noted otherwise, "statistics" throughout this book refers to frequentist statistics. The next two chapters will call on a number of the concepts and issues discussed in Chapters 5 and 6.

Notes

1. Gerd Gigerenzer, *Rationality for Mortals: How People Cope With Uncertainty* (Oxford: Oxford University Press; 2008); an analysis of how people understand the prediction of a "30% chance of rain."
2. David Kaplan, *Bayesian Statistics for the Social Sciences* (New York: Guilford; 2014), p. 284. Quotes in B. de Finneti, *The Theory of Probability* (New York: Wiley; 1974), vols. 1 and 2.
3. Irving J. Good, *Good Thinking: The Foundations of Probability and Its Applications* (Mineola, NY: Dover; 1983).

4. Kaplan, *Bayesian*, pp. 292–294.
5. What does the chance of rain in a national weather report refer to? See http://wxbrad.com/why-a-50-chance-of-rain-usually-means-a-100-chance-of-confusion/.
6. "Heuristics." In Chapter 11, I will contrast Gigerenzer's views (see Note 1) of heuristics as useful cognitive tools with those of Daniel Kahneman, Paul Slovic, and Amos Tversky, *Judgment Under Uncertainty: Heuristics and Biases* (New York: Cambridge University Press; 1982); Daniel Kahneman, *Thinking, Fast and Slow* (New York: Farrar, Strauss and Giroux; 2011) who see them largely as the source of error and bias.
7. Bayesian terminology is somewhat variable. Another way of expressing Bayes' Theorem in words is that the posterior probability is equal to the *prevalence* of marijuana use, p(M/A) among opioid users multiplied by the *prevalence* of marijuana use, p(M), divided by the *prevalence* opioid addiction, p(A). p(M/A) is also known as the *likelihood*.

   Bayesian methods can cope with far more complex data and models than the simple ones that I've described. Parameter fitting via Bayesian networks is at the heart of analytic methods using *Structured Equation Models* (or *Structural Causal Models*) that we'll encounter in Chapter 15.
8. According to the US Census, the US population as of July 1, 2016, was 323,127,513, of which 77.2% were older than age 18, for a total of 249,454,440 adults. See https://www.census.gov/quickfacts/fact/table/US/PST045216.
9. Number of US adults addicted to opioids: 2.1 million, according to https://www.drugabuse.gov/about-nida/legislative-activities/testimony-to-congress/2016/americas-addiction-to-opioids-heroin-prescription-drug-abuse.
10. CNN poll on the number of US adults who smoke pot. See http://www.cnn.com/2016/08/08/health/marijuana-use-doubles-gallup-poll/index.html.
11. Nate Silver; https://fivethirtyeight.com/politics/elections/.
12. http://www.telegraph.co.uk/news/science/large-hadron-collider/9376804/Higgs-boson-Prof-Stephen-Hawking-loses-100-bet.html.
13. https://www.nytimes.com/2015/12/28/opinion/campaign-stops/250000-a-year-is-not-middle-class.html.
14. http://www.wallstreetoasis.com/salary/investment-banking-compensation.
15. Kaplan, *Bayesian*, p. 286. Kaplan summarizes the views of a number of Bayesians.
16. Technically, the p(D) in the denominator is a *normalizing* factor; it ensures that the conditional probabilities add up to 1.0. But if we're comparing hypotheses, we don't really care about the sum. Indeed, in some circumstances, the probabilities won't add up to 1.0: they are *improper probabilities*.
17. Since p(D/H) is also known as the *likelihood*, the Bayes' factor is sometimes referred to as the *likelihood ratio*. The greater the likelihood that a given hypothesis predicts the existing data, the more probable is the hypothesis.
18. This is just an example of how the reasoning behind model testing might go; in real examples, any of the terms to the right of the equal sign in Equation 6.4 will affect the posterior odds ratio and therefore influence which model we decide is preferable.

19. In this form the notation for the *likelihood* is $p(y/\theta)$; that is, the probability of observing evidence y, given that parameter θ is true. In some texts, the likelihood is written as $L(\theta/y)$. In both cases the meaning is the same.
20. For both boys and girls the growth curve from 3 to 10 years is fairly straight: see www.chartsgraphsdiagrams.com/HealthCharts/height-2-20-boys.html; http://www.chartsgraphsdiagrams.com/HealthCharts/height-2-20-girls.html.
21. See Note 6.
22. David Miller, *Critical Rationalism: A Restatement and Defense* (Chicago: Open Court; 1994), pp. 125–132.
23. Andrew Gelman and Cosma Rohilla Shalizi, "Philosophy and the Practice of Bayesian Statistics," *British Journal of Mathematical Statistics Psychology* 66: 8–38, 2013.
24. Ibid.

7
The Reproducibility Crisis

7.A Introduction

The news is everywhere: biomedical science[1] is in trouble. Scientific investigators cannot confirm (i.e., *reproduce*) published findings. Reports of irreproducibility are proliferating, and the trust that scientists and the public have in scientific knowledge about the world is eroding. According to National Institutes of Health leaders Francis Collins and Lawrence Tabak "the complex system for ensuring the reproducibility of biomedical science is failing and is in need of restructuring."[2] And editorials in prominent scientific journals ("Reducing our irreproducibility," "Rigor or mortis: best practices for preclinical research in neuroscience," "Must try harder") conclude that we are now in a "crisis" of reproducibility.[3–5]

Smoldering concerns about repeatability in science burst into flame after scientists at Bayer HealthCare Pharmaceuticals[6] and Amgen[7] vented their frustrations at going down expensive and time-consuming blind alleys in their search for new cancer therapies. The company scientists reported that they could not repeat the findings of 65–90% of the preclinical (basic) science studies that they evaluated. Normally, commercial laboratories extend and amplify leads provided by preclinical research, develop and test candidate drugs, and eventually bring new medicines to market, so the apparent unreliability of the evidence set off alarms throughout science and attracted the attention of the popular press. Columns in the *Economist* ("Unreliable research: trouble at the lab," October 19, 2013), *Los Angeles Times* ("Science has lost its way, at a big cost to humanity," October 27, 2013), *Wall Street Journal* ("Getting the bogus studies out of science," August 19, 2015), *New York Times* ("Why do so many studies fail to replicate?," May 27, 2016), and many others made the public aware that all was not well in the laboratory.[8]

Can things really be as bad as they seem? There are valid concerns mixed in with a lot of noise. Some accounts imply that it is trivial to determine whether results are reproducible or not, but what do we really mean by reproducibility, and how can we tell if we've achieved it? After pulling apart the various meanings of the concept of reproducibility and isolating the ones most important for scientific reasoning, I will show how a flawed view of experimental science has distorted the debate. What is more important: The specific results of a study or its broad conclusions? Is reproducibility equally vital to all kinds of science? Are reproducible results "truer" than others, or only more reliable?

In addressing these questions and more, I want to make three main points in this chapter: first, the significance of reproducibility varies according to the scientific context you are talking about, it is not equally important in all of them; second, even when reproducibility is appropriately defined, there will always be many good reasons that it cannot be achieved; and third, while reproducibility is required for scientific progress and some concern is warranted, panic over irreproducibility is not—science progresses in spite of failure to replicate individual experiments.

Why is reproducibility important? Science is not about isolated individuals or groups doing experiments and publishing data; it is a collective effort that gains its strength from scientists all around the world investigating phenomena and coming to a kind of general agreement, a *consensus*, on how things are. The consensus forms the basis of what science knows and, therefore, what actions it supports. Consensus depends on the reproducibility of results, and so, if reproducibility is threatened, consensus is threatened and the whole enterprise falls apart. That is the big worry.

7.B What Is Reproducibility?

The frantic tone of the newspaper headlines implies that reproducibility is the highest goal of science—that the main purpose of scientific inquiry is to generate reproducible results and that the more reproducible the result, the better the science. This isn't true. As science tries to learn about the world, its objectives include acquiring and interpreting data, explaining its findings, and making predictions of how future experiments will turn out. Reproducibility is a scientific virtue, a core principle that helps science achieve its goals; it is a highly desirable property of research; it is not the main thing. Let's start by defining "reproducibility."

7.B.1 What Are We Trying To Reproduce?

As the Director of the National Institute of General Medical Services at the National Institutes of Health (NIH), Jon Lorsch has noted, " 'Reproducibility' is short-hand for a lot of problems."[9] In this discussion, I assume that we're referring to the ability of investigators to repeat findings reported in a published scientific paper. Since the immediate practical concern is that irreproducible science is bad science, we'll have to be very specific about what reproducibility means. A task force of the American Society for Cell Biology (ASCB)[10] distinguished four kinds of reproducibility: *analytic* reproducibility, which tries to duplicate original conclusions by reanalyzing original data; *direct* reproducibility, which tries to get the same experimental results using the same experimental conditions as in the original report; *systematic* reproducibility, which tries to get the same results as the original study under different experimental conditions than the original ones; and *conceptual* reproducibility, which uses new experimental approaches and aims "to demonstrate the validity of a concept or finding using a different paradigm."

Analytical and direct reproducibility efforts try to confirm the basic facts of your study; they ask whether or not someone else would observe the same effects that you did. Analytic and direct reproducibility are probably what come to most people's minds in this context, and indeed, these are the sources of the greatest anxiety. In contrast, systematic and conceptual reproducibility tests really seek to extend the applicability of your findings, not to duplicate them. The following example illustrates the distinctions among these concepts.

Imagine that, while investigating the neurophysiological basis of emotion in mammals, you discover a drug, CalmDown, that reduces anxiety behavior in mice. That is, if you give animals CalmDown, they readily venture out into the middle of a brightly lit (i.e., potentially dangerous) open area, which they wouldn't normally do. You report your findings and put forward the hypothesis that CalmDown decreases behavioral stress in mammals. If another investigator tries to check your conclusions by reanalyzing your raw dataset (which you dutifully posted online), she is testing it for *analytic reproducibility*. Analytic reproducibility trials want to see if you did your calculations correctly and that your data support your conclusions. Indeed, one view of scientific "evidence" or "facts" is that they represent *interpreted* data, and reanalysis can help ensure that your interpretations of them are reliable. A major part of the controversy regarding the Reproducibility Crisis hinges on questions of analytic reproducibility. We should also note here that interpretation almost always has a subjective dimension, and two people can disagree about what data mean without implying that somebody did something wrong.

The data themselves are the focus of interest when it comes to *direct reproducibility*. If another investigator gets a group of mice and a supply of CalmDown and duplicates your exact procedures as faithfully as possible, he is doing a test of direct reproducibility. Direct reproducibility is the most critical for the validity and integrity of the scientific record; it causes the biggest problems when it fails because the reasons for its failure may be subtle and elusive.

If yet another investigator were to test CalmDown on hamsters instead of mice, as you did, and find that it didn't affect hamsters, it would be a failure of *systematic reproducibility*. Basically, she would have tested a *prediction* of your hypothesis that CalmDown would affect all mammals and falsified it. Although this result would be disappointing if you'd dreamed of having discovered the next Valium, it would not show that your finding was irreproducible. After all, the new experiments did not duplicate your conditions: hamsters are not mice, and it is entirely possible that the drug would affect one species and not another. Your hypothesis would have failed a test of generality, but there is nothing blameworthy in that.

Finally, imagine that, reasoning from your hypothesis that CalmDown reduced mouse stress, other investigators confirmed your behavioral experiments on mice, and then measured the animals' stress hormone levels, inferring that CalmDown should have lowered them. This would test your hypothesis of CalmDown's actions for *conceptual reliability*. If CalmDown did not alter the stress hormone levels, the results would falsify a prediction of the hypothesis, suggesting that the hypothesis was wrong. Again, this would not mean that your study was irreproducible; indeed, in this scenario, your data were actually reproduced. Falsifying your hypothesis would not be a sign that there was anything amiss; rather, it might set off a new inquiry to find out how CalmDown did affect the mice. Maybe it caused stressed-out mice to behave recklessly, without reducing their anxiety, for example. Again, nothing to get upset about.

Science progresses by testing both the generality and the scope of its hypotheses, which is what systematic and conceptual testing do. Because these tests aim to extend original findings by testing hypotheses, their failures probably mean that nature is more complicated than we thought it was. Their failures give us positive new information—our brilliant idea is incorrect!—they point to opportunity, not error, as Stuart Firestein emphasizes (Chapter 10.A.). Thus, there is no contradiction—in fact, no problem—if we succeed in directly reproducing a result while failing to stretch it systematically or conceptually. The distinctions between the various meanings of "reproducibility" have an impact on the debate, and I will follow the ASCB's lead and concentrate on direct reproducibility when trying to decide whether there's a crisis or not.

7.B.2 Reproducibility: A Multipronged Challenge

A failure of direct reproducibility does not imply scientific misconduct,[11] although that distressing issue is always with us.[12] Remarkably enough, irreproducibility can frustrate scientists trying to reproduce their own work.[13] There are many causes of irreproducibility, and learned societies, including the American Association for the Advancement of Science, the National Academy of Science, and the Federation of American Societies for Experimental Biology, have convened panels and proposed strategies to attack the problems.[14] Others have offered technical recommendations for addressing weaknesses in analytical laboratory standards, testing methods, reporting requirements, and so on.[15,16] Not all of the issues are relevant to scientific thinking, which is my main focus, so I propose to group and triage them.

Broadly speaking, the concerns fall into three categories: *material, qualitative,* and *procedural.* The *material* category includes the physical items that science uses and which, if unreliable in consistency, validity, quality, etc., can hamstring reproducibility efforts. Two frequent sources of material problems are antibodies[17] and "cell lines."[18] Antibodies are biological molecules that, ideally, stick with great specificity only to their target proteins and are often used to pinpoint the location or interfere with the function of these proteins in biological tissue. Unselective antibodies stick to the wrong proteins, which leads to incorrect and irreproducible results. Likewise, many biological experiments use in vitro populations of particular cells (cell lines), say kidney cells, which are sometimes obtained from tissue banks or commercial suppliers. Experiments done on contaminated or misidentified cell lines can be irreproducible because, for example, kidney cells and heart cells do not respond in the same way to experimental treatments. If you did experiments on heart cells while believing them to be kidney cells, your reported findings could hardly be replicated. Material-related problems are predominantly quality control and documentation matters that do not affect scientific thinking, and we can set them aside.

Reproducibility is also hampered by *qualitative* issues that include such intangibles as the "culture of science," the "scientific reward system," and cognitive biases, as well as matters of scientific practice governing how we do experiments and publish our reports.[19,20] Even if identifying the remedies for some of these problems were straightforward—and it isn't—putting them into place would require the concerted effort of the scientific community. For example, when investigators do not include enough experimental detail in their reports, their colleagues naturally have difficulty reproducing their data. But, given that scientists have to apply for the money to do their experiments and satisfy editors who publish their papers, granting agencies and scientific journals can bring about change by demanding more information; and, indeed, many are beginning to do so. Scientists, especially mentors, peer reviewers, and journal editors,[21] will exert a major influence on how quickly and thoroughly the community recognizes and corrects well-defined problems like this one.

Other qualitative problems present much greater challenges, either because we do not know much about them or because they are so tightly enmeshed with political and social considerations that they might not be solvable. This category includes various biases—confirmation and publication bias are frequently talked about—that affect how scientists carry out and communicate about science. Bias is unquestionably important, but since it mainly involves psychological and other cognitive factors that are not uniquely linked to the problem of reproducibility, I'll put off discussing them until Chapter 11.

This leaves *procedural* issues, which are closely related to scientific thinking and reproducibility, for us to consider here. These are matters relating to the evidence—the data and theoretical analyses—that can help determine whether reproducibility is a crisis, a problem, or something else.[22] Grappling with procedural issues will give us a better understanding of the Reproducibility Crisis and also illustrate some of the subtleties of science that affect how we think about it.

7.B.3 Is There a Problem?

While there is universal agreement that scientists cannot always reproduce published reports, there is disagreement about the causes of irreproducibility and how common it is. The experiences of the Bayer and Amgen groups are worrisome but might not be typical: confidentiality agreements between the companies and investigators in the original research laboratories meant that all of the details were not disclosed to the public.

How big is the perceived problem? To get a sense of it, I did a PubMed search of scientific literature between January 2014 and March 2018 on "reproducibility crisis or irreproducibility crisis or replicability crisis" and found 156 articles that mention at least one of the terms. The same search on the website of the premier science journal *Nature* (www.nature.com) came up with 261 items (articles, editorials, commentary, etc.). In addition, in May 2016, *Nature* conducted a survey of its readers that drew 1,576 responses.[23] When asked, "Is there a "Reproducibility Crisis?" 90% of the respondents said that there was either a "significant" (52%) or "slight" (38%) crisis, and only 7% felt that there was "no crisis at all." (*Nature*'s definition of "reproducibility" was essentially "direct reproducibility," as defined earlier.) The readers' specialties ran the gamut from physics and chemistry to biology, medicine, and "other." When asked to estimate the percentage of published work in their field that was reproducible, physicists and earth scientists said 80%; chemists and biologists, 70%; biomedical scientists, 50%; and other, 30%. When I posed the identical question to hundreds of biological scientists in an online survey in late 2017, I also found substantial, though slightly lower, levels of concern (Chapter 9). *Nature* is aimed at a scientifically trained audience, and I queried members of scientific societies, so even though neither sample was rigorously controlled, the results probably convey a fair sense of the community attitude: namely, that many scientists feel that a significant fraction of the published work in their field is not reproducible.

The Stanford University statistician John Ioannidis is convinced that 85% of published research is wrong[24] and cites claims that the majority of research dollars are "wasted."[25] Statistician Katherine Button, with Ioannidis and colleagues, analyzed neuroscience studies and concluded that as many as 75% of them reported using procedures that were not rigorous enough to support the reported conclusions[26]; in particular, that the *statistical power* of the studies was much too low. Ioannidis and others argue forcefully that a crisis exists, yet their studies are not above criticism.[27] For instance, Camilla Nord and colleagues[28] reexamined the work that Button et al. had reported on and found that it was not a homogeneous group characterized by low statistical power. Nord et al. concluded that, although "on average" statistical power was too low, there was large and systematic variability in the degree of power, with some studies (notably "candidate gene studies") having quite low power, while "many studies show acceptable or even exemplary power." Nord et al. caution against broad-brush condemnations of scientific findings.

In fact, if you took the direst implications of the work of Ioannidis, Button, and colleagues at face value, you'd wonder how biological science can possibly know as much as it demonstrably does know. Take, for instance, brain studies. Basic neuroscience findings about the brain inform human brain surgery procedures, and, while brain surgery is not a perfected art, it is not as hit or miss as you might imagine it to be from the criticism of basic neuroscience research.

There are those who question the concept of a Reproducibility Crisis,[29] pointing out that it can be extremely difficult to duplicate the exact conditions of a published study. For instance, in fields such as psychology, the precise social context constitutes an amalgam of potentially crucial features—interactions among the individual investigators, the particular environmental setting, etc.—that are probably impossible to duplicate. In general, cutting-edge research must necessarily confront unforeseeable complexity, and it's no surprise to find that scientists at the frontier can't at first guarantee that they've identified all of the pertinent variables. Moreover, an irreproducible result may lead the way to new discoveries, so irreproducibility is not invariably a barrier to scientific advance, as the example in Box 7.1 shows.

In any case, it is scientists, not nonscientists, who discover and rectify the errors that scientists make. This is the self-correcting property of science; finding and fixing mistakes are signs that all is well. Last, complaints about reproducibility are nothing new; they have been around for hundreds of years,[34] and yet nobody denies that we have amassed a huge and rapidly expanding trove of genuine, reliable knowledge about the world. Many commentators think there is nothing to worry about.

So there are two sides to the debate, and we're left with the question, "Is there is a crisis or not?" What actual evidence supports the conclusion that there is one? How is the reproducibility of scientific findings best assessed? The vantage point for approaching such questions is a branch of science that studies itself. This is

Box 7.1 Irreproducibility Can Lead to Better Science

Among many other things, neuroscience wants to know how sensory systems (vision, hearing, touch, etc.) encode and decode information about the world. A group of sensory neuroscientists studies the rat muzzle whisker system. Rats are nocturnal animals. They have poor vision and detect the location and surface properties of objects by sweeping ("whisking") their whiskers around as they explore their environment. This sensory system is accessible, well-organized, and analogous in some ways to the human visual and touch systems. Hence, despite humans' lack of active, controllable facial hairs, we can learn about sensory information gathering and processing by studying rat whiskers. When they are whisked across an object, the whiskers vibrate more or less vigorously according to its roughness or smoothness. The sensory touch receptors associated with the whiskers send information via trains of electrical impulses ("action potentials," somewhat like the 1's or 0's of a computer code) up to a specialized information processing center in the brain (the "barrel cortex),[30] which puts together an image of the animal's surroundings.

Researchers study the impulses that are triggered when the whiskers move, and a key question is, "How does the brain make sense of the impulses?" In particular, how does it decode the information about surface texture that is carried by the impulse trains? One prominent hypothesis[31] held that the brain has a system of miniature neuronal oscillators, basically groups of neurons that act like tiny biological clocks. By comparing the frequency of the incoming impulses to the time kept by the internal clocks, the brain could figure out how fast the whiskers vibrated as they moved over an object's surface and, hence, what its texture was.

Intending to follow-up on the oscillator hypothesis,[32] Asaf Keller and colleagues could not observe the predicted correlations between whisker movements and impulse patterns[33] (i.e., their test of direct reproducibility failed). Keller et al. then did systematic and conceptual tests of the predictions of the oscillator hypothesis. For instance, the hypothesis predicted that the tiny neural clocks should provide timing information whenever the whiskers were moving—not only when they contacted a rough surface. But Keller's group found that whiskers waving freely in the air did not activate the oscillators. This test of a key prediction of the hypothesis was much more informative than the test of direct reproducibility had been, and it implied that the hypothesis was false. This example shows that basic scientists, when confronted with irreproducible findings, are not at a dead end as the applied scientists at Amgen and Bayer had been.

Divorcing specific results (*data*) from a bigger picture (*meaning*) happens frequently in basic science because greater value is attached to being right in a larger sense than in merely making repeatable measurements. Indeed, at times it seems as if opposing groups avoid repeating each other's experiments directly. Perhaps it is more cost-effective not to, or perhaps they think that it will be more productive to get on with new experiments. In Box 7.2, I'll briefly review a famous basic science struggle in which reproducibility was not a deciding factor in the outcome.

"meta-science," the scientific investigation of science. But meta-science is a form of science and, as such, is subject to critical examination ("meta-meta-science!"). How does background knowledge affect the ways in which a meta-science project is carried out or color its conclusions? What unproven assumptions must it take for granted?

The Reproducibility Project: Psychology[35] (RPP) was a systematic attempt to inquire into the pervasiveness of irreproducible scientific findings. If the reports by the Bayer and Amgen scientists ignited the controversy over reproducibility, the conclusions of the RPP and by Ioannidis and colleagues have provided the fuel that sustains it.

7.C The Reproducibility Project: Psychology

The RPP was conducted by an international group of scientists calling themselves the Open Science Collaboration and was led by Brian Nosek. It was a meticulously planned and executed effort to assess the reliability of psychological science. The Open Science Collaboration consisted of numerous teams of researchers who tried to duplicate the experimental findings reported in 100 papers published in top-line psychology journals in 2008. For the most part, the teams stuck closely to the conditions reported in the original papers, checking with the original investigators when necessary; thus, they mainly tested for *direct reproducibility*, although they did allow some variations in procedure, which contributed to the subsequent controversy about the RPP's own conclusions.

An especially importantly methodological feature of the RPP, one that I'll return to several times, was its selection of the particular experiments that the teams would try to replicate. Each paper they reviewed included an average of about 3 "studies,"[36] and each study comprised several experiments that all contributed to the main conclusion of the paper. Each experiment was summarized by a single, statistically derived number (usually either a *p-value* or *effect size.*[37,38]

Obviously, for practical and logistical reasons, it was out of the question for the RPP teams to try to replicate all of the experiments in all of the papers, so the teams focused on only one "key result" per paper and, for the sake of uniformity, took the last one "based on the intuition that the first study in a multiple-study article (the obvious alternative selection strategy) was more frequently a preliminary demonstration." As the story unfolds, we'll see that this apparently sensible strategy limits the applicability of the RPP findings.

What did the RPP teams learn? That precise reproducibility of original experiments was disappointingly hard to come by. For example, the RPP teams had estimated that they should have been able to reproduce 89% of the original experiments,[39] yet they could actually reproduce only 36–47% of them (depending on the particular metric of reproducibility they used). In addition, the replicated effects were, on average, only half as large as the original effects, and 80% of the replicates were numerically smaller than the originals. In the worst cases, an RPP repeat study obtained effects that were opposite to the reported ones (e.g., if the original report stated that X treatment significantly improved memory, the team found that X made memory significantly worse). There were a few bright spots: the more impressive the original results (smaller p-values, larger effect sizes), the more likely they were to be replicated. Nevertheless, the RPP approach was meticulous, and its results appear to constitute a serious indictment of the way science is currently done.

The RPP report is sobering, but does it imply that half of the original research papers came to invalid conclusions? That they were worthless? Bogus? Does the report fatally undermine their credibility or suggest that the money spent on them was wasted? Perhaps not. There are alternative interpretations of the RPP data and of the relationship of reproducibility to science than the one the RPP authors favor.

A published commentary by a group of psychologists led by Daniel Gilbert takes issue with the RPP report's main conclusions.[40] Gilbert et al. looked for *analytical reproducibility*. According to Gilbert et al., the RPP teams overestimated how many studies should have produced positive results; in addition to the number expected to fail because of random sampling error, "infidelities" between the RPP teams' procedures and the reported ones precluded precise replicability in some cases. Furthermore, not all of the replicate studies were overtly testing for direct reproducibility; some were testing for *systematic* or *conceptual* reproducibility which, as I've argued, don't count as forms of reproducibility at all.

Gilbert et al. also point out that when the RPP teams' procedures got the original investigators' full stamp of approval, the rate of reproducibility went up, which hints that slight, perhaps easily overlooked differences in original and replicate protocols may have caused problems. I'll return to this issue shortly, and it will turn out that even apparently insignificant procedural discrepancies can doom efforts to reproduce published results. In summary, Gilbert et al. calculated that if all of the sources of variability, both random and systematic, between the original and replication studies of the RPP teams were completely accounted for, then 34% of the replicates should have failed, which is not far from the number of failures that the RPP actually found (36–47%). For Gilbert et al., the RPP results are no cause for alarm.

The RPP group rebutted Gilbert et al.'s charges,[41] pointing to shortcomings in the critics' own reasoning but, in the end, conceded that the RPP data could support both pessimistic (theirs) and optimistic (Gilbert's) conclusions about the question of reproducibility in psychology research, adding that there is "no such thing as exact replicability." Two subsequent statistical reanalyses of the RPP came to differing assessments about the RPP report: one held that "It really looks like [the RPP's statistical] power was very high,"[42] and the other that "apparent failure of the RPP to replicate many target effects can be adequately explained by overestimation of effect sizes"[43] (i.e., the reproducibility studies themselves were not always above reproach).

There is no disagreement that science needs to tighten up its standards and find ways to improve the reproducibility of its results. It is also the case, however, that reproducibility is not always of transcendent importance, and insisting that it is can generate its own problems. How can this be? A cautionary note from the RPP report provides a clue: "It is too easy to conclude that successful replication means that the theoretical understanding of the original finding is correct. Direct replication mainly provides evidence for the reliability of a result," not that we have achieved good understanding of it.

The significance of any scientific project depends both on what it *has found* (what the data actually show) and what its findings *mean* (what it tells us about the world). Reproducibility is a valuable characteristic of a scientific result, and, ultimately, true results must be reproducible. However, reliability also involves a judgment about theoretical understanding. Basic science tries to learn the truth about the world, and raw reproducibility is at best a leading indicator of the truth value of a scientific result. According to the RPP authors, none ("Zero.") of their positive replications establish the truth of any of the findings that they tested.

We should consider that reproducible results may be unreliable; they may in fact be untrue. Suppose you got a cold and several observations suggested that you contracted it from your cousin at the annual family barbecue: he was there; looked ill; gave you a big, breathy hug; and sneezed frequently. But the sneezing might have been brought on by his hay fever allergy and have had nothing to with his cold. Perhaps, your aunts, uncle, and sister all witnessed your cousin sneezing and drew the same conclusion from it as you did: he's got a cold. Despite the fact that your observation was reproducible, it was unreliable because it was based on your common ignorance of his hay fever allergy. Likewise, the countless reproducible observations that supported Newton's theory of mechanics did not guarantee that the theory itself would be the last word on the subject, as Einstein's theory showed.

What should we make of all this? Two lessons will carry forward: one is about the investigations that the RPP repeated. Given the quantity, quality, and variety of the complaints about irreproducible results, it is likely that there are genuine problems, although blanket claims of a "crisis" seem unwarranted. It is worth recalling in this context the wide range of estimates of reproducibility made by respondents to the *Nature* survey[44]: from 80% (physics) to 30% ("other" sciences, presumably including psychology), which suggests that the extent of the problem varies across scientific fields.

However, another lesson from the RPP controversy is that we shouldn't take its conclusion, any more than we take the conclusion of any scientific report, at face value. How do we decide who is right—the RPP or its critics? This is the kind of intellectual skirmishing that goes on all of the time in science. Science is an inherently adversarial method of knowledge generation; ideas are treated roughly and have to prove themselves worthy of respect by standing up to severe challenges. In empirical sciences, better evidence, rather than nuanced technical debate, determines the outcomes of controversies. A comment supposedly made the physicist Ernest Rutherford, "If your result needs a statistician, then you should design a better experiment," is apropos.

What else can we learn from studying the RPP? In Chapter 1, we saw that all scientific projects have to assume that certain unproven ideas—implicit hypotheses—are true in order to proceed. In the next few sections, I will examine how some of the assumptions that the RPP scientists made could affect the applicability of their conclusions.

7.C.1 The Reproducibility Project as the Subject of Meta-Science

The RPP was a large, controlled experiment, meaning that it attempted to keep as many factors constant as possible to reduce the influence of random variability. Usually, controlled experiments are highly desirable, although their advantages can be lost if they can hide meaningful variability. I mentioned the crucial tactic of the RPP to select a single result from each paper chosen for replication. It is important to stress that the principal "finding" of a multipart scientific paper is hardly ever based on one experiment; typically, the paper pulls together evidence from several experiments into an interpretation that makes sense of all of them.

Note that although the main conclusion of a paper depends on a group of interlinked experiments, one experiment may be absolutely essential and others secondary. The result from this experiment gives rise to the central hypothesis of the paper, and, if it were absent, then the secondary experiments might not stand alone, or at least they would not command the same degree of attention as the whole group would. Indeed, if the crucial result was not repeatable then, depending on its potential impact, many people would lose all interest in the report. And, to turn the argument around, no matter how crucial, a single experiment by itself is scarcely ever granted publication in a respected journal in most fields of biological or social science; the secondary experiments are generally tests of predictions of the central hypothesis, and, in my experience, reviewers and editors can be quite keen to have a few of these predictions tested before they'll approve a piece of work.

The conclusion of each paper selected for replication by the RPP was supported by the outcomes of an average of three multipart experiments. How could the decision to focus on one experimental result affect the relevance of the RPP's conclusions to everyday science? We'll consider some possibilities next.

7.C.2 Reproducibility: All-or-None?

Reproducibility is not a binary condition: a scientific report may not be all right or all wrong. One part of a multipart report may not replicate, while other parts do. Or a replicated result might match the original result qualitatively though not quantitatively, as was the case in many of the RPP experiments. It is often hard to classify reports as "reproducible" or "irreproducible."

"Reproducibility" may also lie in the eye of the beholder. In the blog *Science Based Medicine*, Managing Editor and physician-scientist David Gorski recounts his attempts to duplicate a finding made by the renowned biomedical researcher, Judah Folkman.[45] Folkman and his team had found that a chemical, later named *angiostatin*, was secreted by solid primary tumors and almost completely suppressed the development of secondary tumors, *metastases*, by preventing the outgrowth of blood vessels necessary to feed the metastases. This was a novel and powerful hypothesis, with numerous clinical and theoretical implications. However, Gorski and colleagues observed that angiostatin was "never . . . as potent" as Folkman's group had found. Still, they did find that angiostatin retarded secondary tumor growth. Did they reproduce Folkman's results? Yes and no. For a drug company hoping to discover a new anti-cancer therapy, with a multimillion-dollar R&D project hanging in the balance, the answer could well be "no." Yet angiostatin slowed tumor growth and, as Gorski and colleagues later showed,[46] it augmented the benefits of radiation therapy.[47] This effect of angiostatin might have remained undiscovered had it not been for their *systematic* extension of Folkman's prior work. So, depending on the definition of reproducibility you're using, the answer to the question could be "yes." Science is quite often like this: more nuanced than can be fit into a simple yes-or-no, reproducible or not-reproducible, type of pigeon hole.

7.C.3 Reasonable Assumptions and Unexpected Consequences

Let's take another look at the RPP group's rationale that choosing the last experiment in each paper was safer than choosing the first one. Again, this sounds sensible, but might be problematical. When scientists are just talking, you occasionally hear mutterings about "the last experiment" in a paper they've read, because they're suspicious that it was added at the request (insistence?) of one of the anonymous peer reviewers of the article.[48] (It was once explained to me by a senior luminary in my field that the appropriate posture for an author seeking acceptance of a paper into a top journal is that of "a dog lying on its back, exposing its belly"—i.e., abject submissiveness.) However this may be, it is uncontroversial to say that investigators generally feel that they have little leeway; if one of the omnipotent peer reviewers of their work asks for more information, usually investigators will comply, even it means tacking on an extraneous experiment. It is possible, therefore, that the last experiment in a paper is less essential to the main message than the others. At the other end of the spectrum, some investigators will say that they save the "best for last"; that is, they put their strongest experiment at the end of their paper. Finally, authors sometimes simply add on an experiment that was an interesting tangent from the main stream of the paper, not an integral part of it. My point is that last experiments may not be representative of the rest. Indeed, the RPP authors themselves caution that "it is not necessarily the case that the identified effect [i.e., the one from the last experiment] was central to the overall aims of the article." Still, their conclusion as to whether or not a given report was reproduced was deliberately based on the last experiment.

None of this is meant to cast aspersions on the approach taken by the RPP group, merely to emphasize that their plausible strategy depended on an unproved assumption that could be wrong. A corollary, that all experiments in a paper are equally likely to be reproduced, is also unproved. It remains to be determined whether such assumptions affect the overarching conclusions coming out of the RPP.

7.C.4 One Bad Result May Not Be Fatal

Because the main message of a multipart report rarely hinges on one experimental result, the failure of one result does not completely undermine the report's validity. The fact that your cousin's sneezing was an unreliable sign of his cold does not invalidate your conclusion that you caught your cold from him.

In science, good individual tests of a hypothesis rest on independent background assumptions. Indeed, the primary reason that scientists do multiple tests[49] in the first place is that, while they all have weaknesses, they are unlikely to have the same weaknesses! If the results of several truly independent tests support the same conclusion, you feel more confident that it is not the spurious result of any single test.

While your confidence in a scientific hypothesis based on multiple lines of evidence might be shaken if one of its predictions turned out to be wrong, you might not entirely lose faith in it. Wishing to avoid the negative consequences of basing conclusions on single tests is partly why the RPP authors warn us not to overinterpret their single-result replications. The authors don't emphasize that normally science relies on tests of multiple predictions for strong positive reasons as well.

7.C.5 Do We Know Enough to Be Able to Replicate Experiments?

A critical implicit assumption made by the RPP, and indeed, any scientist, is that you can correctly guess which environmental variables are going to affect your experiments. If you knew all of the factors that affected the original results, then you could take them into account when you try to reproduce those results. Though guessing the relevant variables sounds simple enough, in reality it may be anything but.

For instance, recently, a research group conducting behavioral experiments on laboratory rats found that a male researcher got a different result when he did an experiment than a female researcher got when she did the identical experiment.[50] And they were in the same lab, studying the same animals, with the same equipment! The first hypothesis was that the experimenters themselves somehow influenced the rats; maybe the males were rougher with the rats? But this hypothesis predicted that actual people would have to be present to get the effect, and it turned out that clean, new T-shirts, each worn overnight by either a male or a female and placed near the rodents, were equally good at causing the alterations in rat behavior. What was going on?

The experimenters tested the hypothesis that something transferable to the T-shirts, probably a chemical, affected the animals. They tried various chemical agents, several species of non-human—as well as human—males, behavioral measures of stress, hormone analyses, etc. Only after testing and eliminating numerous hypotheses did they hit on the present explanation—secreted male hormones affect rat behavior by inducing stress responses in the animals. Among the many lessons from this elegant study is one that should make all scientists stop and think: How good we are *really* at intuiting what the important experimental variables are? The lessons for science in general, and reproducibility in particular, are obvious. If, you didn't know about the male-specific effect, and men and women had alternated animal-testing duties, you'd miss detecting the hormonal effect. If two separate labs, one made up of men and one made up of women, had independently conducted the studies, the results of each group would have been irreproducible by the other.

Worse yet, nature can throw up road blocks even we when our do our best to anticipate and avoid them, as a report by Hines et al.[51] reveals. These researchers describe the frustration they experienced in attempting to generate reproducible results. Two laboratories, one in Berkeley, California, and one in Cambridge, Massachusetts, were working on a long-distance collaborative project, and both were thoroughly experienced in all of the technical methods. But they were initially stumped when they could not duplicate each other's observations. Their project required them to identify the various kinds of cells in human breast tissue, and, to do that, both groups had to isolate individual cells from surgically obtained tissue samples. Colored antibody molecules were used to label specific proteins in the cells, and the plan was to use the antibodies in pairs to identify the target proteins in all cells. Each kind of cell had a distinctive combination of the proteins, and the kinds of cells could be sorted into groups according to their color patterns. You could use the same method to identify the different states in the United States according to their ethnic make-up. For example, 46.3% of the population of New Mexico is of Hispanic heritage,[52] while Hispanics constitute only 2.5% of the population of North Dakota. Conversely, German descendants account for 9.8% of New Mexicans and 46.8% of North Dakotans. You could unambiguously identify each state by its ratio of residents of Hispanic heritage to those having German ancestors.

The scientific problem was that the two groups of investigators kept getting completely different proportions of the proteins in their cells despite starting with the same biological tissue. It was as if one group came up with the Hispanic/German ratio typifying New Mexico, while the other got the North Dakota ratio when they were both studying the New Mexicans. After trying for more than a year to pin down the discrepancy, it came down to a technical detail: to isolate the cells in their laboratory, the Cambridge group used a device that swirled their tissue samples in flasks, the Berkeley team gently rocked theirs in a mechanical laboratory shaker. When they both used the same method, they got the same results. Now, just why the difference in devices had such big effects is a mystery; however, the lesson is plain. What looked like a minor difference in technique had major consequences for the results. Had the research teams been working independently, then, again, each team would have obtained perfectly valid results that the other team would have considered irreproducible.

The stories of the cell-sorters and the male and female experimenters are two of countless instances in which subtle and easily overlooked minutiae of the experimental environment could lead to wildly divergent results. Nobody knows how common such dramatic effects are, but probably they're not million-to-one outliers. While more extensive lists of experimental details in scientific papers will alleviate some of the strains caused by irreproducibility, the examples highlight the potential subtleties of the problem. Will we really think to report the genders of our laboratory team members, for example? Even if extreme levels of disclosure were to prove necessary (and societal objections would surely arise at some point), it will be impossible to guess and correct for every conceivably significant variable.

These considerations alone should give us pause before accepting conclusions that most science is irreproducible and that therefore someone needs a scolding. And what about the implication that the replication team's results are more valid than the originals? How do we know, if someone did make a mistake, that it was the originator, not the would-be replicator of the research?

Is irreproducible science invariably bad science? The dataset obtained from swirled tissue is not obviously inferior to that obtained from gently rocked tissue. The experimental examples show that irreproducibility may follow impeccable lab procedures carried out by careful investigators. We must keep in mind that, at first, an irreproducible result is just another observation in search of an explanation.

Finally, the examples raise other deep questions: If a result is so fragile that it is dramatically altered by apparently trivial details, how significant can it be? How can we tell which procedures give us the best insight into the natural systems that we are most interested in? What are the underlying mechanisms of the unanticipated effects—why do male hormones affect rats; how does the swirling or rocking affect human breast tissue cells? Irreproducible results that do not represent error, incompetence, misconduct etc., can be signs of potentially fertile scientific ground, or, in Stuart Firestein's evocative phrase,[53] they may represent areas of "productive ignorance."

7.C.6 Multiple Tests

Intuitively, we feel that conclusions backed up by multiple tests are on firmer ground than those supported by only one test. Let's say that you're trying to improve rats' ability to learn how to run a maze. Your hypothesis predicts that a drug, SmartRat, should be effective, so you give it to one group of animals and a harmless placebo to another group, and you compare how fast the groups learn to get through the maze. If the SmartRat group is significantly faster at the $p < 0.05$ level, you'd be tempted to infer that the drug was beneficial. However, there is a chance that SmartRat didn't do anything and you've been fooled by randomness.[54] You wouldn't want to launch a costly drug development program if there were a 5% chance that you'd lose a lot of money.

How would doing multiple tests change things? Suppose you use SmartRat and a chemically distinct drug, RatHiQ, that has the same predicted effects on the rat nervous system. You give SmartRat to one group and RatHiQ to another and, again, compare them to placebo-dosed control animals. If you find that both SmartRat and RatHiQ boost the rats' performance, you feel more confident than when you had only the one-drug result. Your confidence could still be misplaced, but the odds of getting significant effects with two drugs by random chance surely would be much smaller than they would be with one. The more independent tests that a scientific hypothesis passes, the more well-corroborated it is, and the less likely its successes resulted from chance. (In Chapter 8, I introduce a way of combining individual test results so you can quantify the net probability of getting two [or more] of them.)

The case of the male and female researchers who got different results when they were studying rat behavior is an extreme example of multiple testing. A rough count indicates they did 17 distinct tests before reaching their final conclusion. Even if every result was only significant at the $p < 0.05$ level, the chance of doing 17 tests that would all be consistent with the hypothesis would have been extremely low. While it is true that this was an unusually thorough study, conclusions of most reports rely on multiple tests and are therefore probably more reliable than you would guess from considering only one. Yet, again, much of the concern about reproducibility is based on the examination of one result from multipart studies.

What about the RPP itself? Did the RPP teams try a multiple-tests approach? That is, did they use more than one method to test the generality of their own findings? Not exactly, but another group effort, the Many Labs Project also led by Brian Nosek, did. In a sense, the Many Labs Project was the converse of the RPP. Rather than having many individual teams each test the replicability of a single published result, the Many Labs Project had a group of laboratories all try to replicate the same small number of published results. The results of the two studies were poles apart. Whereas the RPP suggested a reproducibility rate as low as 36%, the Many Labs Project concluded that 77% (10/13) of the studies that they looked at were reproducible. Although there is a dispute about the significance of this difference,[55–57] its existence indicates that meta-science projects are not exempt from the uncertainties that plague ordinary science.

We have been looking broadly at the RPP as a prototypical scientific investigation and, in these last few sections, reviewing background assumptions that the RPP scientists made. When it comes to background assumptions, we always hope that they are innocuous and won't affect the outcomes of our experiments. Science continually faces a choice of whether to devote indefinite amounts of time to keep from being tripped up by an unjustified assumption, or to take a chance, make the assumption, and hope for the best. Science usually manages to strike a balance.

There are one or two other lessons about scientific thinking that we can draw from reviewing the reproducibility studies. They have to do with another kind of implicit assumption that is also quite common in science: that we know what we should do and why we should do it.

7.C.7 Truth, Reproducibility, and the RPP

Although the RPP authors were all well-versed in the principles of scientific thinking, in their report they didn't draw the distinction between seeking truth and seeking reproducibility as cleanly as they could have. The RPP was about data (i.e., direct reproducibility; getting the same numbers as the original report). The RPP authors would agree that the pinnacle of scientific achievement is not spreadsheets packed with reproducible numbers. They say, "Accumulation of evidence is the scientific community's method of self-correction and is the best available option for achieving that ultimate goal: truth" and "Understanding is achieved through multiple, diverse investigations that provide converging support for a theoretical interpretation and rule out alternative explanations." Simply put, what science knows goes beyond mechanical, direct reproducibility to the weighty conceptual issues related to the hypothesis.

7.C.8 Bayes and the Reproducibility Crisis

The RPP group asked a straightforward question—were the original published results of 100 psychology studies replicable or not?—but the answer was not straightforward. Arguments on both sides of the reproducibility debate were based on conventional statistical testing methods, which have their own problems, as we saw in Chapter 5.

Against this backdrop, a paper by Etz and Vanderkerckhove offers a novel perspective.[58] Rather than engaging the topic of reproducibility directly, they asked if the original and replicate studies yielded the same degrees of evidence for or against the hypotheses of interest. Let's suppose that each study, original and replicate, was testing an experimental hypothesis, H_1, against a null hypothesis, $H_{0,}$ where H_0 is that the experimental treatment had no effect. If there was a failure of reproducibility because of experimenter sloppiness, incompetence, or worse, you'd expect the original data to report strong evidence for H_1, and the replicate studies to reject H_1 in favor of H_0. Is this the case? Etz and Vanderkerckhove turned to Bayesian methods to find out.

From a Bayesian perspective, each study in the RPP amounted to a comparison of the strengths of H_0 and H_1. Therefore, you can evaluate each study by its Bayes' factor: that is, the ratio of the likelihoods that H_0 or H_1 accounted for the reported data (D); that is, the Bayes factor is $[p(H_0/D)/p(H_a/D)]$; see Chapter 6, Box 6.2. Etz and Vanderkerckhove calculated Bayes' factors for all of the studies—originals and replicates—and rated a Bayes' factor of ≥10 as strong evidence in favor of one hypothesis over the other and a Bayes' factor of <10 as weak and inconclusive evidence. Their implicit hypothesis was that a systematic tendency toward irreproducibility, the kind that would justify declaring a "crisis," would show up as a marked difference between the Bayes' factors of original and replicate studies. That is not what they found.

Their first remarkable finding was that, although 92% of the original results were significant at the $p < 0.05$ level, only 36% had a Bayes' factor of 10 or greater. In other words, the majority of the original results, which were strong enough to warrant rejection of H_0 by the usual standard, were weak according to the Bayes' factor standard. A greater surprise was that the Bayes' factors of the replicate studies suffered from the same drawback. These were the studies done by the RPP teams that were meant to be the gold standards, yet their conclusions were usually also in the weak range, and none of them provided strong evidence for H_0. This is a stunning outcome. It says that, contrary to our intuition that "irreproducibility" meant H_0 had been wrongly rejected by the initial investigators, neither original nor replicate experiments provided strong evidence either to accept or to reject H_0! And indeed, Etz and Vanderkerckhove found that the original and replicate studies were a lot more similar to each other than had been recognized. Fully 75% of the original and replicate studies gave qualitatively similar results: when the strengths of their evidence for H_0 and H_a were matched up, the weak original studies were associated with weak replicate studies, and the relatively few strong studies were similarly correlated.

What was going on? The explanation is a bit technical. Evidently, the RPP group had overlooked the influence of *publication bias* in designing their replication experiments. Publication bias is the tendency of scientists and journals to want to publish only statistically significant results. However, this tendency inevitably led to overestimation of true effect sizes. The RPP did preliminary power analyses and tried to achieve high power levels in their experiments. Still, because their analyses were based on the overly large reported effect sizes, the replication sample sizes were too small (you don't need as many subjects to detect a large effect as you do a small one). In fact, Etz and Vanderkerckhove noted that the larger the study, the higher the Bayes' factor, regardless of whether it was an original or a replicate. Hence, despite the best efforts of the RPP teams, the replication studies were underpowered, meaning they were insensitive to detecting real effects.

In a final touch, Etz and Vanderkerckhove accounted *quantitatively* for essentially all of the difference between original and replication studies by taking into account publication bias "without recourse to hypothetical hidden moderators." This implies that the RPP doesn't support the more alarmist concerns about questionable scientific practice as a cause for irreproducibility.

Details aside, Etz and Vanderkerckhove's work represents a "good-news, bad-news" scenario. The good news is that science is not heavily tainted by careless error or sleazy behavior. The bad news is that many published studies are neither true nor false, but simply inconclusive. Science is suffering not so much from a "reproducibility" crisis as from a "producibility" crisis; it's not that we can't *replicate* solid findings, it's that we can't *produce* solid findings in the first place. Etz and Vanderkerckhove favor policy measures to combat publication bias, and they call on scientists to pay greater attention to the importance of statistical matters than they frequently do.

And there is still the caveat that I mentioned earlier: the RPP groups only tested one single, statistically significant result in each paper they examined. The analysis of Etz and Vanderkerckhove necessarily suffers because of this defect as well.

7.D Be Careful What You Wish For: Overemphasis On Reproducibility

When it comes to science, reproducibility is not everything, and losing sight of that fact may have unforeseen consequences. In 2015, the US Senate Committee on Environment and Public Works (EPW) passed the Secret Science Reform Act of 2015, which prohibits the Environmental Protection Agency (EPA) from "proposing, finalizing or disseminating" a regulation "unless all scientific and technical information relied on . . . is the best available science, specifically identified, and publicly available in a manner that is sufficient for independent analysis and substantial reproduction of research results."[59]

Given the burgeoning public anxiety about a crisis of reproducibility, you might think this Congressional call for transparency and reproducibility is commendable. If so, you'll be surprised to learn that several Senators and others concerned about climate change and clean air and water are exceedingly worried by the Act and are opposed to it. In holding up reproducibility as a defining characteristic of the "best available science," quantities of valuable, though unavoidably irreproducible, scientific data could be forbidden to the EPA. For instance, the British Petroleum Deepwater Horizon oil spill in the Gulf of Mexico in 2010 was an environmental disaster. It was the largest such spill in history, the subject of intense scientific scrutiny, and the source of enormous amounts of valuable data. It was also a "one-time event" and is, therefore, irreproducible by definition. The Senators who are worried about the Act fear that strict demands for reproducibility would render much of the information gleaned from disasters off-limits to the EPA in making environmental recommendations.

Or consider longitudinal epidemiological studies that last for 40 years[60] and directly inform human health science policy. The 2015 Act could exclude data from such studies unless they are reproducible, and it might take another 40-year study to find out if they are.

The Act also calls for *transparency*, conceivably implying that all of the data used by the EPA, of whatever age, in whatever format, and whether involving confidential patient data or not, must be publicly available for independent reanalysis. Furthermore, much of the old data that continue to guide EPA policy are not in digital or other readily accessible form, and extensive effort would be needed to meet the requirements of the Act. Making everything transparent would be extremely onerous and expensive (the Congressional Budget Office estimated that compliance with the Act's mandates would cost $250 million; the Senate EPW Committee budgeted $1 million for compliance). Finally, satisfying the transparency requirement could be impossible if confidential patient records were involved.

In commenting on the Act,[61] American Association for the Advancement of Science (AAAS) President Geraldine Richmond noted that "while transparency and reproducibility are of utmost importance to the scientific community, this mandate . . . is overly broad and will have severe unintended consequences." It may be too optimistic to expect the US Congress to draw and respect the fine distinction between "utmost importance" and "overly broad" when the political lines are sharp and the stakes are high. If Congress and the public had a better understanding of science, including a realistic appreciation of the roles and limitations of reproducibility, such difficulties might be headed off.

7.E The Variable Value of Reproducibility

The preceding quote from the AAAS suggests that the appeal of reproducibility is not fixed; that it can vary with social context, for example. The fact that a mandate to pursue the goal of reproducibility may be "overly broad" suggests that we need to think about whether the value of reproducibility can vary according to the scientific context as well.

Usually, people who worry about the Reproducibility Crisis do not distinguish between different kinds of science, even though reproducibility might not be uniformly vital in all cases. Recall that the present discussion of reproducibility was started by company scientists at Amgen and Bayer who wanted to develop new cancer therapies and couldn't repeat published results. These scientists needed specific biochemical targets to explore and hoped to pick up directly from where the published reports left off. When their replication efforts failed, they had nothing to work with. They were not trying to understand, say, the basic molecular biology of cancer, but to duplicate enough information to warrant spending time and money on drug development programs.

We've said (Chapter 4) that applied science is practical; it seeks to achieve technological ends, whereas basic scientific research strives for a fundamental understanding of nature, whether or not its practical applications are immediately obvious. The scientists at the commercial laboratories were doing applied science. In the terms introduced earlier, applied scientists are more interested in a report's *data* than its *meaning*. Basic science is more intently focused on meaning, and reproducibility of data is, in a sense, less critical for its advancement. Indeed, at times, it seems as if opposing groups avoid repeating each other's experiments directly, perhaps because pinpointing the reasons for failure of direct reproducibility is less interesting and cost-effective than getting on with new experiments. This suggestion probably seems counterintuitive to many readers, so I'll give an example in Box 7.2.

Divorcing specific results (data) from a bigger picture (meaning) happens frequently in basic science because higher value is placed on finding better explanations (hypotheses) than in merely making repeatable measurements.

I am not saying that reproducibility is unimportant, only that it is not the most important goal for science. We can look at the issue from a logical point of view: "True scientific findings will be reproducible" does not imply that "Reproducible scientific findings will be true." If you argue otherwise, you are *affirming the consequent*, a fallacy that I mentioned in Chapter 1.

7.F Preregistration and the Hypothesis: A Fix for the Reproducibility Crisis?

There is one final issue regarding reproducibility and the hypothesis that we have to go over because it concerns a method that is being widely touted as a solution to the crisis; it is the plan to "preregister" studies before conducting them. Here's the rationale for it and what it would involve.

Box 7.2 When Reproducibility Took a Backseat to Truth: The LTP Wars

From the late 1980s and through the 1990s, rival laboratories studying how we learn and remember facts published a succession of progressively more sophisticated investigations that became known as the "LTP Wars."[62] At stake was an understanding of the molecular basis of memory and a phenomenon called *long-term potentiation* (LTP), which is still the leading contender to be the fundamental neurophysiological mechanism of memory storage[63] Throughout the decade-long contest, the laboratories rarely replicated each other's observations; instead, they executed rigorous conceptual tests of the two diametrically opposed hypotheses that were at the center of the dispute.

For more than 100 years, neuroscientists had suspected that synapses, the submicroscopic[64] junction points where nerve cells in the brain signal to each other, are the sites of memory formation. Suppose that you learn that Lusaka is the capital of Zambia. If we were to zoom in on one of the prime memory formation regions of your brain, the hippocampus, we'd find that the synapses between the signaling cells and their receptive partners had been made "stronger" by LTP while you were learning. The strengthening means that the signaling cells can more easily activate their receptive cell partners in the neural circuit that stored "Lusaka" and retrieve the name from memory when you need it in the future. (Do not worry if you feel that you don't really understand memory much better from this description—no one else does either.) While there is still a lot to learn about memory, elucidating the precise molecular mechanisms of LTP would represent a huge step forward; it could, for instance, contribute to creating a drug to treat memory disorders, such as Alzheimer's disease.[65]

Broadly speaking, two kinds of changes could make synapses stronger: either the presynaptic (sending) cell could send out more chemical neurotransmitter or the postsynaptic (receiving) cell could gain more receptors to the transmitter that was sent out. Think of a synapse as a miniature garden; the transmitter receptors are flowers. The garden becomes more beautiful if you make sure it gets plenty of sunlight ("presynaptic") or, alternatively, if you enrich the soil ("postsynaptic"). The LTP Wars were fought to test the two hypotheses of where synaptic strengthening occurred: on the pre- or postsynaptic sides of the synapse.

The first solid experiments detected more of the chemical neurotransmitter being sent out from the signaling cells after LTP was established.[66] The results were predicted by the Pre-side hypothesis, but the experiment was not always reproducible. Rather than try to resolve the irreproducibility issue, the Post-side researchers took an entirely different tack. The presynaptic hypothesis also predicted that all postsynaptic receptor responses would be strengthened equally, just as all of the flowers grow better if the garden gets more sunlight. The Post-siders found that not all types of receptor responses were strengthened equally by LTP—one kind was greatly enhanced while another was completely unaffected—so the presence of more neurotransmitter alone could not explain the synaptic strengthening. Something on the postsynaptic side, in the "soil" of the synaptic garden, made some of the receptor responses stronger, they felt. The first skirmishes in the LTP wars ended indecisively.

The next round began, again, not with attempts to replicate or reconcile the earlier results, but with the opening of new fronts by testing new predictions. Pre-Siders used high-resolution electrical signaling data and, applying a well-established quantitative theory of neurotransmitter release, concluded that the LTP data was consistent with the presynaptic hypothesis. However, critically, their analysis depended on subtle assumptions that they could not verify. The Post-siders developed a similarly hard-core quantitative conceptual test of the Pre-siders' hypothesis and found that the presynaptic effect was a minor component—the main action was on the postsynaptic side.

And so it went for several more years, with the two sides subjecting the pre- and postsynaptic hypotheses to increasingly severe experimental tests. Because, as Karl Popper emphasized, we must *decide* when to accept that a test falsifies a hypothesis, and because falsification itself can never be conclusive, a struggle for Truth such as this one, between teams of brilliant and fiercely competitive investigators, can be protracted. The experiments were ingenious and intricate; attempts to annihilate the opposing hypothesis at first appeared decisive, only to have it reappear, like the evil robot in the *Terminator* movies, apparently unconquerable. Reproducibility, or the lack of it, no doubt created headaches for the participants throughout the LTP Wars, stimulated controversy, and wasted effort. Yet the headaches also drove the development of increasingly more convincing and telling experiments. In the end, the Pre-siders could not counter the Post-siders sharp attacks, and the Pre-siders essentially conceded. The Post-side hypothesis itself has withstood many challenges and has not been falsified. Thus far.

The story illustrates how a major basic science controversy is settled, not by reproducibility studies, but by intellectual struggle in which specific competing hypotheses are subjected to increasingly rigorous tests until one is eliminated.

If, before you did an investigation, you announced exactly what you planned to do; which data you would collect and how you would collect, analyze, and interpret them; and you were assured that your report would be published regardless of what you found, then, it is thought, you could not be biased in favor of any particular outcome. Your report would be less likely to be tainted by perverse incentives and more likely to be reproducible. The veracity of scientific reports would go up. This is the underlying premise of the program of *preregistration*. According to one scheme, "Before researchers even begin the experiments, they submit a manuscript presenting a clear hypothesis that they plan to test and their proposed experimental methods and analyses." Presumably you could have done a pilot experiment or two, and the concept is that, before doing the major experiments, you submit an outline of proposed work to journals, where, after appropriate review and revision, it is accepted. The journal agrees to publish the eventual data, provided that the authors do the work exactly as proposed. When applicable, the idea of doing experiments and letting the chips fall where they may has much in its favor. The caveat here is that preregistration may not always be applicable.

7.F.1 When Preregistration Can Help

We can start with the term "hypothesis" in the description of preregistration. Does it refer to a genuine, explicit scientific hypothesis, or is it merely a prediction (see Chapter 2)? Preregistration seems well suited to advanced (meaning based on copious prior information) focused studies in which only one or two variables are manipulated, the parameters of the design and analysis can be laid out in detail beforehand, and the experiments are executed almost robotically. This is the essence of a "decision-making" study (see Chapter 4). The archetypal clinical trial of drug efficacy is an ideal case. The study asks, "Does the drug work or doesn't it?" Preregistration is ideally suited for clinical trials, where the important variables—safety, dosage range, statistical procedures, patient pool—have been worked out. We want nothing but reliable data—no new, unexpected twists or discoveries at this stage, please! A preregistered protocol in such a case is just what the doctor ordered.

Pure Discovery Science projects are also compatible with preregistration. In the optimal condition, we simply want to characterize and classify parameters that we encounter in a novel data space. Naturally, although there are no explicit scientific hypotheses, Discovery Science involves plenty of implicit ones that could engage an investigator's conscious or unconscious bias, and preregistration could prevent bias from affecting the results.

7.F.2 When Preregistration Won't Work

Certain types of science are inherently incompatible with preregistration. When basic research scientists conduct an experiment, they engage in an active dialog with nature: you prod the preparation, see how it responds, formulate a hypothesis, test a prediction, do another test, reject the hypothesis, formulate a new one, etc. Yet for preregistration to work, the study must be described in full detail in advance; no changes allowed. Few basic scientists, I suspect, could live with such restrictions.

What would you do if, while conducting a preregistered study, it became clear that your preregistered hypothesis was wrong and that a new one would account for the data much better? Do you stick with the proposed experiments to the bitter end? The answer should be "yes," because that's what you promised to do, and, furthermore, that's what your statistical analysis requires you to do. Still, it might be a big sacrifice to put aside a more exciting idea to carry on the laudable, selfless path of filling in the blanks that you promised to fill in. A glance at many published papers shows that the plot line of a paper often goes through one or more initially promising and ultimately rejected hypotheses. Such studies are precluded by the preregistration protocol, which prohibits modifications to your experimental design. Why? Because this could defeat the whole purpose of the preregistration, which is to prevent you from meddling with the process while it is under way. And it would undercut the promise that you made to the journal to do what you said you would do. Essentially, the journals would being giving you free rein to do anything you like while guaranteeing you a publication! Great for you, not so great for them.

One solution to such problems is to seal the data until it is all in, thus preventing any midcourse corrections. Again, while this could work for certain types of study (e.g., clinical trials), it will be impractical for many Small Science projects where investigators are in constant contact with their results.

Another problem that preregistration poses is how to prevent intellectual "misappropriation" of scientific sights, up to and including intellectual theft. Necessarily, the results of preregistered studies will be unknown before the experiment is conducted: that is, after all, the point. The proposers have to lay all of their cards on the table, their brilliant, original hypothesis along with their experimental design and predicted outcomes. Reviewers, who are also often competitors, must have access to all relevant information in order to review the proposed work. And, unlike a conventional grant application or completed research article, the proposer may have no intellectual protection in the form of established priority (e.g., a meeting abstract or other public communication of the main finding that marks it as hers). Appeals to integrity—stealing ideas is wrong—will often have little effect because blatant intellectual theft is not the main concern. There is a gossamer, nearly invisible line between stealing someone else's idea and having a latent idea of your own take on urgent significance when you become aware of your competitors interest in it. Consider: you've been kicking around a pet hypothesis for years, but some reason haven't gotten around to exploring it in the lab. Suddenly you learn that your competitor is thinking along the same lines! You would be well within your rights to start work on it right away—it was, after all, your idea first! With fearful scenarios like this in mind, many scientists will understandably resist broadcasting their ideas through preregistration.

In other words, preregistration can be problematical, its disadvantages in direct proportion to the novelty and significance of the planned experiment, the ability of others to rip it off, and its value to the preregistrant who might be a young scientist trying to launch her independent career.

There is also a practical question of how ready journals will be to base costly publication decisions on the promised execution of experimental designs. High-impact journals are constantly on the lookout for brand new, game-changing (the current buzz word is "transformative") results. Can they be sufficiently confident that a promising experimental design will deliver the goods without seeing the results in advance? To take the chance that the data that they've agreed to publish will be boring, uninterpretable, or "negative?"

Will the scientific community seek out journals that specialize in pre-registered studies? As it stands, there are lower impact journals that are considered "archival," in which reliable, though not always exciting, work appears. The worth of the journal *impact factor* and similar scientific ratings systems has not yet been resolved, and the availability of scientific rewards—jobs, grants, etc.—to those publishing rigorous, but humdrum, pre-registered studies in less than high-impact journals is also uncertain.

Finally, we may wonder how a pre-registration system could be "gamed" and what could be done about it. It is common lore that researchers sometimes "propose" projects in grant applications that they have already largely completed. You can safely propose to do a seemingly risky experiment when you already know how it will turn out, where the likely pitfalls are, and how to avoid them. Why not collect the experimental results and then, with the data in hand, "pre-register" a plan, complete with eye-catching manipulations, controls, and deep "foresight" into possible problems. If you were clever enough, you'd gain much credibility for a novel design, which will eventually deliver important results and, incidentally, grease the skids for your future projects. Hopes that a pre-registration system cannot be rigged seem faint.

Other, equally commendable experiments in scientific publishing that might have an impact on reproducibility or improve scientific communication have not yet taken hold. It is now possible to publish comments on a publication in an online format linked to an article at a number of journal websites, but this worthwhile-sounding initiative hasn't stimulated much commentary it seems.

Preregistration will no doubt improve the reproducibility of certain kinds of scientific project—Discovery Science and clinical trials being perhaps the prime candidates—but truly hypothesis-based research is unlikely to benefit from the concept, at least without seismic changes in the culture of the basic science that is most closely associated with hypotheses.

7.G Coda

There are three central themes in this chapter. First, examining reproducibility exposes many of the complexities of science. The major conclusion of a typical scientific paper is based on several threads of evidence: its strength comes from weaving them together. The multifaceted structure of a typical study prevents it from being labeled entirely right or entirely wrong, and even its failures can be the source of valuable information. Science is not a unitary field, and not all kinds of science have the same objectives or assess the worth of reproducibility in the same way. Reproducibility is not always the highest goal in science, and, in fact, it might be unwise to insist on a strict goal of reproducibility at all times.

Second, viewing the RPP as a scientific investigation in its own right suggests how its results could have been influenced by its need to make unproven assumptions, including background knowledge. The RPP tactic of investigating just one experiment per report meant that one experiment was taken to represent a report's overall reliability; a questionable move. Although the demand for reproducibility tacitly assumes that original conditions can be adequately duplicated, the subtlety of nature will undermine this assumption at unknown times and in unknown ways.

Third, by distinguishing between the quest for reproducibility and the quest for truth, the Reproducibility Crisis highlighted the difference between the usual recommendations for improving reproducibility—the ones that address problems in the *materials* and *qualitative* categories (see Section 7.A.2)—from the more abstract goal of achieving the truth.

Reproducibility is obviously a multidimensional goal and will not be achieved by any one set of strategies. There is a wealth of expert opinion on rectifying its qualitative and materials-related deficiencies. Far less attention has been paid to possible weaknesses in scientific reasoning. Good experimental design requires clear scientific thinking and skill in using the Scientific Method and explicit hypothesis-based reasoning. Reproducibility is unquestionably an important challenge for science, but how much is too much? How much genuine irreproducibility should we tolerate? "Zero" is obviously unattainable, but what is a reasonable amount? Without good estimates of these factors, the extreme anxiety that has been expressed about a crisis seems at least premature.

Notes

1. Concern about a crisis refers mainly to "basic" preclinical science (i.e., often animal studies), as opposed to "clinical" science involving human beings. Clinical science is much less affected by the reproducibility (or irreproducibility) issues, largely because it does directly involve humans and is therefore tightly regulated already (see Note 2).
2. Francis S. Collins and Lawrence A. Tabak, "Policy: NIH Plans to Enhance Reproducibility," *Nature* 505:612–613, 2014. Senior NIH leaders state that with "rare exceptions" they have no evidence that irreproducibility is caused by scientific misconduct.
3. What is a "crisis?" Originally a crisis was a crucial stage or turning point, especially in medicine. Things could get either better or worse after a crisis point. While "crisis" still implies that a decisive change is coming, it is especially likely to refer to a change with negative consequences: things are bad and will get worse before they get better. This seems to be the sense that most authors writing about the Reproducibility Crisis have in mind. Many essays about the crisis are available at the journal *Nature*; Special Issue on irreproducibility, http://www.nature.com.proxy-hs.researchport.umd.edu/nature/focus/reproducibility/index.html; see also http://phys.org/news/2013-09-science-crisis.html.
4. S. C. Landis, S. G. Amara, K. Asadullah, C. P. Austin, R. Blumenstein, E. W. Bradley, et al., "A Call for Transparent Reporting to Optimize the Predictive Value of Preclinical Research," *Nature* 490:187–191, 2012.
5. Editorials in *Nature* (2015): "Journals Unite for Reproducibility," *Nature*, N515:7, 2014; "Announcement: Reducing Our Irreproducibility," *Nature* 496:398, 2013. See also references under Note 13.
6. C. G. Begley and L. M. Ellis, "Drug Development: Raise Standards for Preclinical Cancer Research," *Nature* 483:531–553, 2012. Former Amgen scientists Begley and Ellis report attempts over the past decade to replicate 53 "landmark" studies in oncology succeeded in confirming "scientific findings" in only 6 (11%) of the cases.
7. F. Prinz, T. Schlange, and K. Asadullah, "Believe It or Not: How Much Can We Rely on Published Data on Potential Drug Targets?" *Nature Reviews Drug Discovery* 10:712, 2011; Prinz et al. systematically studied 4 years (67 total studies) of in-house attempts at Bayer to validate published studies in the areas of oncology (47 studies), women's health, and cardiovascular disease and found that in only about 20–25% of the cases were the data "completely in line with" the relevant published data.
8. Additional articles include J. Lehrer, "The Truth Wears Off: Is There Something Wrong with the Scientific Method?" *The New Yorker*, December 13, 2010. https://www.newyorker.com/magazine/2010/12/13/the-truth-wears-off; G. Naik, "Scientists' Elusive Goal: Reproducing Study Results," *The Wall Street Journal*, December 2, 2011, https://www.wsj.com/articles/SB10001424052970203764804577059841672541590; "How Science Goes Wrong-Problems with Scientific Research:," *The Economist*, October 13, 2013, https://www.economist.com/leaders/2013/10/21/how-science-goes-wrong; E. Yong, "Psychology's Replication Crisis Can't Be Wished Away," *The Atlantic*, March 4, 2016, https://www.theatlantic.com/science/archive/2018/11/psychologys-replication-crisis-real/576223/. There are many more.

9. J. Lorsch quoted by M. Baker, "Muddled Meanings Hamper Efforts to Fix Reproducibility Crisis: Researchers Tease Out Different Definitions of a Crucial Scientific Term," 2016, http://www.nature.com/news/muddled-meanings-hamper-efforts-to-fix-reproducibility-crisis-1.20076
10. Task Force on Reproducibility, American Society for Cell Biology, "How Can Scientists Enhance Rigor in Conducting Basic Research and Reporting Research Results?" White Paper. 2014, http://www.ascb.org/reproducibility/
11. Collins and Tabak, "Policy."
12. This view has been challenged by authors who suggest that misconduct is a much bigger factor than is commonly acknowledged. See, for example, D. S. Kornfield and S. L. Titus, "Stop Ignoring Misconduct," *Nature* 537:29–30, 2016.
13. J. W. Schooler, "Metascience Could Rescue the 'Replication Crisis,'" *Nature* 515:9, 2014.
14. http://faseb.org/Science-Policy-and-Advocacy/Science-Policy-and-Research-Issues/Research-Reproducibility.aspx; http://theoryandpractice.org/2016/05/Reproducibility-Symposium/#.WIjCK9IrIdW; http://sites.nationalacademies.org/DEPS/BMSA/DEPS_153236; https://www.eventbrite.co.uk/e/publishing-better-science-through-better-data-2015-tickets-16695813628; https://aaas.confex.com/aaas/2017/webprogram/Session15112.html; US NIH Principles and Guidelines for Reporting Preclinical Research, http://www.nih.gov/about/reporting-preclinical-research.htm.
15. K. S. Button, J. P. Ioannidis, C. Mokrysz, B. A. Nosek, J. Flint, E. S. Robinson, and M. R. Munafò, "Power Failure: Why Small Sample Size Undermines the Reliability of Neuroscience," *Nature Reviews Neuroscience* 14:365–376, 2013; Erratum in *Nature Reviews Neuroscience* 14:451, 2013.
16. J. P. Ioannidis, "How to Make More Published Research True," *PLoS Medicine* 11, e1001747, 2014; M. R. Munafo, "A Manifesto for Reproducible Science," *Nature*, 2017, http://www.nature.com/articles/s41562-016-0021. Ioannidis and colleagues have been leaders of the inquiry into the veracity of scientific findings, highlighting failures in statistical analyses among other factors.
17. M. Baker, "Reproducibility Crisis: Blame It on the Antibodies," *Nature* 521:274–276, 2015; M. Baker, "Antibody Anarchy: A Call to Order," *Nature* 527:545–551, 2015.
18. M. Baker, "Reproducibility: Respect Your Cells!" *Nature* 537:433–435, 2016.
19. Editorials in *Nature*. See Note 5.
20. O. Steward and R. Balice-Gordon, "Rigor or Mortis: Best Practices for Preclinical Research in Neuroscience," *Neuron* 84:572–581, 2014.
21. Efforts to increase reproducibility by journal policies to encourage full reporting of methodological details: e.g., Nature's Methods Reporting Checklist can be found at http://www.nature.com/neuro/journal/v16/n1/full/nn0113-1.html; see also, http://www.nature.com/neuro/journal/v20/n3/full/nn.4521.html. There are also guidelines for making published data accessible: http://www.nature.com/authors/policies/availability.html#data, another feature for making data reproducible.
22. H. Pashler and C. R. Harris, "Is the Replicability Crisis Overblown? Three Arguments Examined," *Perspectives on Psychological Science* 7:531–536, 2012; W.

Stroebe and F. Strack, "The Alleged Crisis and the Illusion of Exact Replication," *Perspectives on Psychological Science* 9:59–71, 2014; C. Aschwanden, "Science Isn't Broken: It's Just a Hell of a Lot Harder than We Give It Credit For," https://fivethirtyeight.com/features/science-isnt-broken/; M. Baker, "The Reproducibility Crisis Is Good for Science," http://www.slate.com/articles/technology/future_tense/2016/04/the_reproducibility_crisis_is_good_for_science.html; S. Firestein, "Why Failure to Replicate Findings Can Actually Be Good for Science," *Los Angeles Times*, Op-Ed, http://www.latimes.com/opinion/op-ed/la-oe-0214-firestein-science-replication-failure-20160214-story.html; S. Alexander, "90% of All Claims About the Problems with Medical Studies Are Wrong," SlateStarCodex.com, http://slatestarcodex.com/2013/02/17/90-of-all-claims-about-the-problems-with-medical-studies-are-wrong/; and R. Gross, "Psychologists Call Out the Study that Called Out the field of Psychology," http://www.slate.com/blogs/the_slatest/2016/03/03/psychology_study_that_induced_the_reproducibility_crisis_was_wrong.html.

23. M. Baker, "1,500 Scientists Lift the Lid on Reproducibility," *Nature* 533:452–454, 2016.
24. J. P. Ioannidis, "Why Most Published Research Findings Are False," *PLoS Medicine* 2:e124, 2015.
25. I. Chalmers and P. Glasziou, "Avoidable Waste in the Production and Reporting of Research Evidence," *Lancet* 374:86–89, 2009.
26. Button et al., "Power Failure."
27. J. Leek and L. R. Jager, "Is Most Published Research Really False?" *Annual Review of Statistics and Its Application* 4:109–122, 2017.
28. C. L. Nord, V. Valton, J. Wood, and J. P. Roiser, "Power-Up: A Reanalysis of 'Power Failure' in Neuroscience Using Mixture Modeling," *Journal of Neuroscience* 37:8051–8061, 2017.
29. See Note 20.
30. The cortex is the largest and most complex part of most mammalian brains. It sits on top of and surrounds many of the lower brain centers. It is responsible for the highest levels of information processing. The cortex is also the final way station for sensory information coming in from the periphery; for instance the visual and auditory parts of the cortex are the final processing centers for vision and hearing. The "barrel cortex" is the analogous cortical processing center for the information from a rat's whiskers.
31. E. Ahissar, R. Sosnik, and S. Haidarliu, "Transformation from Temporal to Rate Coding in a Somatosensory Thalamocortical Pathway," *Nature* 406:302–306, 2000; E. Ahissar and A. Arieli, "Figuring Space by Time," *Neuron* 32:185–201, 2001.
32. R. Masri, T. Bezdudnaya, J. C. Trageser, and A. Keller, "Encoding of Stimulus Frequency and Sensor Motion in the Posterior Medial Thalamic Nucleus," *Journal of Neurophysiology* 100:681–689, 2008.
33. Attempts to reconcile the discrepancies verbally failed. See F. Ahissar, D. Golomb, S. Haidarliu, R. Sosnik, and C. Yu, "Latency Coding in POm: Importance of Parametric Regimes," *Journal of Neurophysiology* 100:1152–1154, 2008; R. Masri, T. Bezdudnaya,

J. C. Trageser, and A. Keller, "Reply to Ahissar et al.," *Journal of* Neurophysiology 100:1155–1157, 2008.

34. R. Westfall, *Never at Rest: a Biography of Isaac Newton* (Cambridge: Cambridge University Press; 1994). In 1672, Newton described his experiments with prisms that demonstrated that white light was a heterogeneous mix of colored lights in a letter to the Royal Society; hitherto white light had been thought to be primary and the colors derived from it. Two years later, an English Jesuit and college professor, Francis Linus, challenged Newton's results, claiming that he, Linus, had done the experiments and made different observations; i.e., that Isaac Newton's experiments were irreproducible! Newton at first ignored Linus, but when Linus persisted, Newton wrote out detailed instructions and asked the Royal Society members to try to replicate his findings at a meeting.

    In a comment on scientific misconduct (*Nature* 537:29–30, 2016), D. S. Kornfield and S. L. Titus remark that, "The history of science shows that irreproducibility is not a product of our times. Some 350 years ago, the chemist Robert Boyle penned essays on 'the unsuccessfulness of experiments.' He warned readers to be skeptical of reported work. 'You will meet with several Observations and Experiments, which . . . may upon further tryal disappoint your expectation.' He attributed the problem to a 'lack of skill in the scientist and the lack of purity of the ingredients', and what would today be referred to as inadequate statistical power.

    "By 1830, polymath Charles Babbage was writing in more cynical terms. In *Reflections on the Decline of Science in England*, he complains of 'several species of impositions that have been practised in science,' namely 'hoaxing, forging, trimming and cooking.'"

35. Open Science Collaboration (B. Nosek, corresponding author), "Estimating the Reproducibility of Psychological Science," *Science* 349:aac47162015.
36. What psychologists typically identify as a "study" usually consists of several discrete subsections, each one being a mini-study yielding its own result; this naming convention is rarely followed in most areas of biomedical science.
37. Prinz et al., "Believe It or Not."
38. These statistical concepts are reviewed in Chapter 5 and can be found in any introductory statistics textbook.
39. The findings that the Open Science Collaboration tried to replicate were considered "significant" at the $p < 0.05$ level, and, given the replication power, 0.92, of the original studies, only 89 of the 97 should have been replicable.
40. D. T. Gilbert, G. King, S. Pettigrew, and T. D. Wilson, "Comment on 'Estimating the Reproducibility of Psychological Science,'" *Science* 351:1037, 2016.
41. Open Science Collaboration, "Response to Comment on 'Estimating the Reproducibility of Psychological Science,'" *Science* 351:1037, 2016.
42. D. Lakens, "Power of Replications in the Reproducibility Project," 2015. The 20% statistician, http://daniellakens.blogspot.com/2015/08/power-of-replications-in.html.
43. A. Etz and J. Vandekerckhove, "A Bayesian Perspective on the Reproducibility Project: Psychology," *PLoS One* 11:e0149794, 2016. Remember, a Bayes' factor of 3 would imply that one hypothesis is three times more likely to be correct than the

other, but a Bayes' factor of 3 is the barest threshold for thinking that two hypotheses are genuinely different; in fact, it is too weak to support any real conclusion. A Bayes' factor of 3 only means that the probability of one hypothesis is 0.75 and the other is 0.25; you wouldn't want to place a lot of money on a single 3-to-1 bet (even though in the long run you'd be guaranteed of making a profit). This is legitimate if there is no reason to favor H_0 or H_a beforehand; that is, if we can assume that the prior probabilities of H_0 and H_a are equal and there are no other hypotheses on the table.

44. Baker, "1,500 Scientists Lift the Lid."
45. D. Gorski, "Is There a Reproducibility "Crisis" in Biomedical Science? No, but There Is a Reproducibility *Problem*," *Science Based Medicine*, 2016. https://sciencebasedmedicine.org/is-there-a-reproducibility-crisis-in-biomedical-science-no-but-there-is-a-reproducibility-problem/.
46. H. I. Maucer, N. N. Hanna, M. A. Beckett, D. Gorski, et al., "Combined Effects of Angiostatin and Ionizing Radiation in Antitumour Therapy," *Nature* 394:287–291, 1998.
47. The anti-tumor drug *Avastin*, generic name *bevacizumab*, acts similarly.
48. It is the rare paper that gets published without first having to undergo revisions that not infrequently demand doing new experiments.
49. "Multiple tests" refers to a number of distinct experimental tests using different samples, outcome measures, and often different equipment or methods, not multiple repeats of the same experiment, although this is generally a good idea when it can be done.
50. R. E. Sorge, L. J. Martin, K. A. Isbester, S. G. Sotocinal, S. Rosen, et al., "Olfactory Exposure to Males, Including Men, Causes Stress and Related Analgesia in Rodents," *Nature Methods* 11:629–632, 2014.
51. W. C. Hines, Y. Su, I. Kuhn, K. Polyak, and M. J. Bissell, "Sorting Out the FACS: A Devil in the Details," *Cell Reports: Commentary* 6:7790–7810, 2014.
52. German-Hispanic ethnic make-up of North Dakota and New Mexico: https://en.wikipedia.org/wiki/German_Americans; https://en.wikipedia.org/wiki/List_of_U.S._states_by_Hispanic_and_Latino_population.
53. S. Firestein, *Ignorance: How It Drives Science* (New York: Oxford University Press; 2012).
54. N. N. Taleb, *Fooled by Randomness: The Hidden Role of Chance in Life and the Markets* (New York: Random House; 2005). Taleb explores a multitude of ways in which we underestimate the pervasiveness of random chance in numerous aspects of life.
55. Leek and Jager, "Is Most Published Research Really False?"
56. Open Science Collaboration, "Estimating the Reproducibility."
57. See Note 32.
58. Etz and Vandekerckhove, "A Bayesian Perspective."
59. Mr. James Inhofe, from the Committee on Environment and Public Works (United States Congress) submitted the following Report (Calendar no. 124) together with Minority Views [To accompany S. 544] Secret Science Reform Act of 2015, https://www.congress.gov/bill/114th-congress/house-bill/1030/text.

60. D. W. Belsky, T. E. Moffitt, T. W. Baker, K. Biddle, J. P. Evans, H. Harrington, et al., "Polygenic Risk Accelerates the Developmental Progression to Heavy, Persistent Smoking and Nicotine Dependence: Evidence from a 4-Decade Longitudinal Study," *JAMA Psychiatry* 70:534–542, 2013.
61. AAAS President Geraldine Richmon letter to Mr. James Inhofe regarding Secret Science, https:/www.aaas.org/sites/default/files/secret_science_2015_april.pdf.
62. R. A. Nicoll, "A Brief History of Long-Term Potentiation," *Neuron* 93:281–290, 2017; D. M. Kullmann, "The Mother of All Battles 20 Years On: Is LTP Expressed Pre- or Postsynaptically?" *Journal of Physiology (London)* 590:2213–2216, 2012; G. A. Kerchner and R. A. Nicoll, "Silent Synapses and the Emergence of a Postsynaptic Mechanism for LTP," *Nature Reviews Neuroscience* 9:813–825, 2008; S. Choi, J. Klingauf, and R. W. Tsien, "Fusion Pore Modulation as a Presynaptic Mechanism Contributing to Expression of Long-Term Potentiation," *Philosophical Transactions of the Royal Society London B Biological Sciences* 358:695–705, 2003.
63. There are different kinds of memories, memories for facts like the name of the capital of Zambia and memories that you create when you are learning to play the piano or shoot a jump shot in basketball. The latter are sometimes called "muscle memory," but that is just a colloquial label; almost all memories are made and stored in the brain. Memories for facts are also known as "declarative memories"—you can state (declare) exactly what is that you learned, whereas you can't state precisely what you learned (how you control your muscles, how much force to exert, etc.) when it comes to shooting a basketball. The story of the LTP wars is about the cellular basis of declarative memory.
64. The gap between the tips of the signaling (presynaptic) and receiving (postsynaptic) nerve cell is called the "synaptic cleft." The width of the synaptic clefts in the brain is about 20–40 nanometers, or ~1/1,000,000th (one millionth) of an inch. The approximate synaptic area on each nerve cell that forms a synapse is approximately one micrometer, ~1/25,000 of an inch in diameter. Each nerve cell makes ~10,000 synapses with other nerve cells.
65. LTP occurs in essentially all animals including humans, and most experiments were carried out in an in vitro system from the hippocampus of non-human animals.
66. A. C. Dolphin, M. I. Errington, and T. Bliss, "Long-Term Potentiation of the Perforant Path in Vivo Is Associated with Increased Glutamate Release," *Nature* 294:496–498, 1982.

8
Advantages of the Hypothesis

8.A Introduction

What is a hypothesis good for? What advantages does it offer? Previous chapters concentrated on what the hypothesis is and a few of the contexts in which it appears in science, but I haven't said much about why scientists benefit from using it. In this chapter, I'll present two major advantages of the hypothesis: first, hypothesis-based studies are more likely to be reproducible than others, and, second, the hypothesis is a powerful tool for organizing and communicating scientific ideas.

You might have heard something about the second class of advantages, but almost certainly nothing about the first one, so I want to start with it. We'll pick up where the Chapter 7 left off and see how the hypothesis can improve scientific reproducibility and, especially, how it helps avoid certain problems associated with open-ended, non–hypothesis-based investigations. I'll also outline a statistical method that allows you to assess hypothesis-based work realistically, and I'll explain why many hypothesis-based studies are experimentally reliable. After discussing these quantitative advantages, I'll review the qualitative benefits that the hypothesis offers.

8.B Reliably Predicting Reproducibility

Those who are concerned about the current state of science invariably cite the statistical arguments of John Ioannidis, Katherine Button, and their colleagues that science is suffering a Reproducibility Crisis. What gets less attention is that the work of these statisticians also implies that hypothesis-based work should as a rule be more reproducible than less structured research programs.

8.B.1 Positive Predictive Validity

Physically replicating studies to find out if they are reliable can be a formidable undertaking, as we've seen. As an analytical tool for forecasting reproducibility, statisticians introduced the concept of *positive predictive value* (PPV),[1] which is a number that ranges between 0 and 1 (higher is better) and estimates the likelihood that a result will be reproducible. PPV combines three familiar factors: the *p-value, statistical power*, and *probability*. The conclusion that we're going to arrive at is that having a focused experimental hypothesis with only a few alternative hypotheses gives a much higher PPV than, say, gene search strategies where investigators are testing thousands of alternative "hypotheses." This means that the chances of reproducing hypothesis-based experimental outcomes are much higher than reproducing the results of relatively unconstrained searches. I think it will be helpful to outline the reasoning behind the PPV, but if you're only interested in the bottom line, you can skip to Section 8.B.2. and take up the argument there.

Suppose that you've compared the heights of two groups of people and found a significant, $p \leq 0.05$, difference between them (we'll put aside objections to null hypothesis significance testing mode [NHST][2] for the moment). The p-value only tells you the probability that your result is an accident of random chance and that you are *wrong* in thinking that there is a real difference between the experimental groups. However, the 5% ($p \leq 0.05$) chance of being wrong does not mean that you have a 95% chance of being right. That is a major shortcoming. You want to know the truth: Do the two groups differ in height or not? And you can't get that from the p-value alone. Enter the PPV.

To make valid predictions, you want to know if your result is going to be reliable. The PPV depends on the p-value and the statistical power of your test. Power, remember, is the probability that you will correctly *reject* the null hypothesis when it is false and correctly conclude that there is a difference between the experimental groups. The more powerful the test, the more likely you are to correctly decide that one group of women is in fact taller than the other and so on.

In addition to the p-value and statistical power, to calculate PPV you need to have a value of the *pre-study probability* or *pre-study odds* that a result you get will be correct. (In contrast, the PPV itself gives you the *post-study odds* of being correct.) The pre-study odds are analogous to the *prior probability* of Bayesian statistics. As with the Bayesian priors, you can't determine pre-study odds exactly; luckily ballpark estimates are frequently good enough.

To estimate pre-study odds, you can work from informed guesses about the number of "true" (i.e., provisionally true) hypotheses that you might find (T) and the number of all hypotheses that you will evaluate (the universe of all true and false ones, T + F). Assuming that each hypothesis is either T or F, then the simple frequentist probability that your hypothesis is true is T/(T + F). (Yes, Karl Popper would no doubt frown at this whiff of "probable truth," but never mind.) Obviously, when you have a focused hypothesis with one potentially true result and only a few false alternatives, T/T + F will be much greater than when you're doing an experiment where there is only one potentially true result and thousands of false ones. In that case, T + F will be huge and pre-study odds, T/(T + F), tiny. Pre-study odds will be critical because PPV depends directly on them.

There is a slight wrinkle; T/T + F gives the odds of *occurrence* of true results, not the odds of *detecting* true results. What's the difference? Because all tests are imperfect, you cannot detect all of the true results that might occur, but only the fraction that the power of your test allows you to detect. So this is why you need to know about statistical power. If your statistical test had a power of 0.8, which would be very good, you could theoretically detect 80% of the true results, not 100% of them. PPV goes down if power goes down.

8.B.2 Calculating PPV

When you put it all together, the formula for PPV is:

$$\text{PPV} = \frac{\text{Pre-study odds}}{\text{Pre-study odds} + (\text{p-value}/\text{statistical power})} \qquad \text{Equation 8.1}$$

The formula is an arrangement of the three factors we've been discussing. It states that PPV increases toward 1 (the maximum) as statistical power or pre-study odds increase. PPV also increases as p-value decreases because then the fraction, p-value/statistical power, goes toward 0. And, of course, combinations of these factors can occur.

The formula makes intuitive sense. Remember that we're trying to figure out the probability that experimental results will be reproducible. If you know going into an experiment that there is a good chance that you'll detect a true result because one hypothesis is based on a lot of prior information, then the pre-study odds will be high. If the power of your test is high, you'll be likely to reject the null hypothesis correctly. Finally, if the p-value is small, you'll be less likely to mistakenly reject the null hypothesis. All of these factors will make the PPV greater.

Here is a numerical example: suppose that you suspect that a certain disease is caused by a genetic mutation, and you plan to screen 1,000 genes to try to find a genetic marker for the disease (i.e., a gene whose expression is reliably associated with the disease). You have no prior information, and you are blindly sifting through the group of genes, comparing their expression in normal and diseased individuals. You're looking for changes that are significant at the $p \leq 0.05$ level, and, to keep the illustration simple, let's assume there is just one true marker gene for this disease. In a sense, you are evaluating 1,000 distinct "hypotheses" (this is how the statisticians analyze the experiments) and trying to find the right one by eliminating the rest. Notice—and this the crucial point—that the pre-study odds that any particular hypothesis is true are extremely low; just 1/1000 = 0.001!

Now, at the significance level of $p \leq 0.05$, if you screen 1,000 genes, you expect to get about 50 positive hits in total (we'll pretend that the number is exactly 50 to keep it simple). That is, we expect about 50 of our tests to be significant at $p \leq 0.05$. If one of the hits is the right gene, then the rest (49/50) will be false positives; you'll be wrong 98% of the time. The odds of detecting the true gene are heavily stacked against you. What would PPV be in such a case? With pre-study odds of 0.001, p-value of 0.05, and, again to keep the numbers simple, assuming a medium statistical power of 0.5, then PPV, the probability of reproducing your finding, would be 0.01, or 1 in 100. Even if the power of your test were maximal, 1.0, the PPV would only be 0.02. Your finding would stand only a 2% change of being replicated under the best of circumstances.

You could tweak the outcomes a bit by increasing your significance level: The number of false positives would go down if you opted for $p \leq 0.01$, for example, but this would decrease your chances of finding true positives (i.e., your statistical power would go down). And if you were to do a search of 10,000 genes, the PPV would be much lower.

These are the main concepts in working with the PPV. The formula combines all the evidence we have about p-value, power, and pre-study probability and puts them together. The PPV is a driving force for Ioannidis's claim that "most research results are false,"[3] and you can see why he's pessimistic about the reproducibility of this kind of experiment.

8.C Why PPV Is Higher with Hypothesis-Based Research

Surprisingly perhaps, despite the grim impression you get from the gene screen scenario, the PPV calculation strongly supports doing specific hypothesis-testing research. To see why this is, let's look first at pre-study odds. Ordinarily, experimental basic science builds on a great deal of prior information, which greatly reduces the number of realistic alternative hypotheses and dramatically increases the pre-study odds, $T/(T + F)$. In my experience, an experimental neuroscience publication generally evaluates only one or two hypotheses; a report that compares as many as four would be unusual. The fewer the hypotheses, the stronger the argument. To be conservative here, let's assume that you've come up with four hypotheses that might explain the phenomenon you're studying and that you suspect that one of them is correct. Your pre-study odds are 1/4. Already, assuming that p-value and statistical power are the same as in the gene screen case, the odds of hitting on the right hypothesis are 250 times higher. This translates into a big improvement in PPV.

The p-value influences PPV but, since $p \leq 0.05$ is conventional, we'll assume it is constant in this example (in a while, we'll see what happens if we change the p-value). What about the power of your statistical tests? Does the fact that it has often been lamentably low[4] fatally undercut the advantages of the hypothesis? Actually, while low power does reduce PPV overall, the effect is minor when compared with the advantages gained by having a small number of alternative hypotheses. Even with a low statistical power of 0.2, a hypothesis-testing experiment that successfully ruled out 3 of 4 alternative hypotheses would have a PPV of approximately 0.57. That is, the probability that this hypothesis-based experiment would be reproducible is more than *40 times greater* than the open-ended gene screen; we get a 4,000% improvement in PPV just by doing a directed, hypothesis-testing study that evaluates four alternative hypotheses, even if your tests have low statistical power! Improving the pre-study probability of reproducibility of your experimental results is the first quantitative benefit you get from using a hypothesis.

The statisticians would point out that, nevertheless, if your result has a PPV of 0.57, its odds of being reproduced are only slightly better than 50-50, a long way from the level of confidence we expect from science. Obviously, though true, this is not a fatal objection: first, going from abysmally low odds of reproducibility to a 50-50 chance is a giant step forward, despite leaving room for improvement.

Second, and more importantly, the conclusions of an experimental basic science project do not rest entirely on the outcome of a single p-valued test, even though much of the concern about scientific reliability arises directly from this erroneous assumption. As an example, recall that the Reproducibility Project teams that raised the alarm about the reliability of psychology studies (Chapter 7) selected one experimental test from each complex study to try to replicate.[5] If that one study was not fully replicable the authors concluded the study was flawed. Similarly, Ioannidis explicitly posits that "the high rate of non-replication (lack of confirmation) of research discoveries is a consequence of the convenient, yet ill-founded strategy of claiming conclusive research findings *solely on the basis of a single study assessed by formal statistical significance*, typically for a p-value less than 0.05" [emphasis added]."[6] And he explains that, "Research findings are defined here as any relationship reaching formal statistical significance."

While there may be a kind of science that "claims conclusive findings" as a result of "a single study," basic bioscientific research does not operate like that.

Instead of relying on a single p-valued test, bioscientists take the results of numerous experiments into account in drawing their conclusions.

8.C.1 How Multiple Tests of a Hypothesis Can Lead to Higher PPV

The principle that "an overall inference can be more reliable and precise than any of its premises individually" is what the philosopher/statistician Deborah Mayo calls "lift-off."[7] For Mayo, lift-off is related to her "argument from coincidence," which she illustrates with an example: wondering if she's going to gain weight during a trip abroad, she weighs herself on three separate, carefully calibrated scales before and after the trip. When she's back, all three scales report similarly higher weights only for her, not her calibration standards. Since this procedure is surely capable of rejecting the null hypothesis that she didn't gain any weight, it constitutes a genuinely "severe" test, thus illustrating how she'll flesh out the definition of severity, which Popper failed to do. More importantly for us, however, is that her procedure illustrates the power of her "argument from coincidence." Her procedure is especially convincing, she says, because it would be "preposterous" to assume that she'd get the results that she did because all three scales just happened to be off by similar amounts when she weighs herself. Therefore we believe the scales are right and reject the null hypothesis that she didn't gain weight.[8]

Without saying so exactly, this is the same reasoning that scientists employ when they do multiple tests of a given hypothesis: if the conclusions from all the tests agree, then we mentally *combine* them when we interpret the validity of the hypothesis because it would be preposterous to think that they all just happened to agree. We reasonably expect conclusions based on multiple tests to be more reliable than those based on single tests.

Although this informal reasoning process is roughly correct, it is vague and doesn't convincingly counter the quantitative arguments of the statisticians. It would be helpful if we had an uncomplicated, formal method of combining the results of a cluster of experiments that test predictions of the same hypothesis. This method would give us an objective estimate of the probability of observing the group of results that we could use to test the hypothesis. Unfortunately, Mayo does not provide a way of calculating the odds of overall inferences of this kind, and it will take more than a judgment of "preposterousness" to satisfy scientific readers.

Fortunately, there are methods for combining results, and they could help solve our problem, although they haven't been used much for this purpose so far. I'll devote the next section to describing one of these methods.

8.C.2 Multiple Testing of Hypotheses: Fisher's Method for Combining Results

We want to calculate how having a collection of different experimental results, all bearing on a single hypothesis, alters our confidence in the conclusions relating to that hypothesis. I'll start with a nonscientific example to give a better picture of what we need: Suppose a carnival has come to town and offers games of chance. You notice a shady-looking character who seems to be winning a lot, and you hypothesize that he is a shill—someone in league with the carnival owners. For a shill, the games are rigged in his favor, and he only appears to be a legitimate player: his job is to create the illusion that anybody can win and lure onlookers to play (and lose their money because, for them, the games are rigged the other way).

Your hypothesis predicts that the shady guy will win far more games than you'd expect from chance alone. You discretely follow him around and take notes. At one booth, a player has to roll a sequentially numbered 16-sided die and, if an 8 comes up, he wins. The suspect rolls and wins, but his chance of winning, 1/16, is not that unusual. He next tries a game where a player has to fish a green ball out of an opaque urn having 10 green balls scattered among 190 other balls of different colors. The suspect manages to pull out a green ball on his first try. Once again, the odds of that happening, 1/20, are not extremely small. Finally, the suspect goes to the magician's booth. The magician puts a pea under one of five inverted cups, moves the cups around very quickly, and right away, the suspect correctly identifies the one hiding the pea. His chance of guessing correctly was 1 in 5. Is he part of a crooked scheme or an honest, though lucky, guy? He's won every time, which certainly seems suspicious, and yet his chances of winning any single game were not particularly small; he could legitimately have won each of them. On the other hand, he won all three. What are his chances of doing that by luck alone? You can find the odds of his string of successes by multiplying the probabilities of winning all three games: that is, $p(\text{winning all three}) = 1/16 \times 1/20 \times 1/5 = 1/1{,}600$. His run had a chance of only 1 in 1,600 of occurring. It is not impossible for that to happen randomly, but your suspicion that he was a part of a scam looks as if it could be justified.

A scientific hypothesis makes many independent predictions, and scientists test more than one to test their hypothesis. Ideally, we'd like to be able to combine the statistical results, the p-values from our tests, more or less the way we did at the carnival. For technical reasons,[9,10] we can't simply multiply the p-values from a group of, say, t-tests, together and come up with a global p-value for the probability of getting them all. However, we can effectively do the same thing with a method invented by R. A. Fisher: *Fisher's Method for Combining Probabilities*.[11]

According to Fisher,[12] "When a number of quite independent tests of significance have been made . . . the aggregate gives an impression that the probabilities are on the whole lower than would often have been obtained by chance. It is sometimes desired, taking account only of these probabilities, . . . to obtain a single test of the significance of the aggregate, based on the product of the probabilities individually observed." This is exactly what we want to do.

Here is his formula:

$$X^2 = -2\sum_{i}^{k} \ln(pi)$$

Although it might look daunting, it's not (its derivation is available, if you're interested.)[13] You begin by testing several independent predictions and get a p-value for each test, p_i. If you did k different tests, then you have k p-values. You take the natural logarithm (*ln*) of each p-value, add the logarithms together (the capital Greek letter Σ, sigma, tells you to add what comes after it). Finally, you multiply the sum by −2. The resulting value is a statistical construct called a *chi-square* variable, and to evaluate it you consult a chi-square table (they are available online or at the back of many statistics textbooks) with $2k$ *degrees of freedom*. I'll go through an example shortly. There are some caveats,[14,15] the most important being that your tests *must be strictly independent*—different methods, subjects, dataset, etc. If they are independent, then the number you get from the chi-square table tells you the probability that you'd get that value or a smaller one by chance alone.

As was the case with the carnival games, the probability of getting a cluster ("aggregate") of experimental results will generally be much lower than the probability of getting any single result. As usual, the smaller the probability, the less likely it is to be a random event, and the more confident you can be in the conclusion it supports. Fisher's Method let's you calculate a parameter which, although it is a chi-square variable, I'll call "p_{FM}" because it is like a p-value.

8.C.3 Example of Fisher's Test to Combine Results

Assume that you're a neuroscientist, and you have a hypothesis that a group of neurons in the mouse hypothalamus is inhibited by endogenous cannabinoids, which makes the mice more relaxed and less fearful (i.e., "mellower"). Your hypothesis predicts that endocannabinoid activation of cannabinoid receptors on these neurons is crucial for the mellowing response. You test four predictions of the hypothesis in four groups of mouse hypothalamic tissue samples: one group is from mice that were genetically engineered to lack the receptor, one group is treated with a drug that activates the receptor, one group with a drug that inhibits the receptor, and one normal group in which you collect the fluid around the activated neurons to see if it contains endocannabinoids. Of course, you do all appropriate controls and use recommended group sizes.

The p-values for the four tests are 0.032, 0.042, 0.049, and 0.058. While the last test would be considered insignificant under the conventional, $p < 0.05$ standard, Fisher would approve of including all values because he felt that standards should not be rigid (Chapter 5), and, moreover, in the combined test individual, test results don't matter: we are only interested in the aggregate probability.

First, you take the logarithms of the p-values: −3.44, −3.17, −3.02, −2.85

Second, you add them together and get: −12.45

Third, you multiply by −2 and get: 24.9

This is your chi-squared value, and you evaluate it by consulting a chi-square table with 2k (i.e., $2 \times 4 = 8$) degrees of freedom.

The table tells you that the probability of this value is $p < 0.005$; that is, it is the probability of getting the group of p-values for your four independent tests, what I'm calling "p_{FM}." This value is much lower than any individual p-value and implies that getting the whole collection of your results was very unlikely. You're entitled to feel that your hypothesis has been well-corroborated. The ability to take advantage of Fisher's Method to obtain this parameter, p_{FM}, is the second quantitative benefit of using a hypothesis. But what, besides a greater sense of confidence, can we get from p_{FM}? We can use it to improve our estimate of PPV, the future predictive value of the results.

8.C.4 Using the Significance Level from Fisher's Method to Reassess PPV

How does the p-value affect PPV? Let's assume that, instead of $p \leq 0.05$, we used $p \leq 0.001$ in the example of the 1,000-gene screen, and, since statistical power generally decreases as p-value decreases, let's also assume that power is only 0.25 rather than 0.5. In this case, PPV would increase from 0.01, which we calculated earlier, to 0.05; thus, decreasing the p-value increased PPV five-fold.

Once again, because you're a basic scientist who wants to assess the validity of your entire hypothesis, not merely the validity of any one test of a prediction, you should take into account the outcomes of all the experiments that you did to test it. It obviously doesn't make sense to use the p-value from a single test to calculate PPV for the overall hypothesis. You could, instead, use a combined estimate, such as p_{FM}, in the PPV formula. Since p_{FM} will generally be much lower than a typical p-value, then PPV will be higher for the hypothesis than for the single test, as we just saw in the gene screen example.

How will p_{FM} affect PPV? For the four-part endocannabinoid experiment discussed in the previous section, let's compare PPV calculated from one test with a conventional significance level of $p < 0.05$ and PPV calculated using the p_{FM} that we calculated in the preceding section. Assuming you have four alternative hypotheses in mind, your pre-study odds are 1/4 = 0.25. We'll use the estimate from Button et al. that the average power of neuroscience experiments is only 0.2. With the $p < 0.05$ case:

$$PPV = 0.25/[0.25 + (0.05/0.2)] = 0.5$$

(i.e., there is a 50% chance this single result would be reproducible)
Using p_{FM} from the aggregate of the four tests we get:

$$PPV = 0.25/[0.25 + (0.005/0.2)] = 0.91$$

That is, the PPV of the study goes from a meager 50% chance of reproducibility to a very respectable 91% chance when the p-value equivalent goes down. Confidence in your overall conclusion appears to be on much firmer ground than it was when you assessed PPV from only one test. Getting a more realistic estimate of PPV is the third quantitative benefit you get from using the hypothesis.

In summary, use of the scientific hypothesis can enhance the statistical reliability of scientific results quantitatively in three ways. First, a research plan that is focused on testing only a small number of alternative hypotheses has much higher predictive validity (PPV) than less focused investigations because the pre-study odds are higher. Second, combining the results from testing multiple predictions of a single hypothesis with Fisher's Combined Method gives you a new measure of statistical significance, p_{FM}, that will in general be lower, and hence less likely to be explained by random chance, than a single p-value. Third, you can use p_{FM} to estimate a PPV for the hypothesis that will generally be higher than it will for p-valued tests of individual predictions.

8.C.5 Other Benefits of Fisher Method for Combining Results

Does the PPV calculation have practical utility for science or is it merely a tool for various forms of meta-analyses ? Could the Fisher Method have any impact on science more generally? I think that there several potential additional benefits of the combined method:

1. Decreased emphasis on the single "p-value" per se and a shift in attention to the overall study, its rationale, and design. Since only tests of predictions of one hypothesis can be included in the Fisher Method, authors would have to consider their arguments and analyses carefully.
2. Since the Fisher Method is not constrained to a specific "significance level" for any of the constituent tests (i.e., authors would include a result with $p = 0.071$ as readily as one with $p = 0.011$). This, too, would reduce inappropriate focus on p-values and thereby the decrease the incentive for selectively reporting only the most significant results.
3. The method would make it easier for others to evaluate an entire study and might even make it possible to compare and evaluate competing hypotheses that purport to explain a given phenomenon.
4. More broadly speaking, some of the theoretical benefits of a much lower p-value (e.g., $p < 0.005$) (we touched on this proposal in Chapter 5.E.2.d) could be had by using p_{FM} obtained with Fisher's Method, while at the same time avoiding some of the pitfalls of switching to a much more stringent standard.

Taken together, these quantitative benefits constitute a strong argument in favor of hypothesis-based science. The hypothesis also offers numerous qualitative advantages as a thinking tool, and we'll review a few of these next.

8.D Cognitive Advantages Offered by the Hypothesis

Historically, scientists wrestled with the concept of the hypothesis, trying to balance its advantages with its real or imagined disadvantages; partly because of the many meanings the word had (Chapter 2). Take, for example, Santiago Ramón y Cajal, who shared the Nobel Prize in 1906 for his meticulous descriptions of the cellular structure of the brain. Cajal was a passionately dedicated experimenter who held observation and the discovery of facts in the highest possible regard. In *Advice for a Young Investigator*,[16] he record his musings about scientific thinking. He has no patience for the grand though futile theorizing that had characterized "the Aristotelian principles of intuition, inspiration and dogmatism" that "involves exploring one's own mind or soul to discover universal laws." He seconds the advice of the nineteenth-century chemist, Justus Liebig: "Don't make hypotheses. They will bring the enmity of the wise upon you," and believes that the unknown is the most important stimulus to future scientific progress. Scientific theories, for Cajal, were elevated, abstract things made by theorists, those "wonderfully endowed minds whose wills suffer from a particular form of lethargy" such that, "when faced with a difficult problem they feel an irresistible urge to formulate a theory, rather than to question nature."

The comments seem unambiguous, but before placing Cajal in the anti-hypothesis camp, we read, "The hypothesis is an interpretative questioning of nature" and that "observation, explanation or hypothesis, and proof" are key elements in scientific discovery. He feels that "A hypothesis is necessary; without it phenomena cannot be explained," and he quotes Jacob Christoph Le Blon, "he who refuses to accept hypothesis as a guide is resigned to accept chance as a master."

Cajal resolves these contradictory points of view by distinguishing "between working hypotheses . . . and scientific theories." He finds the working hypothesis to be "an integral part of the investigation." The hypothesis, in other words, was useful provided that you didn't confuse using it with creating theories that were distant from data. Similarly, long before Cajal, the English philosopher John Locke (1632–1704), who was often critical of the hypothesis, also found much to like about it. "Hypotheses," Locke says, "are great helps to the memory and often direct us to new discoveries."[17]

Karl Popper, at least, was unambivalent about the virtues of the hypothesis: in the preface to *Conjecture and Refutations*,[18] Popper writes, "The essays and lectures of which this book is composed are variations upon one very simple theme—the thesis that we *can learn from our mistakes*." His work establishes the hypothesis as the preeminent practical tool for making and learning from mistakes.

In the remainder of this chapter, I will expand on the themes introduced by Cajal, Locke, and Popper.

8.C.1 Working with Your Built-In Drive to Understand the World

The psychological literature is full of examples of our being under the influence of our own unconscious interpretations of the world. The influence of the unconscious can be very hard to overcome, especially if we're unaware of it. These concerns touch on cognitive matters that we'll explore in greater detail later, but an example from the work[19] of Daniel Kahneman and Amos Tversky helps illustrate the point. Take a minute to think about the infamous "Linda problem" if you've never encountered it:

> "Linda is 31 years old, single, outspoken, and very bright. In college she was a philosophy major and was deeply concerned with issues of discrimination and social justice. She participated in antinuclear demonstrations." Which of the

following two statements about Linda would you say is more likely? (A) Linda is a bank teller. (B) Linda is a bank teller and is active in the feminist movement.

What was your impression of Linda? Which is the correct answer? Would you believe either (A) or (B) might be correct? It's true. Depending on what you mean by "likely" and "correct"—depending on how you interpret the context of the question, you can make a plausible case for either one. The case in favor of answer A, "bank teller," is that it is the larger class, it includes the smaller class, and, therefore, logically, it must be the right answer if by "more likely," we mean "more statistically probable." The case for answer B, "bank teller who is a feminist," is that we're being asked for our best guess about Linda's societal identity—whether she is a "feminist" or not—given our knowledge of the world.

Thousands of people have considered and evaluated the Linda Problem, and most of us preferred answer B, the socially aware but logically weaker one. There is much more to say about the Linda Problem and its interpretations that we'll cover in Chapter 11 because it has several lessons about scientific reasoning. Here I use the problem to illustrate our universal tendency to spin a narrative when we're trying to understand the world. Working from the skimpy information in the initial description of Linda, we instinctively perceived her in a social context and constructed a story, a coherent narrative, about the kind of person she is.

A narrative is an account of a sequence of events or an arrangement of elements that reveals connections among the parts. A narrative makes sense. Our subliminal internal story about Linda is essentially an implicit stereotyped hypothesis about her. It puts all of the pieces of the evidence together and clarifies her character. Our hypothesis tells us what she's like and predicts that she must be a feminist.

The hypothesis works well in science in part because of this predisposal to create interpretive organizational structures and to make predictions.[20] But, at the same time, this drive to understand creates superstitions, gives rise to certain religious beliefs,[21] and explains why we are so good at seeing shapes in the clouds or more generally imagining patterns in randomness.[22] We impose orderly interpretations on nature as much as we discover them in nature.

Given that our in-born drive to understand leads both to science and superstition, the big question is how do we, as scientists, cope with it? How do we reap the benefits of our tendency to seek hidden relationships, which is the core of the scientific urge to explain, while avoiding its pitfalls? The physicist Henri Poincare expresses the danger this way: "Some hypotheses are dangerous: first and foremost are those which are tacit and unconscious."[23] Where do tacit and dangerous hypotheses come from? From the never-ending saga, the ongoing series of stories, that we tell ourselves.

Instead of pretending that we can easily escape the traps set by our automatic, hypothesis generating minds and be innocently led by pure questions or curiosity, we should consciously recognize and go with our innate tendencies. We should take charge by formulating and testing our hypotheses explicitly.

As I've suggested, the difficulty in dealing with the Linda Problem stems from the narrative that we spin about her and the implicit hypothesis about her life that the narrative suggests[24]. In contrast, when you have a problem that lacks any evocative narrative subtext, you are more likely to reason about it logically. If you were to asked to guess whether it is more likely that I have an "apple" or a "green apple" in my lunch box, you'd immediately realize that the category of "apple" is the larger one and you'd have no trouble getting the logically correct answer. Formally, of course, the apple problem is identical to the Linda Problem (we'll return to these nuances in Chapter 11). Scientific thinking often profits from objectifying problems, turning them into questions about neutral, visualizable elements so that you can think clearly about their relationships.

If you deliberately make your hypothesis conscious and explicit, you can also take strategic advantage of two major safeguards against bias and self-deception: rigorous skepticism and multiple hypotheses.

As a Popperian Critical Rationalist, you are always skeptical about hypotheses; your own and others'. You look for flaws in them, not for ways to confirm them. You break the grip of the narrative of the hypothesis and look at it in the way a literary critic looks at a novel: skeptically, analytically, trying to determine whether the book achieves its stated purpose or how it fails.

8.C.2 Multiple Hypotheses as a Defense Against Narrative Bias

John Platt's program[25] of Strong Inference extends and amplifies elements of Conjectures and Refutations. According to Platt, the best defense against the infiltration of bias into your thinking is more, not less, hypothesizing; any phenomenon has more than one conceivable explanation, meaning that more than one hypothesis can, conceivably, explain it. Although it's infrequently used, the technique of multiple hypotheses is invaluable. Since we inherently want explanations and interpretations, we should be actively trying to create more of them. Try to develop several, quite different, possible explanations to account for the same initial observations. If there is a danger of forming too close an attachment to one hypothesis, do not have one. Have two! Even three or four, if your imagination and diligence are up to it. See what you've been taking for granted, or ask if relaxing your assumptions would lead to a new hypothesis. Once you have multiple hypotheses, you can take an Olympian view of them. They are all your babies, and you can be justly proud of each one. And, if having one hypothesis is bad because you want it to be right, then having multiple hypotheses gives you more chances to be right.

Best of all, having multiple hypotheses not only encourages you to maintain a proper distance from your pet theory, it also helps you design better experiments. An ideal experiment acts like a razor, neatly separating results into classes that are either compatible or incompatible with a hypothesis. You can simultaneously falsify one and corroborate others.

Here's an example of how this works. As noted in Box 7.2, neuroscientists have long been fascinated by the problem of how memories are stored in the brain. Our best guess is that a phenomenon called *long-term potentiation* (LTP) is responsible, and the current consensus is that the synapse, the junction between signal sending cells (presynaptic) and signal receiving cells (postsynaptic) is where LTP occurs. Somehow the synapse becomes "strengthened" by LTP, meaning that the presynaptic cell becomes more effective in stimulating the postsynaptic cell and either of two hypotheses could explain LTP: an increased release of the chemical neurotransmitter, glutamate, from the presynaptic cell, or an increased number of glutamate receptors (and there are two kinds of glutamate receptor) on the postsynaptic cell.

I want to look at one of the classic experiments from a slightly different perspective. Remember that investigators studied the postsynaptic response that was caused by glutamate. If LTP was caused by the release of more glutamate (regulated by the presynaptic cell), then the responses caused by both glutamate receptor types should have gotten bigger. If LTP was accompanied by an increase in the response caused by only one kind of receptor, it would mean that LTP was regulated by the postsynaptic cell. In fact, the investigators observed a highly selective increase in only one kind of receptor-mediated response, thereby simultaneously falsifying the presynaptic hypothesis and corroborating the postsynaptic one. Cutting just like razor.

You have to know what the alternative hypotheses are to design good experiments. Of course, even key experiments can be asymmetrical at times: one outcome could be fatal for one hypothesis, but an alternative outcome would be inconclusive. This is not a problem; normal science advances unevenly, and if you can falsify one hypothesis by an experimental outcome, that's progress.

Encouraging you to consider competing ideas and develop multiple hypothesis is a major advantage to using hypothesis. Hypotheses can also help organize your scientific thinking in a number of less specific ways as well.

8.C.3 The Hypothesis as an Aid to Scientific Thinking

You can be a committed hypothesis-testing scientist without being busy testing hypotheses every minute of the day, and you don't have to swear a loyalty oath to do hypothesis-based science either. Most experimental investigations, including the determinedly hypothesis-testing ones, meander at times; an experimenter may draw a tentative conclusion, try out a new idea, and, after some pilot experiments, return to the original or a modified design. The hypothesis is a cognitive tool, a device that helps you think clearly, and, in the following sections, I'll briefly review few of the ways in which it helps.

8.C.3.a The Hypothesis as Blueprint

A blueprint is a detailed architectural drawing; it is a definite plan with specific details—the doorframe is to be X inches wide and Y inches tall. If a blueprint were imprecise and vague, the builder wouldn't know how to proceed. Likewise, a scientific hypothesis provides a blueprint for an investigation, and it must be specific if you, its author, as well as the scientific community, are to know how to proceed.

As an explanation of a phenomenon, a hypothesis must include specific details about what it says and doesn't say for several reasons: you can only test a hypothesis if you know what it says. The hypothesis *entails* predictions, and you use deductive reasoning to bring them out. This is critical: by making predictions, a hypothesis tells you in no uncertain terms what you must do to see if your explanation is false. If your plan is fuzzy and indistinct, then neither you nor anyone else knows how to test it.

In this way, the framework of the hypothesis supplies built-in answers to perennial experimental questions: "What's next?" "What have I overlooked?" Once formulated, your hypothesis crystalizes what your investigation is about, then it indicates a course of action.

The LTP experiment constitutes a textbook example of how hypotheses provide a blueprint for the investigators. The presynaptic hypothesis *entailed* not only the prediction of a greater release of glutamate, but also a constellation of biochemical and biophysical predictions that could cause more glutamate to be released. Similarly, the postsynaptic hypothesis entailed an analogous constellation of predictions having to do with glutamate receptor regulation and expression. As always, numerous known and unknown biological and other complexities conspire to keep us from being absolutely certain about the result of any one experiment, no matter how elegant and persuasive. Having done an experiment, experimenters go back to the blueprint, derive a new prediction, and test it, repeating the process until they and their colleagues decide to let "the investigation rest."

Together, the hypothesis and the principle of falsification triage the experiments; they indicate which are the more important ones to do. You can't do everything, and there is little point in doing "confirmatory" experiments where the results are hardly in doubt and your hypothesis is left unchallenged. If you can carry out a severe test that puts the hypothesis at risk, then you're sure to learn something. And you must do it because your arch-competitor definitely will. The blueprint makes it plain to everybody what to do.

Of course, it's important to stress that blueprints are drawn on paper not engraved in granite. A hypothesis is not a rigid set of steps to follow. Just as a prospective home builder might look at an architect's blueprint and request another window in the bedroom, a hypothesis is a set of guidelines, an intended plan. It is flexible and can be changed according to need or circumstance. Which tests you do will depend on practical as well as scientific or logical reasons, and they may reflect such imponderables as your esthetic sense and intuition.

8.C.3.b Hypotheses Define Success and Failure in Research

Everyone agrees that failure is an integral part of science.[26] Often experiments do not work (e.g., the results are obscured by noise or are uninterpretable) the first (or second, or third) time. You have to be willing to try, fail, and learn from your mistakes. Hypothesis-based reasoning is a systematic way of making mistakes and extracting learning opportunities from them. Because your hypothesis makes definite predictions, you know how to falsify it; hence, failure of a hypothesis is a well-defined end-point. In contrast, if you go about science in a loose, undirected way, you can't reap the same benefit from the lessons of failure. You can't even fail if you haven't tried to succeed. If you seriously respect the importance of making mistakes in science, you will inevitably gravitate toward the hypothesis.

But we should be careful in how we talk about failure. A falsified hypothesis is not a failure. Eliminating a hypothesis is progress; going backward or going nowhere is failing. In Chapter 10B, I'll review the classic example of the Michelson-Morley experiment, which falsified the prevailing "ether" theory of light propagation, and argue that it was a brilliant scientific success. Why does it matter how we label experimental outcomes? Calling successful falsification "failure" sustains the mistaken notion that only experiments that confirm predictions should be considered successful. This notion, in turn, contributes to *publication bias* that favors publishing only "positive" results.[27] Encouraging you to think objectively about your results and what they mean is, accordingly, another advantage of hypothesis-based science.

8.C.3.c Hypothesis Testing as a Conscious Cognitive Process

You might decide to do an experiment or launch an investigation because you have some conscious or unconscious curious thoughts or questions. We can't say where hypotheses come from, but, even if we could, it would hardly ever matter. Having a hypothesis is not a prerequisite for starting an investigation, and once you've started, you quickly become absorbed in the "what" and the "how" of the plan, and you don't worry about the "why."

Having a hypothesis keeps you intellectually involved in the experiment. This is the "interpretive questioning" of nature that Cajal was referring to. You cannot fully carry out an investigation in a purely passive mode, and hypothesis-based science encourages active engagement at every step. One consequence of increased automatic data collection is that you can become disengaged from scientific reasoning, which has potential upsides and downsides. The ability to gather and analyze massive quantities of high-quality data mechanically and impartially could definitely be a good thing. However, no matter what your technical resources are, you need to decide which data to collect and how to analyze them. Designing and shepherding a hypothesis-testing experiment through to completion is an interactive process; you do something, see what happens, puzzle out what that means, and continue, or do something else.

Obviously, a scientist asking a series of questions may be as intellectually involved as a hypothesis tester is; intellectual involvement is a personal characteristic that, one hopes, all scientists share. My point here is that, by its nature, rigorously testing a hypothesis demands constant, conscious interpretative interaction in a way that other methods, question-answering, for example, may or may not do.

8.C.3.d Self-Organization

The writer Norman Mailer reported that doing the morning *New York Times* crossword puzzle each day was how he "combed his brain"[28] to get ready for the tasks ahead. Forming a hypothesis is a way of combing your brain and creating order from the occasional tangle of thoughts or the mass of unclear or inconsistent data that can pile up. Thinking about your hypothesis helps straighten things out.

You need to exert determined efforts to make sense of observations, and yet the course of an experiment is not always (i.e., is rarely) as straightforward as you imagine it will be. Consideration of possible test outcomes and their interpretations is invaluable in making course corrections as required when a failed prediction falsifies your hypothesis. And making an explicit hypothesis makes it harder to for you to "fool yourself" in Feynman's often-quoted words. In other words, the *cognitive act of formulating* a hypothesis, independent of the hypothesis per se, is one of the advantages of working with hypotheses.

8.C.3.e Memory and Narrative Structure

Organized information is the easiest to recall, and the hypothesis organizes information into a narrative structure that fosters memory. A dramatic nonscientific illustration of the power of structure to facilitate memory comes from competitions where the participants, "Memory Athletes"[29] try to recall large numbers of items. Most of the athletes rely on the "method of loci," a technique of mentally placing images of items that they need to recall in various locations in a familiar physical structure, also called a "memory palace." Mentally revisiting a location triggers retrieval of the image that was placed there. Recall of words arranged into meaningful sentences is much better than recall of random lists of words. Cognitive psychology also assures us that, when we're given complicated information, we tend to remember broad meanings not specific details.[30] And we recall stories better than lists of facts, meaningful sentences better than random word lists. Nature is complicated: biological sciences often deal with large numbers of entities interrelated in complex ways.

The narrative structure provided by the hypothesis helps organize scientific facts into logical, coherent chunks of information. We see the world as an orderly arrangement of thematically linked mental images.

8.C.3.f Communication

Not only do you, the individual scientist, gain a good sense of what is involved in your hypothesis-based inquiries, but so do your colleagues who read your reports and want to follow-up on them. The hypothesis is an exceptionally efficient mechanism for conveying the big picture because it puts your results into a logical, easy-to-follow context. Indeed, as a general rule, one really can't overstate the importance of readily-grasped communication in all aspects of science. The quality of your scientific publications, grant proposals, job talks, lectures, etc. are all directly related to how easily others can understand them. Superb communication skills on their own do not produce excellent science, but deficient communication skills can most certainly obscure scientific excellence. And there is no necessary correlation between scientific expertise and communication skills. Brilliant thinkers are not necessarily brilliant communicators.

You may protest; surely a good idea will be recognized, even if it is not ideally expressed! Yes, if you are a surpassing genius, it may well be that the significance of your revolutionary contribution will instantly be perceived and rewarded no matter how inarticulately it is expressed. On the other hand, the history of science is replete with tales of key discoveries being neglected in their day and with disputes among scientific giants that often focused minutely on who said precisely what. And there is even the possibility that one of your competitors will simply appropriate you poorly articulated discovery, do a few related experiments, and rebrand it as his own (this sort of thing does, unfortunately, happen). In any event, the great majority of scientists will almost certainly benefit from paying close attention to what they want to say and how they say it.

Science is a social enterprise and having, or anticipating, the critical comments of others will also help improve the soundness of your thinking and uncover your unconscious biases. Stating your hypothesis is a public declaration of your thoughts and reasoning; reviewers and editors of scientific journals and others routinely provide valuable feedback, and the quantity and quality of their feedback depends on their ability to understand what you're trying to say. You cannot benefit as much if your ideas are not readily intelligible. Many of us are familiar with the global condemnations—"diffuse" "unfocused," and "poorly organized"—that spell doom if they appear in the review of your National Institutes of Health grant application.

Efficient, informative communication is the overriding objective of a scientific report, which, therefore, is not a veridical historical accounting of the actual course of an investigation. Readers expect to learn about the new information you've found, not contorted descriptions of your thought process, no matter how fascinating you believe them to be. You can assemble your report around your hypothesis to maximize its logical flow and scientific cogency. You might have become aware of the killer control experiment, the one that you should probably have done first, only after your project is well under way. This is not a problem: do the experiment and report the outcome in the first paragraph of the Results section if that makes the most sense. The purpose of a scientific publication is to present your data and communicate your ideas, not to build suspense or impress the reader with your literary skills. (After you've won the Nobel Prize, historians of science may find the exact sequence in which you did your experiments to be deeply informative, but probably not before then.)

Finally, the explicit hypothesis is especially useful for sciences such as neuroscience and other biological sciences that rely to a large extent on natural language for communication. Hard sciences (the rigorously quantitative ones—e.g., physics, chemistry, etc.) use mathematics as a common language; chemistry also has unambiguous sets of symbols for organizing, expressing, and communicating important concepts. Neuroscience (much of which is somewhere on the continuum between hard and soft sciences—it could be considered a "firm" science) uses English as the de facto official language (the *lingua franca*) of most large international conferences and science journals, in part because English readily assimilates new technical terms and is notoriously flexible. However, these advantages are partly offset by a tolerance of imprecision and lack of rigorous standardization; no "Academie Englaise" stands guard over its usage. Hence, any field that relies heavily on English for communication, evaluation, etc., on an international scale may be more in need of uniform linguistic conventions to foster communications than more quantitative fields. Devices such as the explicit hypothesis help reduce ambiguities that can otherwise arise.

8.D Coda

Hypothesis-based reasoning has many advantages for scientific thinking and practice. The range runs from quantitative applications to qualitative principles.

Statistical theory suggests that hypothesis-based work can avoid many of the more egregious pitfalls that lead to irreproducible science. This chapter also explored several cognitive advantages that using the hypothesis offers, including its ability to combat bias and create an organizational structure that facilitates experimental design, self-organization, and scientific communication.

In the next chapter, I present data from two surveys that I conducted to find out how scientists themselves view the hypothesis and how they use it in their written work. In Chapter 10, I review and critique the work of three authors who disagree that the hypothesis has any advantages and argue that science can dispense with it.

Notes

1. Positive Predictive Value (PPV) is sometimes given other names, but this is the one that Ioannidis and colleagues use. Wacholder et al. calculate an analogous factor, the False Positive Report Probability (FPRP), that is basically the inverse of the PPV. S. Wacholder, S. Chanock, M. Garcia-Closas, L. Elghormli, and N. Rothman, "Assessing the Probability that a Positive Report Is False: An Approach for Molecular Epidemiology Studies," *Journal of the National Cancer Institute* 96:434–442, 2004.
2. Null hypothesis significance testing (NHST) and objections to it were discussed in Chapter 3.D.
3. J. P. Ioannidis, "Why Most Published Research Findings Are False," *Public Library of Science/Medicine* e124, August 2, 2005. (The paper has been cited well over 5,000 times—by Google Scholar—so it has had a big impact.)
4. K. S. Button, J. P. Ioannidis, C. Mokrysz, B. A. Nosek, J. Flint, E. S. Robinson, and M. R. Munafò, "Power Failure: Why Small Sample Size Undermines the Reliability of Neuroscience," *Nature Reviews Neuroscience* 14:365–376, 2013.
5. Open Science Collaboration (B. Nosek, corresponding author), "Estimating the Reproducibility of Psychological Science," *Science* 349:aac4716, 2015; "An Open, Large-Scale, Collaborative Effort to Estimate the Reproducibility of Psychological Science," *Perspectives in Psychological Science* 7:657–660, 2012.
6. Ioannidis, "Why Most."
7. Deborah G. Mayo, *Statistical Inference as Severe Testing: How to Get Beyond the Statistics Wars* (New York: Cambridge University Press; 2018).
8. Mayo argues forcefully that evidence that severely tests and persuades us to reject, for example, H_0, constitutes evidence *in favor* of H_1, a formulation that Popper, but not John Platt (Chapter 2.H), would have resisted.
9. Fisher's Method (Fisher's Combined Probability Test); https://en.wikipedia.org/wiki/Fisher%27s_method.
10. R. A. Fisher, *Statistical Methods for Research Workers: Biological Monographs and Manuals Series* (Edinburgh: Oliver and Boyd; 1925), gives the formula, referred to as Fisher's Method, for calculating the joint probability of multiple independent

p-valued tests that all bear on the same hypothesis, as a chi-square value where p_i is the probability of the *i*th test, and k is the number of independent tests.

11. Brainder, "The Logic of the Fisher Method to Combine Pvalues," posted on May 11, 2012, https://brainder.org/2012/05/11/the-logic-of-the-fisher-method-to-combine-p-values/.
12. R. A. Fisher, 1932 (as quoted in https://brainder.org/2012/05/11/the-logic-of-the-fisher-method-to-combine-p-values/):

    "When a number of quite independent tests of significance have been made, it sometimes happens that although few or none can be claimed individually as significant, yet the aggregate gives an impression that the probabilities are on the whole lower than would often have been obtained by chance. It is sometimes desired, taking account only of these probabilities, and not of the detailed composition of the data from which they are derived, which may be of very different kinds, to obtain a single test of the significance of the aggregate, based on the product of the probabilities individually observed."
13. See Note 11. Brainder gives an easy to follow derivation of Fisher's formula.
14. R. A. Fisher (1932), quoted in "The Logic of Fisher's Method," https://brainder.org/2012/05/11/the-logic-of-the-fisher-method-to-combine-p-values/.

    "The circumstance that the sum of a number of values of X^2 is itself distributed in the X^2 distribution with the appropriate number of degrees of freedom, may be made the basis of such a test. For in the particular case when $n = 2$, the natural logarithm of the probability is equal to $\frac{1}{2}X^2$. If therefore we take the natural logarithm of a probability, change its sign and double it, we have the equivalent value of X^2 for 2 degrees of freedom. Any number of such values may be added together, to give a composite test, using the Table of X^2 to examine the significance of the result."

15. Caveats to and extensions of Fisher's Method: https://en.wikipedia.org/wiki/Fisher%27s_method; https://en.wikipedia.org/wiki/Extensions_of_Fisher%27s_method. See also Brainder, "Non-Parametric Combination (NPC) for Brain Imaging," posted on February 8, 2016, https://www.brainder.org; Anderson M. Winkler, Matthew A. Webster, Jonathan C. Brooks, Irene Tracey, Stephen M. Smith, and Thomas E. Nichols, "Non-Parametric Combination and Related Permutation Tests for Neuroimaging," *Human Brain Mapping* 37:1486–1511 2016.
16. Santiago Ramón y Cajal, *Advice to a Young Investigator;* translated by Neely Swanson and Larry N. Swanson. (Cambridge, MA: MIT Press; 1998).
17. J. Locke, quoted in Laurens Laudan, *Science and Hypothesis: Historical Essays on Scientific Methodology* (Boston: D. Reidel;1981), *Science*, p. 64.
18. Karl Popper, *Conjectures and Refutations: The Growth of Scientific Knowledge* (New York: Routledge Classics; 2002).
19. Daniel Kahneman, *Thinking, Fast and Slow* (New York: Farrar, Straus and Giroux, 2011).
20. M. Bar, *Predictions in the Brain* (New York: Oxford University Press; 2011).
21. Pascal Boyer, *Religion Explained: The Evolutionary Origins of Religious Thought* (New York: Basic Books; 2002).

22. N. N. Taleb, *Fooled by Randomness: The Hidden Role of Chance in Life and in the Markets (Incerto)* (New York; Random House; 2005).
23. H. Poincare, *Science and Hypothesis* (Create Space Independent Platform. Original publisher, New York: Walter Scott; 1905).
24. Kahneman, *Thinking*, p. 159. Kahneman cites the frustration that the naturalist Stephen Jay Gould had with the Linda Problem. Despite knowing the answer, Gould wrote: "a little homunculus in my head continues to jump and down, shouting at me—'but she can't just be a bank teller; read the description.'"
25. J. Platt, "Strong Inference: Certain Systematic Methods of Scientific Thinking May Produce Much More Rapid Progress Than Others," *Science* 146:347–353, 1964.
26. Stuart Firestein, *Failure: Why Science Is So Successful* (New York: Oxford University Press; 2016).
27. S. C. Landis, S. G. Amara, K. Asadullah, C. P. Austin, R. Blumenstein, E. W. Bradley, et al., "A Call for Transparent Reporting to Optimize the Predictive Value of Preclinical Research," *Nature* 490:187–191, 2012. See also references therein.
28. N. Mailer (from the documentary movie, *Wordplay*, directed by Phillip Creadon, 2006).
29. Memory athletes. https://www.psychologytoday.com/us/blog/brain-waves/201703/how-train-your-brain-memory-champion
30. Teachings of cognitive psychology regarding memory. A good basic text is John R. Anderson, *Cognitive Psychology and its Implications, 8th Ed.* (New York: Worth Publishers, 2015), Chapters 5–7, pp. 97–180.

9
What Scientists Think About Scientific Thinking

9.A Introduction

Philosophers and policy makers, with an occasional scientific author chiming in, have dominated the conversation about science and the hypothesis. Trying to understand the role of the hypothesis in science by listening only to these folks would be like trying to figure out why people vote the way they do by listening only to political scientists and media pundits. If you want to know about voting behavior, you have to talk to voters. To find out how scientists feel about the hypothesis, I carried out two different investigations: an email survey that solicited their opinions and a scientific literature review. This is an exploratory, Discovery Science, effort; I didn't have an explicit hypothesis when I conducted it. I report percentages, rather than the results of statistical tests, so you, the reader, can determine their significance.

Caution: This chapter is for those who, for reasons of interest, curiosity, or masochism, want to get down into the weeds and wrestle with data; though brief, it is full of numbers and graphs. If you'd rather skip the details, here is the executive summary:

- Scientists get only a minimal amount of formal education in scientific thinking.
- Scientists would welcome such formal instruction.
- While hypothesis testing is the most widely used mode of doing science, all scientists engage in multiple modes.
- The advantages of the hypothesis testing outweigh its disadvantages.
- The Scientific Method and the hypothesis are still essential for scientific progress.
- There is no widespread enthusiasm for major changes in grant or journal policies relating to the hypothesis.
- The scientific hypothesis of a paper is usually implied, not stated explicitly.

9.B What Do Scientists Say About the Hypothesis? A Survey

I sent a survey via SurveyMonkey[1] in late 2017 to 3,813 randomly selected members of biological science societies.[2] The survey, mainly substantive multiple-choice questions, brought in responses from 444 individuals.[3] Most replies (83%) were from academics, while the rest were from scientists in government or industry laboratories (3% and 5%, respectively), or held other kinds of positions or were retired (9%). My first question, which was included in the introductory email, was neutral—it merely asked about the respondent's educational experience with the hypothesis—to try to avoid selecting only for those having strong opinions one way or the other about the hypothesis. Not every respondent answered every question, in part because certain sections (e.g., those relating to National Institute of Health [NIH] research grant applications) were inappropriate for graduate students and postdocs who hadn't applied for those grants. The numbers of responses for each question are listed in the figure legends. Because I couldn't control who replied to the survey or answered individual questions, the survey wasn't a rigorous study; it would be interesting to see detailed follow-up.

In 2014, *Nature* asked its readers about the Reproducibility Crisis.[4] To see how my sample compared to theirs, I used one of their questions verbatim: "Do you believe that there is a Reproducibility Crisis?" Our results (Figure 9.1) are grossly similar, although fewer of my respondents felt that there is a "significant" or "slight crisis" (70%, my survey vs. 90%, theirs); more of my respondents

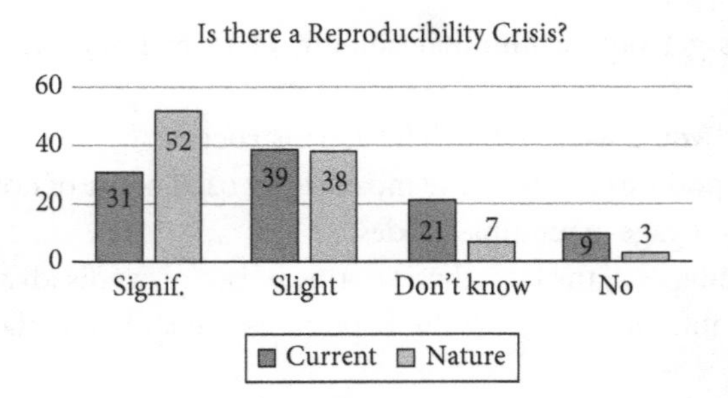

Figure 9.1 Is there a Reproducibility Crisis? Question taken from *Nature* readers' survey; *Nature* results in light bars ($n = 1{,}576$); current survey results in dark bars ($n = 210$). The groups differ ($p < 0.001$) according to a chi-squared test with one degree of freedom. In all figures, the captions summarize the gist of the survey questions; the number of respondents (n) to each question is in parentheses.

doubted that there is any crisis at all (9%, mine vs. 3%, theirs), or they "didn't know" whether or not there is one (21%, mine vs. 7%, theirs).[5]

Based on these numbers, it appears that the two groups are similar but not identical, which is something to keep in mind when thinking about extrapolating the conclusions from either survey.

9.B.1 Scientific Training Regarding the Hypothesis

How do scientists acquire their skills in scientific thinking? Answers to this question can shed light on the diversity of opinion about the hypothesis that I've alluded to earlier.

For the most part, scientists don't learn about the hypothesis through formal training. Survey respondents, 70% of them, had either no formal instruction or a minimal amount (≤1 hour) of it (Figure 9.2A), and yet a total of 92% said that they thought that formal instruction in hypothesis testing would be "very" or "moderately" useful for science trainees (Figure 9.2B). Despite their lack of formal training, 89% of respondents were either "very confident" or "confident" in their knowledge of the Scientific Method and the hypothesis, with only 11% admitting that they were only partially or "not at all confident" in their knowledge (Figure 9.3).

If scientists don't acquire their knowledge of the Scientific Method from formal training, where does the knowledge implied by their high confidence levels come from? Apprenticeship methods: exposure. Most respondents, 81% (not shown), had gotten either a "great deal" or a "moderate amount" of informal experience in evaluating hypothesis-based research through journal clubs or lab meetings. In these venues, participants think out loud, challenge and criticize each other, and students learn through experience how their colleagues

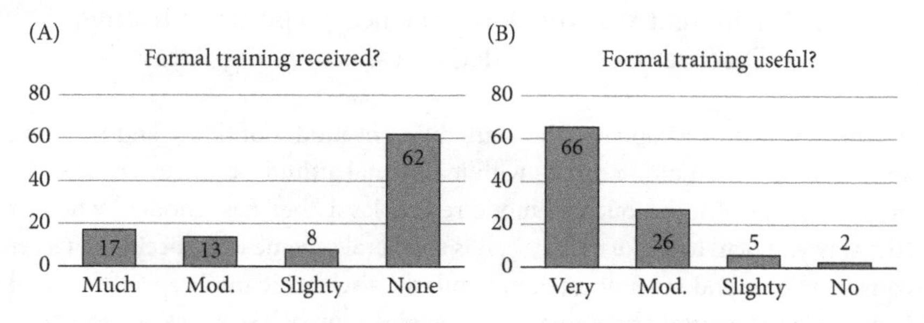

Figure 9.2 How much formal instruction on the scientific method, including the hypothesis did you receive? (A) During your doctoral or postdoctoral training, how much formal instruction on the scientific method, including the hypothesis, did you receive ($n = 444$)? (B) How useful do you think formal instruction in hypothesis testing would be for science trainees ($n = 348$)?

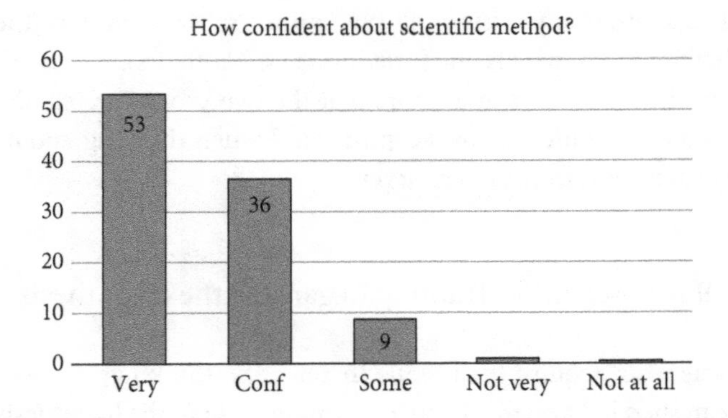

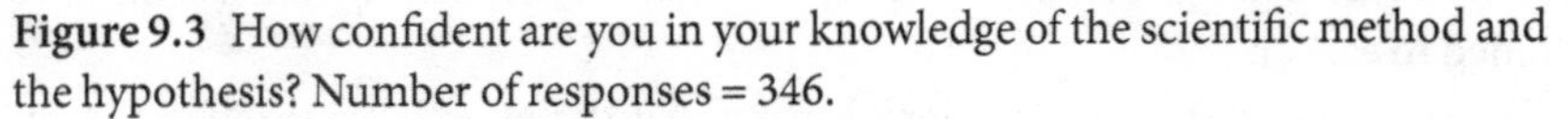

Figure 9.3 How confident are you in your knowledge of the scientific method and the hypothesis? Number of responses = 346.

reason scientifically. The confidence that scientists express in their knowledge of the Scientific Method would seem to suggest that the apprenticeship methods are extremely effective. Alternatively, the replies to the received and wished-for formal instruction (Figure 9.2A, B) convey the impression that the informal methods are not fully satisfactory. If they were, you'd expect the graphs to have similar shapes, with few people wishing for more formal instruction. Indeed, it is a little hard to believe that, if scientists were getting adequate training in scientific thinking in relaxed, informal settings, they'd want to sit through formal lectures on it. Most can't wait to escape the classroom and get into the lab, and the fact that the vast majority of respondents think that additional formal instruction would be beneficial suggests that there is room for improvement in how we're teaching them about scientific thinking.

9.B.2 Different Ways of Doing Science: Hypothesis Testing, Discovery, Open-Ended Questioning, Big Data

As we reviewed in Chapter 4, there are different modes of doing and thinking about science. Each mode carries with it a mental attitude as much as a specific method or set of techniques. A mode resembles a "business model," which is "the way you plan to make money"[6]; it is a general scheme that specifies a target consumer base and revenue stream. Similarly, a scientific mode specifies a goal (e.g., explain a natural phenomenon), a general approach (e.g., test a hypothesis by doing experiments), and so on. Unlike business models, however, scientific modes are not mutually exclusive; an individual scientist typically goes back and forth between different modes. Sometimes a scientist is, say, trying to develop a catalog of the kinds of play behavior that male rat pups engage in (Discovery Science), sometimes she wonders if a drug would affect the pups without having any idea of what it might do (open question), and sometimes she is trying to explain why a drug affects the pups the way it does (hypothesis testing).

The survey asked participants to estimate how important each mode had been for them at different stages of their careers (i.e., during their PhD thesis work, postdoctoral training, their current research, or during the writing of their most recent NIH grant proposal). The subgroups of respondents for each stage are not identical—not everybody had postdoctoral training or had written an NIH application, for instance.

The four-bar clusters in Figure 9.4 show the percentages[7] of respondents who found a particular scientific mode to be highly significant for them at a given career stage. A scientist might say that, during her PhD thesis work, both hypothesis testing and Discovery Science had been very important, open-ended questioning slightly important, and Big Data not at all important. My hope was to find out how scientists rated the different modes in terms of their perceived importance, not to see how they accounted for their effort, which is why the totals do not add up to 100%. Thus, 77% of respondents said that hypothesis testing was very important for their PhD work, 48% said that Discovery Science was, 34% said that open-ended questioning was, and 16% said that Big Data was.

The overall pattern of the rankings was consistent: at each career stage the hypothesis testing mode was cited as the most important, followed by Discovery

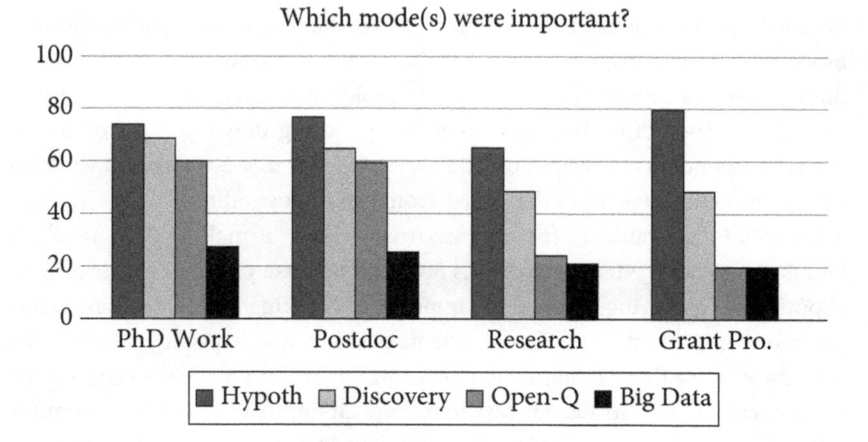

Figure 9.4 How important were different modes of science to you at different stages of your career? Respondents were asked to rank order the importance of different modes at the indicated stage of their careers ($n = 339, 288, 294, 255$ for the stages from left to right. The rating categories were qualitative and were not mutually exclusive, so the totals in each cluster of bars add up to more than 100%.) Large fractions of the group (not plotted) said that they had nothing to do with Big Data at the four stages: 56%, 54%, 40%, and 56%, respectively, from left to right.

Science, open-ended questioning, and Big Data modes. There was no significant (chi-square test) change in the proportions of responses for either the hypothesis testing or Big Data modes across the four stages. However, there were large and significant differences in the proportions citing Discovery Science and open-ended questioning as highly important across the stages. These two modes became much more influential during the postdoc and current research periods than they had been during the PhD years. A curious result, from a sociology-of-science point of view, is the comparison of the rankings during the "PhD Work" and the grant proposal "Grant Pro." The chi-square test shows that, at these two stages, the rankings of the modes are not significantly different. Maybe when scientists go to write grants, they revert to what worked during the PhD years, or, alternatively, they believe that the grant reviewers expect hypothesis-based work to be front and center again.

Probably the main implication of Figure 9.4 is that scientists are not stuck in any one mode of doing science; most of us operate in multiple modes. Note also that Big Data methods have not, so far, made great inroads into biological science even though Big Data was defined quite broadly as "data-mining"; not only was Big Data least frequently cited, but about half of the respondents said that they had nothing to do with it.

9.B.3 Advantages and Disadvantages of the Hypothesis

We can't assume that just because scientists say they're in the hypothesis testing mode most frequently, we know what they really think about it. Maybe hypothesis testing is simply the most familiar choice and therefore they use it but, deep down, they think it is worthless. To try to dig down and get opinions, I asked respondents to select potential advantages or disadvantages of working with hypotheses, as many as they liked, from lists of possibilities.

Potential disadvantages (paraphrased) were that, "it makes you biased," "it blinds you to other alternatives," "it is pointless because you can only disprove a hypothesis," "gathering data is more important," "making precise measurements are more important," and "most scientists don't use hypotheses." Potential advantages were that, "it helps you to organize your thinking and to know what experiments to do," "its logic is easy to follow," "it improves scientific communication," "it advances science," "you can test your ideas with it," "the hypothesis is the main message of a scientific paper."

Substantial minorities felt that disadvantages were that a hypothesis made you biased in its favor (picked by 48% of respondents) and that it blinded you to other ideas (35%); no other disadvantage was selected by more than 18% of the group (Figure 9.5A). The top advantages were that a hypothesis helps organize

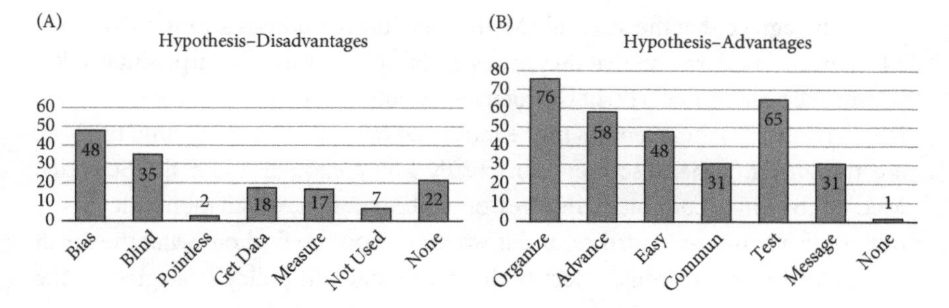

Figure 9.5 What the advantages and disadvantages of having a hypothesis? Disadvantages (A) and advantages (B) of basing scientific work on hypotheses (select all that apply); see text for descriptions. Totals: 440 disadvantages selected ($n = 331$); 1,024 advantages selected ($n = 345$).

your thinking (76%), allows you to test your ideas (65%), and advances science (58%). Other potential advantages were cited by healthy minorities of 31–48% (Figure 9.5B).

Interestingly, respondents found much more to like than to dislike about the hypothesis: the survey participants freely selected a total of 1,024 advantages and only 440 disadvantages (i.e., each respondent found an average of 3.0 advantages and only 1.6 disadvantages). And, whereas 99% of them thought that a hypothesis offered at least some advantages, only 78% thought that it had any disadvantages. We might infer that scientists do a lot of hypothesis testing because they find it beneficial; in any case, the results indicate that many scientists hold favorable views of the hypothesis.

The hypothesis and the Scientific Method are considered "essential" for scientific progress by 94% of the survey participants (Figure 9.6). However, the result may not be exactly the ringing endorsement that it seems at first glance. Only 60%

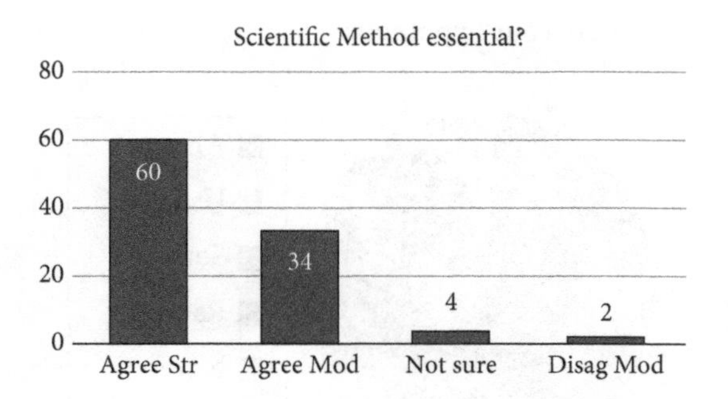

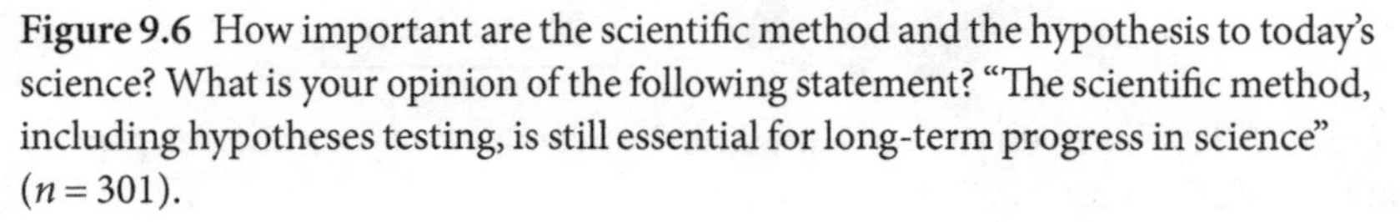
Figure 9.6 How important are the scientific method and the hypothesis to today's science? What is your opinion of the following statement? "The scientific method, including hypotheses testing, is still essential for long-term progress in science" ($n = 301$).

"strongly" agreed that the Scientific Method and the hypothesis are essential, while 34% agreed "moderately" that they are essential. To me, this was surprisingly lukewarm; like hearing that 34% of US voters were only moderately in favor of free and fair elections. Do the scientists in the moderate camp favor a competing method, are they hedging because they don't really know enough about the Scientific Method to want to commit themselves, or is there another explanation? The survey did not probe this area further, and it would be good to find out what the result means as the answer could affect both educational and policy strategies for the future.

9.B.4 The Hypothesis at Large

How does the hypothesis affect scientists' thinking? For one thing, the generally high favorability rating that respondents gave to the hypothesis probably explains why so many people (75%) claim that they "always" or "usually" state their hypothesis in their papers (Figure 9.7). A small group said that they don't state their hypothesis even though they have one (n = 71, data not shown), and most of this group said that their hypothesis was too obvious to need stating (42%) or that it would seem "artificial" to do so (28%). Although these numbers were too small to support firm conclusions, the responses may hint that elements of social awareness shape the ways in which we present our scientific work.

How do scientists evaluate the influence of the hypothesis beyond the walls of their own laboratories? That is, how do they perceive its influence on a broader scale? One survey question asked respondents to name the

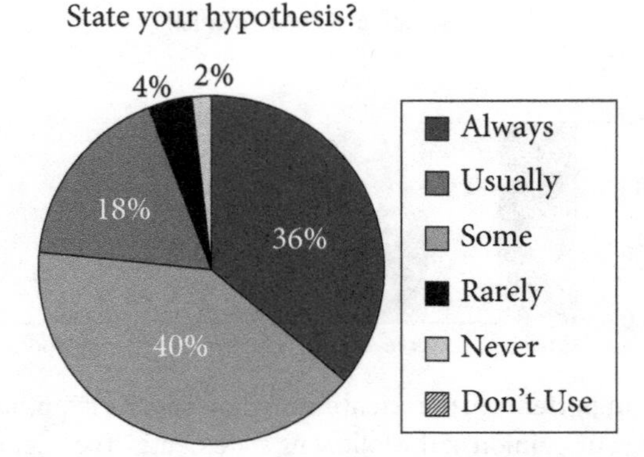

Figure 9.7 How often do you explicitly state the hypothesis of your research papers? Number of responses = 295.

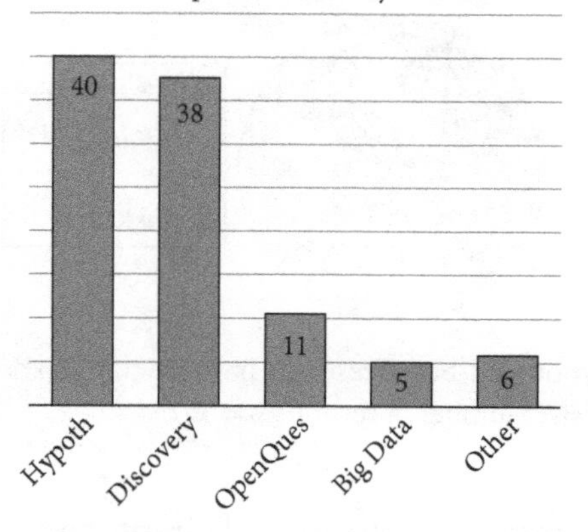

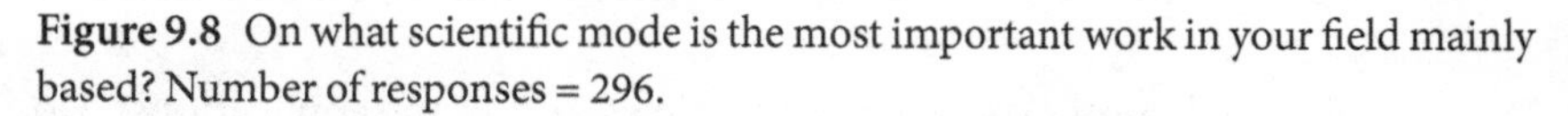

Figure 9.8 On what scientific mode is the most important work in your field mainly based? Number of responses = 296.

scientific mode that produced "the most important work being done in your field": hypothesis-based work was the nominee of 40% of the respondents and Discovery Science was the top choice for 37% (Figure 9.8) (i.e., co-equal in importance; cf., Figure 9.4). It seems that, even though most people say they're primarily engaged in hypothesis-based science, they recognize that major advances also come from Discovery Science. The replies confirm that scientists do distinguish between different modes of science and appreciate the unique contributions of more than one. In the future it would be interesting to follow-up and find out how the community perceives the relationships of "Curiosity-Driven Science," "open-ended questioning," and even Big Data to hypothesis-based and Discovery Science.

As I've noted, a commonly expressed concern about the hypothesis is the danger of becoming so wedded to it that you can get into trouble—that having a hypothesis can lead to bias, and the like (Figure 9.5A). Do scientists believe that using a hypothesis has "anything to do with the 'Reproducibility Crisis?'" Survey-takers were either divided on this issue or openly skeptical: 17% of them said "yes" because there is "too much" emphasis on the hypothesis, and 17% said "yes" because there is "too little" emphasis on the hypothesis. The greatest fraction, 44%, said that the hypothesis had nothing to do with the crisis; the remaining 22% didn't know enough about a crisis to express an opinion (Figure 9.9). At least it seems safe to conclude that there is no consensus that overreliance on the hypothesis should be blamed for the purported Reproducibility Crisis in science. Perhaps because scientists move fluidly between scientific modes, they don't want to pin such a complex problem on any one source, or perhaps they feel other factors are to blame.

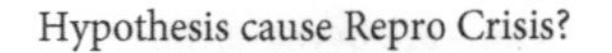

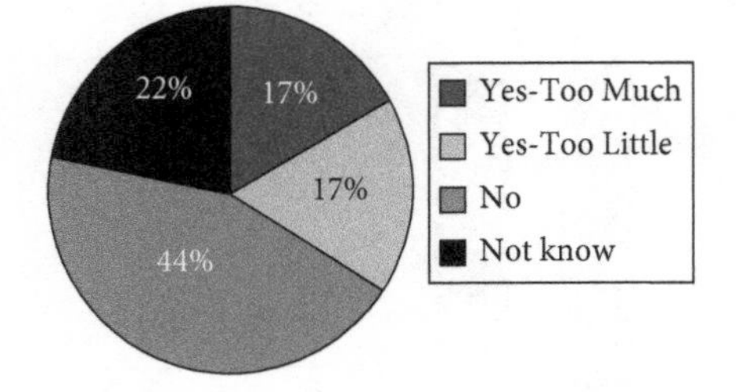

Figure 9.9 Does hypothesis-based research have anything to do with the Reproducibility Crisis? Number of responses = 302.

9.B.5 The Hypothesis in Reviews of Grants and Journal Articles

What about the hypothesis in journal articles and grant applications? Do journal reviewers and editors place too much, too little, or the right amount of emphasis on the hypothesis of a research report? For journal publications, the results were remarkably balanced (Figure 9.10A), with 19% of the respondents saying there was too much emphasis on the hypothesis, 21% that there was too little emphasis on it, and 60% that the current emphasis was "appropriate."

How about the all-important grant application? To get the money that they need to run their labs, academic scientists apply for grants and hence are at the mercy of the application review process. Do grant reviewers focus on the hypothesis of a grant proposal? Confirming what most people have assumed, most survey-takers, 92% (data not shown) said "yes," grant reviewers did indeed

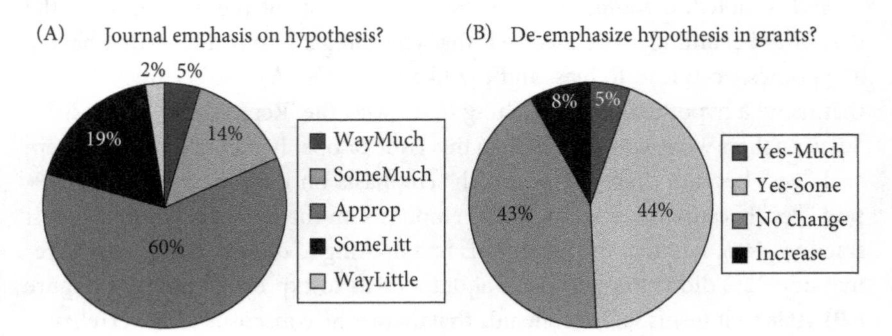

Figure 9.10 Is the hypothesis weighted too heavily in scientific publications and grants? (A) How much emphasis do you feel that journal reviewers and editors place on it ($n = 292$)? (B) Should the National Institutes of Health and similar granting agencies de-emphasize the hypothesis ($n = 292$)?

take the grant hypothesis into consideration in their reviews; only 8% felt that reviewers did not focus on the hypothesis of their applications at all.

Should the NIH and similar agencies de-emphasize "hypothesis-driven" research? The sample group was split, with 5% and 44% favoring great or moderate de-emphasis, respectively, 43% favoring no change in current practice, and 8% wishing for an increased emphasis on the hypothesis (Figure 9.10B). How should we interpret this array of answers? From one point of view, the group sentiment (57% vs. 43%) was for a change in the prominence of the hypothesis—Let's do something!—with the large majority who want change urging de-emphasis (49% vs. 8%). However, from another perspective, the data say that 95% of the group did not want to see a major de-emphasis of the hypothesis. Be careful, the baby's still in the bathtub!

9.C How Do Scientists Use the Hypothesis? Analysis of Neuroscience Literature

The survey tells us what scientists say about the hypothesis. What they say isn't necessarily what they do, however. Part of what scientists do with the hypothesis is reflected in their research papers. A research paper is not like a diary or a stream of consciousness, but it is also not a robotic printout of data. In style, it is somewhere between a literary short story and a medical diagnosis. The factual content of a journal article is often woven into a narrative in which the author is trying to make her story interesting enough to keep you reading and yet not so colorful that she'll turn off the traditionalists in her audience or come across as an intellectual lightweight. In their journal articles scientists describe the experiments that they did and what they found. They also report why they did the experiments and what they believe that the results mean. Above all, a scientific author wants to come across as being a good scientist: observant, insightful, clever, well-read, and logical.

By examining how scientists integrate the hypothesis into their research reports, I wanted to get a sense of how it figures in their reasoning or, at least, what they want us to think about their reasoning. Because scientists, like any professionals, are prone to breaking out their most arcane jargon when they're addressing other experts in their narrow areas, I avoided highly specialized technical journals in favor of the high-profile, generalized science journals, *Science* and *Nature*, plus two broad-based, top-notch journals within neuroscience, *Neuron* and *Nature Neuroscience*.[8]

I searched each article for key words related to hypothesis testing and the other scientific modes and tallied the numbers of times that they appeared, initially hoping to use the search function in Adobe DC exclusively, I nevertheless frequently had to read and score the text by hand to sort out ambiguities. For instance, in published reports the word "hypothesis" overwhelmingly refers to a statistical hypothesis, and, since I was interested in scientific hypotheses, I had to weed the statistical ones out of the results manually. Figure 9.11 shows the literature survey results.

I found that scientists rarely state their scientific hypothesis directly ("Our hypothesis is that . . ."): Only 16% of the papers were so overt. Then, assuming (see Chapter 2) that a scientific "model" (i.e., a conceptual model, not a "model system" or "animal model"—I weeded them out as well) is effectively the same as a "hypothesis," I counted another 12% of the papers, which explicitly tested "models," as being "hypothesis-based." When authors used "hypothesis" as a synonym for "prediction" (17%), I didn't count them as *explicitly* hypothesis-based. Even allowing for the possibility of alternative judgments in a few cases (we can be surprisingly creative, or opaque, in presenting our ideas), I estimate that less than approximately 33% of the papers stated their hypothesis in so many words, which is not what you'd expect given that roughly 75% of the scientists responding to my survey (Figure 9.7) said that they always or usually stated the hypothesis of their papers. This suggests that a gap exists between what they say and what they do (assuming, of course, that survey respondents and authors do not represent independent groups). How could we explain such a gap? The

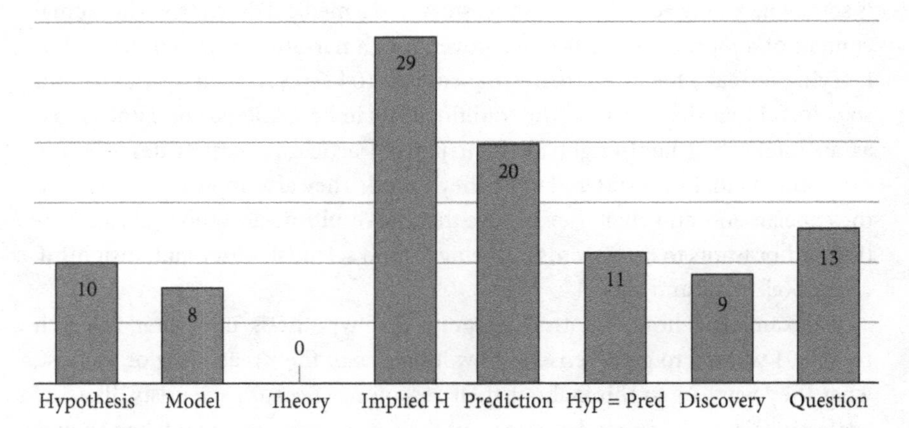

Figure 9.11 What do scientific papers reveal about how (neuro)scientists think about the hypothesis? A pdf search of the articles counted the times that "hypothesis," "model," "theory," and "prediction" appeared in neuroscience articles in *Nature, Science, Neuron*, and *Nature Neuroscience* during 2015. To make decisions about whether the papers were based on "implied hypotheses," "discovery science," or "open-ended questioning," or when the word "hypothesis" really meant "prediction," I read and evaluated the article's text. (See Note 8 for details.)

answer may be that authors do not distinguish between their own "implicit hypotheses" and explicitly stated ones.

I found that, while roughly 45% of the papers did not state a hypothesis, they are plainly based on hypothesis testing procedures, and they do attempt to explain a particular phenomenon. That is, these are not "discovery" projects and are not asking "open-ended," "what-if" kinds of questions. Moreover, though they lack an explicit hypothesis, the experiments laid out in the papers follow a logical sequence. The papers are also often sprinkled liberally with free-floating "predictions" which, though unmoored to a clearly stated hypothesis, let the reader in on why the authors went from one experiment to the next. The papers carried through a theme and reached concrete conclusions, usually proposing a conceptual model to give the reader a physical picture of what the authors have in mind.

I did classify some papers as Discovery Science (9%) or open-ended questioning (13%), so it wasn't as if these kinds work weren't represented in the sample. Rather, the main reason that an explicit hypothesis is missing from most research papers is that the authors prefer to avoid directly saying that they had one. It seems that scientific authors expect readers to sense its presence in the way that an art connoisseur senses the harmony of a painting or a chess master senses a dangerous position on the board. Why we adopt this round-about way of presenting out work is not self-evident.

You might object that, "If the only things missing are stated hypotheses and predictions, then there is no big deal; we, at least the experts, can figure them out." Indeed, you can frequently ferret out what's going on in a paper but then, with effort, you can figure out what "sntncs mssng vwls mn, bt ts nt lwys sy r bvs"; "bvs" is not obviously "obvious," for example. The mere fact that readers can often decode such writing doesn't mean that writers *should* write that way. Especially if the goal is to facilitate communication. What if it isn't? What other goals are there?

Why not speak plainly? If you have a hypothesis, why not come right out and say what it is? My survey participants who didn't always state their hypothesis said that they felt it was too "obvious" to need stating or that it seemed "artificial" to state it. Socially sensitive factors such as these are undoubtedly important, but there may be others. We'll come back to this topic in later chapters, where we'll cover a number of reasons for problems in communication, together with a few suggestions for solving or at least alleviating them.

9.D Coda

This chapter reported scientists' opinions of the hypothesis and how it enters into their written work. It was a preliminary attempt to add to the discussions focused on the hypothesis which, thus far, have been remarkably data-free. Although my investigations are only pilot studies, they are consistent with the conclusions that scientists consider that the hypothesis continues to occupy a central place in scientific thinking, practice, and policy.

While there is ambivalence and disagreement among respondents, the hypothesis testing mode remains dominant in both scientific thinking and scientific practice. The survey respondents identified twice as many advantages as disadvantages with it, implying that they see hypotheses in a favorable light. They acknowledged its pervasiveness throughout their careers and rated its contributions to science highly. The respondents also saw Discovery Science as an important complement to hypothesis-based science, but I found no evidence that scientists feel that they have to choose between them; indeed, all four modes of scientific procedures that I asked about were well-represented in modern practice, although thus far Big Data has not made a major impact on the bioscientists that I queried. There was consensus that major changes in official policies, either publishing or governmental, are unwarranted.

Two important findings from my review of the neuroscience literature in top journals was that most work is hypothesis-based, but that the hypothesis is rarely explicitly stated. Instead, the hypothesis is left unstated in an implicit form that reveals itself in the organization, structure, and conclusions of the work. The fact that the hypothesis is so often left unstated partly explains the confusion about it and its roles in science.

Notes

1. https://www.surveymonkey.com/
2. I used email addresses from membership lists of the Society for Neuroscience and Federation of American Societies for Biology society, selecting email addresses associated with US institutions to try to limit complications caused by answers coming from respondents with very different scientific educational backgrounds. I did not try to determine where respondents had actually received their training or ask for more fine-grained information, such as their scientific specialty, age, or gender. I did ask about their current academic position, which ranged from "graduate student" to "laboratory supervisor" or "other." About half of the recipients did not open the introductory email—the subject line of the email invitation was "Community Hypothesis Survey"—and I don't know how the self-selection might have affected the results. As an inducement to participate in the survey, recipients' email addresses were entered into a lottery that offered a chance to win one of two $50 Amazon gift cards, which were awarded.
3. There was a 12% overall rate of return from the initial mailing; of the 2,004 recipients who opened the email, the return rate was 22%.

4. https://www.nature.com/news/1-500-scientists-lift-the-lid-on-reproducibility-1.19970.
5. Although the absolute differences are fairly small, the response patterns differ (chi-square test, $p < 0.001$), meaning that the underlying populations may not be the same. It could be that the steps taken to curb irreproducibility in the past few years have begun changing community practice and opinion, that more people are now genuinely confused about whether a problem exists or not, that the differences between the surveys represent sampling error, or that dedicated *Nature* readers—which included physical scientists, whereas mine did not—are truly more alarmed than the group that I sampled. In any case, I can't say how well the conclusions from my survey generalize to the world of science. My survey went out in two phases, and this question was included only in the second phase; it generated 210 answers.
6. Andrea Ovans, "What Is a Business Model?" *Harvard Business Review* January 23, 2015, https://hbr.org/2015/01/what-is-a-business-model.
7. The wording of the questions for the different modes was, regrettably, slightly different; nevertheless the rating categories were essentially the same, and they were the basis for calculating the proportions illustrated in the graph. I used the Excel ChiSqTst function to make the comparisons reported in the text. Given my inability to control the sample group—who responded to the survey, who responded to each question, etc.—the survey results can only be taken as suggestive, and the statistical tests were done purely for descriptive purposes.
8. Literature analysis. I examined all non-human animal studies in cellular, molecular, and behavioral neuroscience appearing in these journals during the calendar year 2015. I narrowed the focus partly so that I could be reasonably sure of understanding the papers and partly to reduce the variability caused by including scientific fields that use different conventions. For example, psychologists frequently refer to "theories" and rarely to "hypotheses" in their research papers, whereas neuroscientists do exactly the opposite; I suspect the underlying concepts are not very different, though. I did automated searches on "hypothesis or model or theory or prediction" and then examined the context of each hit in the text. When necessary, I read the abstract and enough of the paper to get the gist of the reasoning; this, for example, is how I distinguished papers built around an implicit hypothesis from those in which non–hypothesis-driven "discovery" or "open-ended questioning" seemed to guide the work. I ignored all key word hits in the references and supplemental data.

PART II
OPPOSITION AND COGNITIVE CONCERNS

10
Opponents of the Hypothesis
Stuart Firestein, David J. Glass, and David Deutsch

10.A Introduction

"I hate hypotheses," is how one well-known neuroscientist sums up his feelings about the subject of this book. In earlier chapters, I've noted that there is a debate about the role of the hypothesis in modern science and mentioned that there are people who are not in favor of the hypothesis, though I haven't gone into their arguments. It's time to go into them. In this chapter, I'll review the work of three prominent scientists who are opposed to the hypothesis. They oppose it with different degrees of intensity, for different reasons, and to promote different alternatives. They are united only in thinking that the hypothesis is no longer necessary for science. It is impossible to gain a full appreciation for the hypothesis today without evaluating their criticisms.

The critics are Stuart Firestein (*Ignorance: How It Drives Science*[1] and *Failure: Why Science Is So Successful*[2]—the quote that opens this chapter is his), David J. Glass[3] (*Experimental Design for Biologists*, 2nd ed.), and David Deutsch (*The Beginning of Infinity: Explanations that Transform the World*[4]). Their programs are called *Curiosity-Driven Science, Questioning-and-Model-Building* (QMB), and *Conjectures and Criticism*, respectively. Each shuns the hypothesis and advances another way of understanding and conducting science. Both Firestein and Glass accept the philosophical principle that *empiricism* (Chapter 1) is the ultimate standard for scientific truth. David Deutsch rejects empiricism. Deutsch is a theoretical physicist who is very sympathetic to the philosophy of Karl Popper (Chapter 2), but whose unique agenda supersedes Popper's ideas.

My plan is, first, to sketch out the key points of each program, trying to represent it in a light that its author would be comfortable with and then to analyze his reasons for spurning the hypothesis. Firestein's ideas are the most radical and the purest—he's against almost everything you learned about how to do science and what it's good for—and we'll start with them.

10.B Curiosity-Driven Science (Stuart Firestein)

Stuart Firestein puts forward the case for Curiosity-Driven Science[5] in his books and a TED talk.[6] He emphasizes that a scientist doesn't need a hypothesis to start a scientific inquiry and shouldn't wait for one to pop up. He thinks the Scientific Method is a sham. This is strong stuff. However, as an eminent scientist (he is Professor of Neuroscience and former Chair of Biological Sciences at Columbia University), Firestein knows his way around the issues of scientific thinking. For him, scientific investigations are not motivated by hypotheses, but by curiosity: an interested but puzzled individual asks, "What if . . .?" and begins doing experiments to find out. The urge to try something, to turn over a rock to see what's under it, is universal among nonscientists and scientists alike and has always been a fundamental motivator of science, although Firestein pushes the idea beyond its usual boundaries.

10.B.1 Ignorance

Ignorance is the core principle of Firestein's philosophy. He stresses that science is not a collection of well-established facts and that scientists are not people with fact-stuffed heads. Science is not about amassing information; it is about trying to shine light into dark places, to explore the unknown for the sake of exploring, with no guarantee that you'll find anything. Scientists are path-finders, discovering questions and inventing tools to answer them. The words of the two-time Nobel Prize winner Marie Curie express one of Firestein's main themes: "One never notices what has been done; one can only see what remains to be done." Well-understood things are of great interest to society and technology, but, to a basic research scientist, established knowledge is merely raw material for new investigations. What is not known—ignorance—alone is truly important, and facts only lead basic science to new areas of ignorance. From the insight that ignorance drives the curiosity that drives science, it follows that ignorance is the most important product of an investigation. "Creating ignorance" is Firestein's catchy phrase for the scientific enterprise.

Firestein likens scientific progress to the widening ripples on a pond, with their increasing circumferences representing the expansion of the known into the unknown. We can imagine that the pond is the human grasp of nature. In prehistoric times, before anything that we could call knowledge existed, the pond was the "unknown unknown"; we were unaware, not only of what was in it, but that it existed. Knowledge acquisition began when somebody dropped a pebble into the water, and knowledge increased as the rings of ripples spread. Our knowledge is in the center, within the rings. As knowledge expands, we get a glimpse of what we don't know, the ripples of the "known unknowns," and our ignorance focuses on them. The ripples form the border between the known and the unknown unknown; they are where science takes place. Ripples represent the *productive ignorance* that science seeks because it enables investigation to begin.

Still, the world is complicated and more than simple ignorance of facts is at stake. There is profound and impenetrable ignorance, "*unknowable* unknowns" (big rocks sticking up in the pond!)—Heisenberg's Uncertainty Principle in physics and Kurt Gödel's Incompleteness Theorems in mathematical logic are two examples of facts that reveal absolute limitations on our knowledge and thus redirect science to productive avenues of inquiry. Thus, as science widens the breadth of knowledge, it increases productive ignorance.

Here's an example (not Firestein's) from neuroscience of how productive ignorance spreads from a small splash of new knowledge. For thousands of years, people have used cannabis plants for all sorts of purposes, for treating pain or depression, for religious rituals, or just for getting high. Nobody had any idea of how cannabis worked or what its active agent was until two biochemists, Raphael Mechoulam and Yechiel Gaoni, isolated Δ^9-tetrahydrocannabinol (THC)[7] from cannabis plants in 1964. Before then scientists could not investigate how THC affected the body because they didn't know it existed; the problem was deep in the unknown unknown. The discovery of THC sent out ripples of productive ignorance. Because of THC's origin, it was called a "cannabinoid." One ripple involved THC's chemical properties; what was it like? The investigation turned up a surprise and a wealth of productive ignorance. THC was an oily lipid, a fatty chemical, when many people had guessed it would be water soluble, like molecules that were already known to affect nerve cells. Attempts to find out how THC affected nerve cells led to new surprises—it clamped tightly and specifically onto a kind protein molecule, a receptor—and therefore new ripples of known unknowns to explore. The receptor seemed tailor-made for THC, so it was called the "cannabinoid receptor." THC came from plants, which allowed researchers to investigate why animal brains have receptors for a chemical made by plants. And so it went, with every discovery expanding the area of knowledge into unguessed at areas of ignorance about the cannabinoid system. There were natural chemicals in the body that, because they also clamp onto cannabinoid receptors, were named "endogenous cannabinoids" (shortened to "endocannabinoids") and a growing list of things that endocannabinoids do (dull pain, control weight gain, regulate mood, and more).

Firestein's imagery is in line with that of earlier writers:

> It has been well said, that the more we enlarge the diameter or sphere of light, the more, too, do we enlarge the circumambient darkness—so that with a wider

> field of light on which to expatiate, we shall have a more extended border of unexplored territory than ever.[8]

(Or, in plain English, "As our circle of knowledge expands, so does the circumference of darkness surrounding it."[9])

10.B.2 Curiosity, the Hypothesis, and the Scientific Method

When I first read *Ignorance*, I found myself agreeing with most of what it had to say: the disorderly nature of research, its rewards and pleasures, the descriptions of life in the lab, were all familiar—until the topic of the hypothesis came up. When my colleagues and I came across an unforeseen observation, we'd would dream up a rough-and-ready hypotheses, see what it predicted, and toy with schemes about how to test it, poking holes in the schemes as they came up, and, if we hit on a plausible one, take it into the lab and try it out. The process was inelegant and informal, but we found making up hypotheses and experiments to test them to be intellectually stimulating, useful, and fun. I assumed that Firestein felt the same way. Not so.

In Firestein's view, the Curiosity-Driven and the hypothesis-based ways of doing science are unalterably opposed and as we've seen, he doesn't mince words: he hates hypotheses. Such passion is startling; he is not off-handedly dismissing a procedure that he has long since outgrown or expressing a lack of enthusiasm. "Hate" is committed antagonism.

Does Firestein perhaps have something else in mind than what others would call a hypothesis? No, his is a standard textbook definition: It is "a scientist's idea of how something works," "falsifiable," and "a statement of what is not known and a strategy for finding out." But this is the rote answer; his real opinion is that a hypothesis is "imprisoning," "biasing," and "discriminatory." Because it is *our* hypothesis, we become unreasonably attached to it—we want it to be right and defend it against all comers. He believes that if you have a hypothesis you cannot be open-minded, as you must be to follow your best instincts. Hypotheses are also "trendy" and "faddish" and attract entrenched cliques that support them.

These forces can destroy a true curiosity-driven investigation, which is nurtured by an accepting state of mind that is willing to consider any avenue of research and to welcome any outcome from it. Hypotheses filter and distort our perception of and appreciation for the data, which are *the* concrete products of an experimental investigation. Everything else is *interpretation*, which Firestein insists must be left up to the consumers of the published data, the scientific audience. The worst part of hypothesis-based science is that, if you're devoted to a hypothesis, you overlook experimental results that don't support it; you may even fudge (fake, alter, misrepresent) data that threaten it. Firestein warns that the hypothesis is not merely useless and unnecessary; it is "a real danger."

What does Firestein say about the Scientific Method itself? It is a "calamity," a rigidly confining, simplistic recipe; an empty concept; a fake process that no one actually uses. A major shortcoming is that the Method does not tell you how to generate a hypothesis, and yet it supposedly requires hypotheses! An absurd situation, for Firestein. Science is about good explanations, and searching for an explanation is divorced from any Method. Creativity is the key. We have no real idea how creativity works, yet it and passionate curiosity, not sterile reasoning, are the most important attributes for a scientist. A scientific search begins with animated personal interest, not cold logic. Ideally, a scientist is like "a guileless, intent field biologist collecting bugs in a remote ecological niche." Science is trying things out, staying as unbiased as possible, and not having a definite plan; the main thing is to be looking. None of what's most important for science has anything to do with the Scientific Method, and therefore, in Firestein's vision, science could dispense with it without losing a beat.

10.B.3 The Importance of Failure

In *Failure: Why Science Is So Successful*, Firestein again flips common wisdom on its head by insisting that failure is a more significant stimulus for scientific research than is success. Both risk-taking and failure are vital to the advancement of knowledge. We should try to venture outside the safe confines of what we know to tackle bigger experimental challenges and ask more penetrating questions than we ordinarily would.

And failures are usually more informative than successes, says Firestein. A failure of an experiment shows that you've reached the limits of your knowledge and gotten to the edge of productive ignorance. Failure breeds creativity. He calls on a classical example—it was the failure of Newton's theory of gravity that led to Einstein's dramatic reconceptualization of space and time. A scientist should always seek to push the envelope of knowledge to the point of failure. Indeed, good science is all about failure; the uncertainty of knowledge guarantees that we are continually immersed in failure and should therefore embrace it. The vast majority of all scientific theories in history, like the vast majority of species in evolution, have gone extinct; their apparent successes were transitory, the lessons of their failures lasting. We can increase our respect for failure by, for example, preferring research projects that have no better than, say, a 50-50 chance of succeeding. We need to know more about failure and should study and teach it at least as much as we study and teach success.

10.B.4 Summary

At times, grasping Firestein's message demands the mental agility and stamina of a Zen Buddhist acolyte wrestling with a koan: "Forget the answers, work on the questions"; "Trying to succeed . . . is driving science into a corner. . . . The alternative? Fail better." His discussions are full of penetrating, often biting and witty insights, and the outrageousness works: it makes you think. The importance of ignorance and failure in science is frequently overlooked, and calling attention to them is unquestionably a good thing. Wholesale rejection of the scientific hypothesis is not a necessary consequence of his argument, but Firestein acknowledges that he is not after tight, logical coherence. Curiosity-Driven Science is less a defined program than a state of mind; it is best understood in motivational terms.

Firestein has a lot more to say, but this brief overview covers his fundamental messages. While there is much to admire about his maverick stances, you might wonder if he goes a bit too far to make his case. I'll look at a few outstanding issues in the next sections and focus on those parts of Curiosity-Driven Science that touch on the hypothesis.

10.B.5 Critique: Is Curiosity Enough?

Hearing that curiosity is a major instigator of scientific investigation might come as a surprise to hard-working, tax-paying nonscientists who think that science should be about formal procedures and rigorous control experiments, not about satisfying curiosity. However, Firestein is right; curiosity and ignorance are as influential as he says they are. Good scientific ideas come from aimless mental meandering, and curiosity is a great personal trait for a scientist. Still, is curiosity enough to guide scientific thinking? What happens after the curious thought has sprung up?

While Firestein's Curiosity-Driven scientists spend time "testing their ideas," it is not clear what *testing ideas* means for them. When traditional researchers talk about testing their ideas, they are referring to their (explicit or implicit) hypotheses, their predictions, plus the experimental designs that could falsify them. Curiosity-Driven scientists want to their ideas to be correct, but, because they don't hypothesize anything, they can't use falsification as a tool to rule out the incorrect ones. What tests do they actually do, and why do they do them? Curiosity alone is not enough to sustain science, but they obviously can't look to the Scientific Method for guidance, and Firestein steadfastly declines to put forward a Method of Ignorance or a Method of Failure to take its place. This refusal is entirely in keeping with his philosophy; it may not be enough for students or others hoping to understand science.

The Curiosity-Driven program emphasizes the accumulation of observations unencumbered by interpretations and the baggage that goes with interpretations. After quoting Newton's proud, if flexible, reluctance to form a hypothesis (see Box 10.1, for a different take on Newton's position), Firestein remarks, "Just the data, please." Give us the results and we'll make up our own minds about them, he seems to say. In this he anticipates the recent push to make scientific data widely available for checking and reanalysis (see Chapter 7), which is surely a good thing. But we can't act as if data interpretation doesn't take place, even in Curiosity-Driven Science.

Firestein suggests that data morphs naturally into facts; that observations, measurements, and so on "accumulate and at some point may gel into a fact." On the contrary, nothing as free from human agency, as outerwordly, occurs. Perhaps he is alluding to the nebulous but real phenomenon of *consensus*; if enough scientists seem to accept an observation as true, then it is accepted as "fact," keeping in mind that "a fact is where the investigation rests." However it is that consensus emerges, accepted facts are nonetheless *interpretations* of data, and active human minds generate interpretations. Just because we don't understand how minds interpret data doesn't mean that they don't interpret them; cognitive science assures us that we are continually, reflexively engaged in trying to understand the world.

Stated plainly, we do not have the option of leaving evidence in a perpetually uninterpreted state. We can either try to suppress our innate urge to make sense of evidence—in effect denying our nature—or be aware of this urge and work with it. Most importantly, we mustn't forget that uncertainty guarantees that our interpretations are hypotheses, *provisional* truths that are subject to revision.

The Curiosity-Driven program officially declines to sanction summary statements, models, and interpretations along with hypotheses, which means that its inquiries leave us little beyond new questions and uninterpreted data. I say officially, because, of course, despite preferring to remain aloof from data interpretation, Firestein realizes that interpretation is necessary at some level; science, he acknowledges, does generate facts and facts can come in handy. Still, he tolerates facts as he tolerates success in discovering them: generating facts is "not a bad thing" and can even lead to accomplishments, "if by that you mean publishing papers and getting grants." In effect, discovering facts is not exactly disreputable, but you get the distinct impression that the best people don't do it.

This probably sounds a little confusing, and it is because Firestein's overarching goal is to shake up hidebound, complacent thinking; paradox serves his purposes. Though he cares little for knowledge, he knows that we can't disregard it altogether because we can only appreciate ignorance, the driver and product of scientific investigation, against a backdrop of knowledge. Knowledge is more than a guide to ignorance; it defines ignorance. Just as darkness is defined

Box 10.1 Isaac Newton and Francis Bacon: Not Arch- Foes of the Hypothesis After All

In his magnum opus, *The New Organon*[10], Bacon generally refers to hypotheses as *axioms* and distinguishes several kinds of them; how they're used is their crucial characteristic. He condemns thinking that "flies from the senses and particulars to the most general axioms, and from these principles, the truth of which it takes for settled and immovable, proceeds to judgment and to the discovery of middle axioms." It's a mistake, in other words, to leap from simple observations to grand general principles and then use the general principles to justify intermediate-level explanations. Above all, you don't start with the most general axioms and reason deductively about nature from them. That's what made Aristotle's natural philosophy "useless." "The true way" says Bacon, "derives axioms from the senses and particulars, rising by a general and unbroken ascent, so that *it arrives at the most general axioms last of all* [emphasis added]." If you develop axioms from observations and use them gradually to discover larger, more general truths, they're fine.

Bacon champions "induction" but disparages "simple enumeration"[11] (i.e., enumerative induction. True induction "must analyze nature by proper rejections and exclusions," and only after "a sufficient number of negatives, come to a conclusion." Moreover, when evaluating axioms "the negative instance is the more forcible of the two"[12]. Searching for rejections, exclusions, and negatives is the essence of Popperian falsification 300 years before Popper. Bacon wants to guide science onto a course that begins with observations, proceeds through stages of understanding[13] represented by formulating and testing intermediate hypotheses (axioms), until it achieves general principles of nature. Readers who focus only on isolated words—hypothesis, induction—and ignore their context miss his main argument.

What about Isaac Newton? Wasn't he dead-set against hypotheses? Not really. In the *Principia Mathematica*, Newton's "laws of motion" showed that a force, gravity, that decreased with distance between the sun and the earth could account for the earth's elliptical orbit. Newton had no idea what gravity might be, and he declined to guess at this point, saying, "I form no hypotheses!"

His genuine feeling toward hypotheses was by no means clear-cut, however. In his notebooks, Newton first called his laws "Hypotheses" and even years after publication of the *Principia*, he still referred to his "hypothesis" of gravity[14]. During a dispute with Robert Hooke, Newton remarks to Edmond Halley[15], who got the *Principia* published, that "there was an hypothesis of mine published in your books wherein I hinted a cause of gravity towards the earth." He adds, "I hope I shall not be urged to declare, in print, that I understood not the obvious mathematical condition of my own hypothesis." (Hooke! Back off! It was *my* hypothesis; I know what it says!)

And Newton was not finished with hypotheses; he later published a formal one. He had discovered that white light was not pure and unadulterated, and colored light complex, but the other way around. White light was what you got when you combined, using prisms, all of the colored rays into one beam. He described his experiments and his interpretation of them in *An Hypothesis explaining the Properties of Light, discoursed of in my several Papers*[16]. Since he had publicly despised hypotheses, Newton realized the title was a trifle awkward. He begins a letter to Oldenburg, the head of the Royal Society in 1695, well after the *Principia*: "Sir, I had formerly purposed never to write any hypothesis to engage me in vain disputes"; however, Newton now feels that "such an hypothesis would much illustrate the papers I promised to send you and having a little time this last week to spare, I have not scrupled to describe one" because it will make them "more intelligible," presumably, to less-elevated minds.[3] We should not, he implies, dream that he, Isaac Newton, would ordinarily stoop to a hypothesis, but he believed one would aid others in following his thoughts. So, having a little extra time on his hands, he dashed one off.

In fact, Newton had much more to do with the hypothesis. Nobel Prize-winning physicist Frank Wilczek remarks that "in [Newton's] vast notebooks one finds hypotheses galore about all sorts of things."[17] Though, often, says Wilczek, Newton used the "charming trick" of "putting a question mark at the end of statements. For then they are not assertions or Hypotheses, but only Queries." He gives as example, "Do not Bodies act upon Light at a distance, and by their action bend its Rays; and is not the bending greatest at the least distance?"

Wilczek sees this as a "suggestion for research"; in other words, as a hypothesis (and not just any hypothesis, but the first hypothesis that gravity could actually cause the path of a light ray to bend; i.e., the one that Eddington famously tested in the context of Einstein's Theory of Special Relativity more than 150 years later).

Newton, like Bacon, thought intensely about the nature of science and scientific reasoning. Despite the bad reputation that the hypothesis had acquired prior to the Enlightenment, both men gave it a central place in their thinking.

by light and makes no sense without it, so is ignorance defined by knowledge. Science is more than peering out at the universe through innocent eyes and doing experiments. Scientists do try to figure out why things are the way they are and why they work the way they do.

Because his enthusiasm for acquiring knowledge is tepid, he doesn't delve deeply into the relationship between ignorance and knowledge. The metaphor of the pond, with its ripples of productive ignorance spreading out from a central circle of knowledge, is misleading. The image subtly suggests that, once established, facts are, pretty much, facts; they remain in the calm, settled waters, and Firestein warns against lingering there, bobbing around. He is well aware that this is not an accurate image, of course, but he doesn't explore the consequences of the notion that, because facts are indistinguishable from hypotheses, they are always susceptible to testing, modification, or rejection. Science is not always about pushing on into uncharted regions of ignorance for its discoveries.

Here's another picture of how science works. We live in a low, watery, boggy place, susceptible to intermittent flooding, say the Netherlands or the city of New Orleans. We need dry, stable land for houses, schools, roads, so we put up a levee, pump out the water behind it, secure the land (now called a "polder"), and build. Then we advance into more wet land, put up another levee, pump out the water, and repeat, moving steadily outward into newly dried areas. But the levees are never perfectly secure; they need constant tending or they leak; yearly storms weaken them, and they must be repaired; periodically they must be replaced. And every once in great while a major hurricane strikes, bursting levees and inundating swaths of formerly dry ground. After the disaster, the city regroups, builds new levees, and, maybe, abandons some territory in favor of higher ground. Thomas Kuhn called big intellectual storms *scientific revolutions*,[25] and they require major reconstruction of the foundations as well. Like levees, scientific facts, no matter how solid they look, are never entirely secure. A group of them may seem very good, and we take them for granted while we move out into new areas. However, we need to monitor our facts even as we are on the lookout for novel productive ignorance because as our facts go, so goes our ignorance. We need to treat our facts as the tentative concepts that they are and always be willing to revisit and retest them.

Firestein's decision to focus on data rather than interpretations has additional consequences. Practically speaking, there is the genuine danger that, if a scientist does no more than collect and publish data, then she can be, and probably soon will be, replaced by a robot that will do the job faster and better. Curiosity will be replaced by a computer algorithm for detecting gaps in data, and programs will direct machines to collect and upload these data into colossal databases in the cloud. Human scientists will become technicians (see Chapter 15). At the moment, we hold the edge over computers in creative thinking, including data interpretation. Our lead is shrinking though, and we ought to work hard to maintain what we have.

Human science is communal, not solitary. Gathering data for personal satisfaction is not enough for us; we need to spread our information around.

If the scientific community and society at large are going to benefit from an individual's investigations (and why else pay for them?) that information must be available to all. Your fellow scientists want context and interpretation; while they want the numbers, the data, they also want to know what you think they mean. Hypotheses (and their alter egos, theories and models) are excellent vehicles for summarizing and communicating scientific thoughts and reasoning.

Finally, I believe there is a serious ethical issue at stake. Scientists have benefited from years of education and advanced training and have acquired a great deal of unique knowledge and mental skills. They understand the significance and limitations of data; they can think in novel and fruitful ways about problems. Their intellectual input adds value to their numbers. Thinking is therefore an integral part of scientists' jobs, and sharing the benefits of their thinking is an ethical responsibility. Scientists who decline to contribute to communal data interpretation are shirking their responsibility.

10.B.6 Similarities and Differences Between Hypothesis Testing and Curiosity-Driven Science

Is curiosity freer, more authentic, less artificially constrained than hypothetical thinking? This seems dubious. We don't know how the mind concocts a hypothesis any more than we know how it comes up with a curious thought. The enigmatic motivation to formulate a hypothesis resembles the ping of wonderment that signals curiosity: both are responses to mysteries. This reality creates a subtle problem for Firestein who rejects the Scientific Method partly because it doesn't tell us how form a hypothesis; that despite the orderly connotation of a Method, "you just sort of passively make one up, sometimes out of thin air." In contrast, he tells us that curiosity springs from "night science" our "intuitive, inspirational" side. Now of course night science does not take place only at night, but its name evokes the imaginative, romantic side of science. Good. The awkward truth for Firestein is that the answer to the question, "where do hypotheses come from?" is the same "night science."

Hypothesis-based and the Curiosity-Driven modes of doing science *are* different—curiosity is an emotion; a hypothesis is a plan—but they are not in conflict. Driven by curiosity, a scientist may remain engaged with a problem, or she may go on to ponder about something else entirely. Curiosity does not offer guidelines about how to carry on, what questions to pursue, how to pursue them, or when to let an investigation rest; it does not summarize, interpret, account for observations, or suggest the next step. This is where a hypothesis comes in; it helps to make plans, draw conclusions, and organize and communicate thoughts.

Firestein says that real scientists that he knows don't follow the Scientific Method or use hypotheses. But other scientists do (Chapter 9), and sometimes quite explicitly. Here's an example from neuroscience. In the early 1900s, neuroscientists were investigating how brain cells communicated with each other at the synapses where they come close together. There were two camps: one camp (the "spark boys") thought that electrical signaling in one cell directly caused electrical changes to occur in another, and a second camp (the "soup boys") held that communication between neurons required a chemical messenger—a *neurotransmitter*—to carry the signals by diffusing from one cell to the other. Prior to the development of the electron microscope or the identification of chemical neurotransmitters in the brain, both hypotheses did about equally well in explaining the existing data. The soup boys based their arguments on the well-established finding that chemical transmission took place at synapses in the heart, and, hence, a parsimonious hypothesis was that all synapses operated on the same principles. The spark boys countered that the brain was not the heart and, besides, the speed of communication between nerve cells (~5/10000th of a second) was too fast for transmission by chemical diffusion.

Eventually, the spark boys, led by John Eccles, devised methods for measuring the electrical changes taking place inside a nerve cell during synaptic transmission to test their hypothesis. They had predicted that the intracellular electrical change would be in the "plus" direction and were surprised to find that it was actually in the "minus" direction. This observation was completely incompatible with their electrical hypothesis and could only be explained by the chemical messenger hypothesis. The authors concluded that "[their former] hypothesis is thereby falsified."[19] Eccles instantly switched sides, became an outspoken proponent of the soup theory, and eventually won a Nobel Prize (and a knighthood) for his efforts. The story shows not only that scientists truly do use hypotheses and hypothesis testing, but that there are benefits to changing your mind when facts fly in the face of your opinions.

Firestein's instincts about scientists and the hypothesis are certainly correct in at least one respect, however. Scientists nowadays, at least the neuroscientists whose papers I reviewed in Chapter 9, generally avoid saying that they have "falsified" anything. I did not find one occurrence of the term, even in those studies that tested hypotheses explicitly or implicitly, used models, or made predictions. Researchers seem to have a greater aversion to rejecting a hypothesis than to having one, a state of affairs that probably contributes to *publication bias* (Chapter 11). In any case, we should keep in mind that the apparent prejudice against "rejecting" hypotheses is a convention, not an indictment of the hypothesis.

10.B.7 Calling It "Curiosity"

"Curiosity-Driven Science" isn't precisely defined, which makes sense because curiosity itself isn't precisely defined. For Firestein, a scientist's curiosity is not an idle, casual thing; its essential element is passion, and a scientist has a burning desire to satisfy it. Curiosity is a drive to explore or to know without necessarily having a fixed objective. At the same time, calling it "curiosity" spotlights the human side of science and reinforces the perception that scientists are people with feelings like everyone else. Science done by regular, curious people seems less forbidding than science done by super-rational clones of *Star Trek*'s Mr. Spock. This emphasis has the positive side effect of demystifying science and making it more approachable.

On the other hand, overdramatizing its emotional allure can misrepresent science. For example, one commentator describes the multibillion dollar Large Hadron Collider project in physics (and subject of the documentary film *Particle Fever*) as "Curiosity-Driven research."[20] A major objective of the collider project was to search for the Higgs boson, a hypothetical entity that the Standard Model of physics says is key to the mechanism that gives subatomic particles their mass.[21] While individual physicists were definitely curious about its outcome, the project itself was as rigorously hypothesis-bound as anything you could imagine. Curiosity alone did not prompt thousands of physicists to spend decades collaborating in building, testing, and operating that immense machine. All the scientists involved wanted to know if the highly quantitative theoretical predictions were right or if the theory would be falsified and they would get a glimpse of a world of new physics. Calling the Large Hadron Collider project "Curiosity-Driven" obscures its theory-bound significance; outside of the context of the Standard Model, evidence for or against the Higgs boson might have been merely one more naked observation.

10.B.8 Curiosity Satisfying, Hypothesis Testing, or Both

Life in a Curiosity-Driven laboratory as portrayed by Firestein is engaging, lively, and deliberately unstructured. The scientists spend lots of time "screwing around with things to see what will happen" and bouncing ideas off each other. I think that all labs are like this at times, and Firestein's books accurately capture the atmosphere, although his hostility toward the hypothesis evidently blinds him to the complementary functions that Curiosity-Driven and hypothesis testing research serve; scientists slip into and out of both modes all the time. An experiment will produce unexpected results that spark curiosity that leads to a hypothesis and an experiment to test it and more surprises, etc.

Any scientist can come up with dozens of examples of the blurred lines. Here is one of mine. While recording electrical responses from a nerve cell to test a rather boring hypothesis, a colleague and I noticed some small electrical signals coming from a nearby cell; technically they are called *inhibitory postsynaptic currents* (IPSCs). Remember that these cells are microscopic in size, and synapses are submicroscopic, so we couldn't tell immediately what was going on. The diagram in Figure 10.1 shows the set up; we were recording the activity of the P (for pyramidal) cell, while the small signals were coming from synapses made by the I (for inhibitory) cell. We had studied IPSCs in the past so they were familiar and, since our experiment at the moment was not focused on IPSCs,

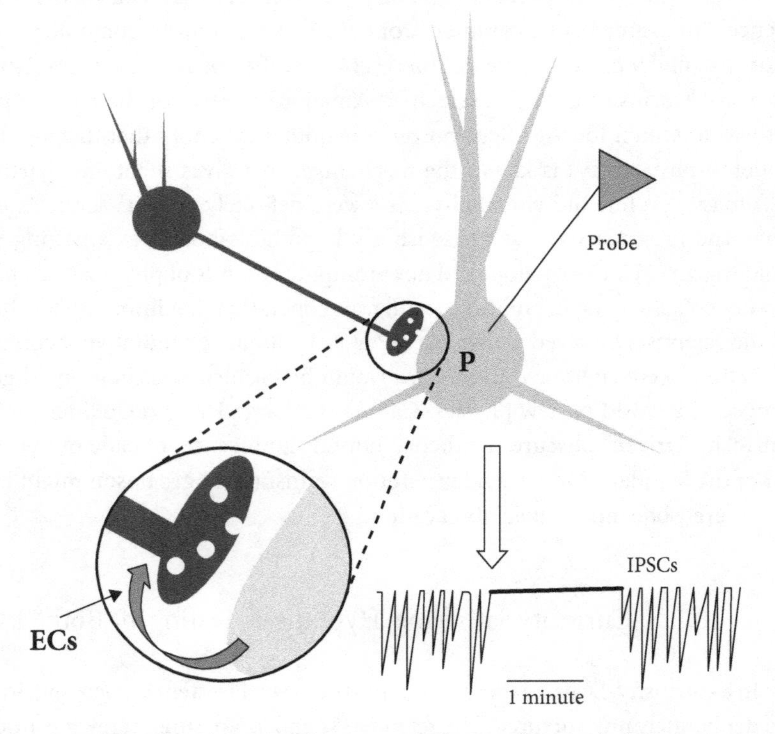

Figure 10.1 Schematic drawing of experimental set up (see text). A probe inserted into a pyramidal cell (P) recorded inhibitory postsynaptic currents (IPSCs; lower right) that originated from the inhibitory cell (I) and occurred spontaneously and continuously until the P cell was activated (*open arrow*). For about 1 minute after activation of the P cell the IPSCs disappeared and spontaneously reappeared. In the blow-up of the synaptic region (lower left) the gray arrow indicates that endocannabinoids (ECs) were released from the P cell when it was activated. The ECs' effect, which only lasted for a short time, caused the transient IPSC disappearance by preventing their release from the synaptic terminal of the I cell.

we ignored them; they were a bit like gentle static coming through on a radio. Then we got a surprise: when we activated the P cell, the IPSCs disappeared and reappeared a minute or so later. And every time we activated the P cell, they vanished and then came back. Now, ordinarily, the IPSCs are monotonously predictable, and their sudden capriciousness was new. We mentally filed it away, and, after we finished the boring project, we thought about the odd behavior of the IPSCs again.

Our first hypothesis was that the P-cell properties temporarily changed and prevented us from detecting the IPSCs, even though they were still there. Yet we couldn't find any evidence for such changes. The P-cell hypothesis was obviously false, and we were even more curious. We then hypothesized that something interfered with the IPSCs after they had been sent from the I cell to the P cell. This hypothesis, too, was wrong. In desperation, we hypothesized that activating the P cell itself prevented the IPSCs from being sent out from the I cell. It was a crazy idea; let me explain why. The P cell and the I cell are completely independent cells, and there was no physical connection between them. Thinking that activating the P cell could prevent the I cell from sending IPSCs would be like thinking that you could stop radio static by yelling at the radio. Strange though it was, we tested this hypothesis in many ways over the next several years and could never disprove it. It turned out that whenever we activated a P cell, it sent a new kind of "do not leave!" message to the I cell that kept IPSCs from leaving. Equally remarkably, as others eventually discovered,[22] the new message going from the P cell was carried by the marijuana-like endocannabinoids in your brain that I mentioned a while back. When, driven at first by curiosity, we started our investigation, nobody knew that backward signaling with endocannabinoids could happen. Chance, curiosity, and hypothesis testing all were necessary. You can be a Curiosity-Driven hypothesis tester, or vice versa, going from one to the other and back. Nobody checks up on you, nobody cares.

10.B.9 Failure and the Hypothesis

Besides exalting *ignorance*, Firestein also promotes *failure* as a crucial ingredient for science. He quotes the physicist Enrico Fermi, "If your experiments succeed in proving the hypothesis, you have made a measurement; if they fail to prove the hypothesis, you have made a discovery." While Firestein's pronouncements are deliberately jarring, it is hard to argue with his main point, and yet something seems wrong. To "fail" means not to succeed, and to think about failure, you must know about success, in the same way that you have to know about "knowledge" to understand "ignorance." We can't even define failure without knowing what success is, so what counts as success for Firestein? Surprisingly, at this point, he balks. He won't go there. Not only does he not explain or explore "success," but he inverts our intuitive understanding of it in his discussion of "failure."

In a characteristic quip, Firestein tells us that Albert Michelson (of Michelson-Morely fame) won a Nobel Prize "for an experiment that didn't work." This seems odd; why would anyone get a Prize for that, we wonder? A bit of background: in the late 1880s, physicists thought that light was a wave in a mysterious stuff called the "luminiferous ether" ("ether") that permeated all of space. The hypothesis was that ether was necessary for the propagation of light waves somewhat like water is necessary for the propagation of water waves.

Although there was no direct evidence for ether's existence, it was a reasonable hypothesis that led a physicist, Augustin-Jean Fresnel, to predict that, if the earth was constantly plowing through ether-filled space as it orbited the sun, the ether would exert a "drag" on a light beam generated on earth. If you stick your hand out of the window of a moving car on a windless day you can feel the drag of the otherwise stationary air because the car is traveling through it. Analogous drag caused by ether should slow the speed of a light beam as it traveled from one place on earth to another, depending on how the beam was oriented with respect to the direction of the earth's movement.

The light beam should be slowed more if it was heading against the direction in which the earth was heading than when it was traveling at right angles to the earth's path. To change the metaphor, imagine you were swimming in a river, either going some way upstream or the same distance across it. It would take you longer to go upstream than across, and the same is true for the light. Remarkably, despite making unprecedentedly precise and sensitive measurements, Michelson and Morley observed no difference in the speed of light regardless of its direction of travel. Firestein classifies the experiment as a "failure" because the result predicted by the ether hypothesis was not obtained. And this, I think, is where he goes overboard.

Let's look again at what Michelson and Morley did. They rigorously tested a specific prediction of a dominant hypothesis and discovered that it was false. They found no evidence for ether, although light was undeniably propagated through space. Since its critical prediction was false, the hypothesis was rejected: light propagation occurs in the absence of ether. (Naturally, no one test is decisive, but numerous replications of the experiment have come to the same conclusion.) In falsifying a critical hypothesis, the experiment became one of the first major pieces of evidence for Einstein's theory of Special Relativity. (And, as a side note: Michelson got his Nobel Prize for inventing the interferometer, the device that he and Morely used, which is applied for making highly precise measurements in many branches of physics. His Nobel Prize Award citation[23] does not mention the famous experiment.)

Michelson and Morley would have *failed* if their experiment had not tested the prediction that it set out to test. The significance of the experiment was that it paved the way for a better explanation. The Michelson-Morely experiment was a success under any ordinary definition of "success," given any ordinary definition of "science". It is a red flag that Firestein's dictionary is not the same as yours. You have to beware when decoding his messages.

10.B.10 Fishing Expeditions

Is Curiosity-Driven Science as unstructured as it seems to be? Sometimes Curiosity-Driven research is criticized as being a "fishing expedition," and, in the arcane lingo of scientific criticism, this is always a pejorative. No real scientist, it is thought, wants to be caught dead on one of those, lackadaisically drifting along with the currents, free from the cares afflicting serious, no-nonsense types busily grinding away on their hypotheses. Advocates of Curiosity-Driven Science vigorously reject this criticism: philistines may not understand or appreciate fishing, but that's their problem; there is nothing wrong with fishing. Here again, Firestein is on target: if you don't know exactly what you'll find or where you'll find it, then fishing is what you do; it is the epitome of an open-minded unscripted search. And if you can't know in advance what shape the information will take, you can't very well have an explicit experimental hypothesis about it. There is no serious doubt that scientific fishing expeditions count as real science, even if convincing a grant review panel to give you money for them is hard to do.

We shouldn't forget though, that fishing is not an entirely random activity, and Curiosity-Driven searches are not utterly unguided. As Firestein points out, there are tricks to fishing successfully: you need to know "where to fish, and what's likely to be tasty and what not." Indeed, you need to know more; if you want to catch fish, you have to know what gear to use and how go about fishing. To catch the tasty fish, you need to know what bait or lures they will go for. Background information, assumptions, lore, and so on come under the heading of *implicit hypotheses* (Chapter 2). Moreover, if fisher-folk take their gear to a likely spot and don't catch any fish, they might try again, vary their equipment, etc.; however, at some point, they'll give up and not go there again. Their implicit hypothesis about fishing will have been falsified.

A final thought: Firestein opens his book, *Ignorance*, with an old proverb that symbolizes the state that scientists are frequently in, "It is very difficult to find a black cat in a dark room. . . . Especially when there is no cat." This is "the best description of science" that he knows. The object of your search is often elusive, and worse, you might be wrong about its very existence. But let's think about it: if for some reason you really did have to find a black cat in a dark room, a random search would never do—cats are quick and clever and quiet; you'd better have a plan. You'll want to know a lot about cats, the equipment you might need to catch one, and a scheme for how to go about it. You'll need a hypothesis. Especially if there is no cat.

I'll save a few more general comments on Curiosity-Driven Science until we've considered an alternative approach that, while having similarities to the philosophy of Curiosity-Driven Science, differs from it in striking ways. The name says it all: Questioning and Model-Building. In contrast to the free-form nature of Curiosity-Driven Science, QMB is a tightly organized.

10.C Questioning and Model-Building (David J. Glass)[24]

QMB, the method championed by David Glass, also rejects the hypothesis. In QMB, a question comes first. Investigations progress by asking and answering more questions. Scientists reason inductively from experimental results and create *models* that explain the results. Generally speaking, a scientific model is a data-based idealization that takes various forms—diagrams, such as Figure 10.1; physical structures, such as the ball-and-stick molecular models of high school chemistry labs; and narrative explanations, flow charts, or, if sufficient quantitative information is available, systems of mathematical equations. Many biological models are qualitative descriptions, accompanied by more or less elaborate schematic diagrams of presumed underlying cellular or molecular mechanisms that scientists use as explanatory devices.

While you might think that, with its emphasis on questions, QMB is very similar to Curiosity-Driven Science, nothing could be further from the truth. Let's look at Glass's definition of science to see why. Glass posits that "scientific research is the process of determining some property *Y* about some thing *X*, to degree of accuracy sufficient for another person to confirm the property *Y*."[24] Note the requirement for accurate and reproducible measurement. For Glass, the world is sufficiently regular that inductive reasoning leads to models that we can count on to predict the future. Explanation is defined as successful prediction and "verification of [the model] is the explicit goal of the experiment."

Science, in other words, is primarily an observational and descriptive activity; later efforts may determine cause-and-effect relationships, but the core value is accurate description. For Glass, scientists do experiments because they need to acquire data to answer one or more of seven types of questions: Does *X* exist? What does *X* look like? What is *X* made of? What does *X* do? What causes *X* to do *Y*? By what mechanism does *X* do *Y*? And, finally, what can perturb *X* to perform its function differently? In comparison with the flexible, unstructured

Curiosity-Driven program, QMB is sharply defined and heavily prescriptive. There are other differences.

The *question* is the logical starting point for QMB science because scientists are in a state of ignorance. A crucial difference between QMB and Curiosity-Driven Science is the end product of an investigation which, for the Curiosity-Driven scientist, is the generation of new data and new ignorance and for the QMB program is another question or a model. For QMB, whichever has priority—question or model—depends on the circumstances. If your data are insufficient to support a model, pose a question; if you have enough data, construct a model. Once you have a model, you gather the "negatives and affirmatives," the contradicting and supporting observations needed to verify or refine it.

Glass places high value on description, so he puts inductive reasoning—the process of extrapolating and generalizing from past experience—at the center of QMB. He believes that models arise from induction and, although he concedes that inductive reasoning does not allow for air-tight conclusions, he argues that the virtues of inductive reasoning outweigh its drawbacks. Moreover, he sees inductive reasoning as unavoidable: we necessarily rely on it to decide whether one model is a better predictor of the future than another. And, if things keep happening the way a model predicts they should, the model gains "proof of verification." This process depends on both induction and probability: when you've done a sufficient number of experiments, your data will reliably predict future outcomes. Glass admits that reliance on probability doesn't solve Hume's problem of induction (Chapter 1) but says that a probable outcome "shifts the burden to the critic." He feels that there are "built-in preferences for particular kinds of models" and that a "workable" system should be satisfactory for science, even if it is not provable.

In QMB, the concepts of model and hypothesis are poles apart: (1) a model is based on data gathered during an investigation, rather than on the "prior unproven assumptions" on which a hypothesis, in his view, must be based; (2) unlike the hypothesis, which is "established to be falsified," the model must be held up for *verification* via inductive reasoning; and finally, (3) an unsuccessful model need not be scrapped, the way a falsified hypothesis must be, but can be used as a "starting point for suitably refined successors."

Ideally, a scientist is an open-minded questioner. Glass gives an example: If you want to know the color of the sky, you ask, "What is the color of the sky?" This is much better than being forced to adopt a narrow, falsifiable hypothesis. If you collect data objectively to answer a question and construct a predictive model to account for it, rather than to test a hypothesis, then you won't be biased in favor of the model. Glass suggests that scientists operating inductively to build a model "might be insulated" from an impulse to defend an unproven premise. Most importantly, once constructed, a model that makes additional correct predictions gains *inductive power*, and the greater the predictive ability of a model, the more inductive power it has. This approach is practical because "sufficient experience is reproducibly predictive of the future—within certain probabilities."

To illustrate the importance of induction in biomedically relevant reasoning, Glass has us envision the ethical morass that we'd be in if we abandoned inductive reasoning. Clinical trials are used to determine how much inductive power a particular medical treatment has. We must assume that the inductive conclusions from the study are valid if we're going to offer the clinical treatment to the public in good conscience. Glass acknowledges that the inductive power of conclusions drawn in biological sciences might fall short of the "absolute predictability" associated with physical laws but asks where we'd be if past experience was irrelevant to evaluating a treatment or a model. He considers whether falsification or verification should weigh more heavily in the evaluation and concludes that it is verification, which is provided by the inductive method and measured by inductive power, that is key.

10.C.1 Rejection of the Hypothesis

Glass stipulates that, for him, a hypothesis is not derived from data, that it is equivalent to the "fictitious and unrealistic" construct that was rejected by Francis Bacon and Isaac Newton. Glass cites Newton's definition of a hypothesis as "whatever is not derived from phenomena, an unproven premise, advanced without evidence, as a tentative explanation." (See Box 10.1 for a discussion Newton's and Bacon's views.) Glass agrees and argues that a corollary follows: if an idea is well supported by data then it is, *by definition*, a model, not a hypothesis. He dismisses hypothesis-based science: "The notion that a scientist need not actually do an experiment to derive the answer is . . . quite seductive." (This peculiar remark follows from Glass's idiosyncratic definition of a hypothesis "in the strictest sense" as having no empirical basis.)

For Glass, hypotheses are aligned with "ideologies" and are dangerous because they appeal to vanity. Furthermore, a scientist working with a hypothesis will "filter data through the lens of that hypothesis, rejecting contradicting evidence in favor of validating evidence." In his favorite example, Glass suggests that an unenlightened scientist seeking to know the color of the sky would formulate and test a narrow hypothesis such as "The sky is red." To do so, the scientist would have to construct a special filtering device, a "redometer," to distinguish between red and non-red colors and, as a consequence, would miss making any non-red observations. In another example, Glass considers a scientist testing the hypothesis that "caffeine increases blood pressure." This person, he says, will try to set up conditions in which caffeine does increase blood pressure, rather than simply asking, "what is the effect of caffeine on blood pressure?" The very statement of the hypothesis compels the scientist to look for an increase, as opposed to a decrease, and establishes a filter that "forces methodology to determine an increase in particular."

Additionally, Glass argues that a hypothesis sets up a "dysfunctional positive/negative binary" that dictates that a scientist's experimental objective is negation, whereas no one can doubt that hypothesis verification, not falsification, is the real goal of science. He wonders why anyone would spend hundreds of millions of dollars if their purpose were to disprove the hypothesis?

And insistence on a preexisting hypothesis stifles innovation. Glass believes that the National Institutes of Health (NIH) grant review system is failing because of a misguided reliance on hypotheses. For instance, large systems biology projects that aim to characterize organisms comprehensively cannot be usefully framed by a hypothesis. Moreover, the fact that NIH distinguishes between "hypothesis-generating" experiments and hypothesis-testing ones proves that hypotheses are not required by big science. If a hypothesis is not required to sequence a genome, why then, he asks, should one be required to study a particular gene?

Finally, in case the reader might suppose that Karl Popper's proposal to seek potentially falsifying evidence to test the validity of an idea would solve some the problems that Glass identifies, he waves away Popper's program as "too problematic," inconsistent on philosophical grounds, and harmful to the process of discovery. And there is the problem of inconclusive falsification: if a hypothesis cannot be unambiguously falsified, then nothing has been gained. Glass concludes that hypotheses are "unhelpful, disadvantageous, inconsistent, and unworkable"; it is better for science to ask good questions.

10.C.2 Summary

Unlike the Curiosity-Driven approach, QMB enumerates detailed directions about how scientists should carry out investigations and stresses the value of precise and reproducible measurements. QMB is philosophically compact: there are a few core principles and specific questions that guide its inquiries. Glass's examples of scientific projects focus on detecting experimental regularities and deriving general principles to make models. He considers the process of model refinement as the reshaping of existing conceptual structures rather than aggressive attempt to probe their weaknesses or overthrow them. Glass sees science as a smooth, gradual, and increasingly accurate approach to truth; a vision that has as its goals practical measurement and workable solutions to existing problems, such as you'd find in applied science or industry.

10.C.3 Critique: QMB, Open-Ended Questioning

The model is at the heart of Glass's program, and his dismissal of the hypothesis in favor of the model rests on questionable assumptions about both concepts and their relationship. There is a lot to disentangle, so let's start with his characterization of the hypothesis. Glass posits that in "its strictest definition," a hypothesis is a data-free construct that "is held up for falsification." The first part of the description is a caricature of the hypothesis as it was 400 years ago, and the notion that a modern scientific hypothesis is entirely divorced from data is, frankly, silly. He offers no evidence that any scientist today considers it valid and admits that "in actual practice" data are often used to justify a hypothesis. The definition of a hypothesis as being an empirically empty proposition is a straw man that enables him to perceive distinctions between hypotheses and models. It doesn't make for a convincing argument.

What about the second part of Glass's characterization of the hypothesis, that it is "held up for falsification"? As I discussed in Chapter 2, this kind of comment stems from a misreading or misunderstanding of Popper. To review briefly, you conjecture a hypothesis as a *true* explanation for some phenomenon that you don't understand. You do not "hold it up for falsification" as if you're trying to prove that it is false; instead, you try to find out whether or not it is false. If your tests don't falsify it, then you continue to act as if it's true.

Look at Glass's test statement, "The sky is red." As it stands, it's not a hypothesis, and that's part of the problem. Let's recycle it as a hypothesis: you are in a room late in the day, and you notice that light on the wall opposite the window has a warm, orange-red cast to it. You conjecture the hypothesis that "The sky is red," to account for the color. Alternative hypotheses might include someone shining a light into your window or a fiery blaze in a nearby building. Your hypothesis predicts that, if you look out the window, you'll see that the sky is red and also that you won't see a light directed at your building or a fire next door. You look out the window to test your hypothesis. If the sky is indeed red, then you have learned a good deal more than that your hypothesis is "not falsified"—which is all that Glass will allow you. You have corroborated your hypothesis; that is, you've confirmed its prediction and falsified the predictions of alternative hypotheses.

Corroboration, once again, is not confirmation of your hypothesis about the sky, which could still be wrong; a neighbor might have been shining a red light into your window even if the sky was also red, but you do feel more confident in your explanation of the color on your wall.

The process of testing "The sky is red," when it is a hypothesis that is based on data is not ridiculous or counterintuitive; it's the kind of thing we do all the time. Answering an open-ended question, such as "What color is the sky?" is a different matter because it does not suggest, or call for, a hypothesis; it is a matter for Discovery Science (Chapter 4). Keeping the distinctions between them in mind helps avoid confusion.

Unexpectedly, given the decidedly anti-hypothesis tenor of his book, at one point Glass suddenly acknowledges that hypotheses *can* be useful when you have adequate "prior knowledge of the system," and he distinguishes between times when hypotheses are "practical" and when they are not. For instance, a hypothesis such as "gene 278 is a receptor tyrosine kinase" is "perfectly testable" and is within "the critical rationalist framework." Now, he says, we are "back in the realm of useful hypotheses" (a surprise since we hadn't known that such realm existed for him). While in the realm, Glass sees hypotheses as scientists ordinarily do: as explanations whose purpose is to explain actual data. What changed, you wonder; can we arrange it so that we don't leave the realm again? Rather than explain, Glass, in a final wrenching twist of the argument, asserts that this association between the hypothesis and data is, nevertheless, a "problem for the critical rationalist framework" and segues back to his original position that, "strictly speaking," hypotheses are not based on prior knowledge! It remains unclear why he fixates on a definition of hypothesis that no one has taken seriously for centuries.

In Chapter 2, I discussed the background information that scientists, including QMB scientists, have to rely on as a system of deep implicit hypotheses about nature. I'll return to this problem in Section 10.C. Before that, I want to explore the distinction between models and hypotheses that is so crucial for QMB.

10.C.4 Models and Hypotheses: What's the Difference?

In QMB the model summarizes the results of an investigation, reveals relationships among experimental variables, and, at least implicitly, makes predictions about how it can be tested. I don't know how widely accepted his position is. Stuart Firestein, for example, states that[17] "A scientific model is more or less synonymous with a *theory* or a *hypothesis*." Glass firmly rejects this conclusion, however, Firestein's comment does raise the question of whether there are substantive distinctions between a hypothesis and a model. Glass proposes that models are *summary* statements (or diagrams, etc.) that are created to account for data presented in a scientific paper and therefore are necessarily found at its end, rather than at the beginning where, he implies, the hypothesis must appear.

Is it mandatory that a model be introduced only at the end of an investigation? Working scientists do not seem to think so. If a model is at all significant, somebody will challenge it and focus on it at the beginning of a new investigation. Indeed, in support of this surmise, I found (Chapter 9) that in 20 of 158 scientific reports the investigators introduced their paper by stating that their goal was to test a model; for comparison, 41 of 158 explicitly said that they were testing a hypothesis. Hence, the evidence supports Firestein's intuition that scientists don't recognize a significant difference between models and hypotheses.

Why does any of this matter? One reason is that, in claiming that there is a conceptual difference between hypotheses and models, Glass's interpretation obscures the iterative, recursive nature of science. As Chapter 2 emphasizes, hypotheses are developed, tested, and, in response to experimental outcomes, modified ("recycled"), retested, etc. The model resulting from one investigation is the raw material that gets fed into the investigational hopper in the next cycle. It becomes the new hypothesis.

Glass's distinction between models and hypotheses includes the notion that a model is not subject to falsification. A major benefit of a model, allegedly, is its ability to make good predictions; when their predictions are inaccurate, models need "refinement," or they can be "improved," or "suitably revised." Not falsified, though; only hypotheses are falsified in his scheme. Let's pause to assess this extraordinary position. For one thing, it easily accommodates the reasoning that sustained the Ptolemaic model of the heavens, which could always be "suitably revised" but not falsified. When their model made inaccurate predictions, the Ptolemaic astronomers refined it by adding more epicycles until the latest unpredicted wiggles of planetary motion were accounted for. Its inexhaustible flexibility made the Ptolemaic scheme unfalsifiable, and it has served as an archetype of bad or, to be fair, premodern, science. QMB holds the door open for such an approach.

But, Ptolemy to the side, I see no major distinction between a model that needs refinement and a hypothesis that has been falsified. You'd only need to refine a model if it failed in some way, and, when you correct its deficiencies, you replace it with a new version and get rid of the old one. Although "refinement" and "improvement" may sound kinder and less judgmental than "rejection," trying to draw a meaningful distinction between "refinement" and "rejection" seems unlikely to get us anywhere. We could make a similar argument about "revising" and "rejecting" a hypothesis. I am reminded of the story of the axe that had been in one man's family for more than 200 years: the axe head had been replaced twice and the handle three times. He treasured it as *the same axe* that his great-great-great grandfather had used. However, hypotheses are not like axes; when you change part of a hypothesis, it becomes a new hypothesis. It's the same with models: a model that has been replaced by another is a model that has been falsified, even if you call it something else. What this means is that the results of testing don't distinguish between models and hypotheses.

QMB suggests that models can be verified, whereas hypotheses can only be falsified. This issue is not really about the properties of models and hypotheses per se; it gets back to the well-worn philosophical dispute about verification versus falsification (Chapter 1), which we don't need to rehash. Scientific models cannot be conclusively verified, and QMB does not claim that they can be; according to QMB, verification is always incomplete and conditional. This sensible position leads immediately to a conundrum: Glass rejects Popper's hypothesis testing program in part because falsification cannot be complete. We might ask, if partial or incomplete falsification is a big problem for Popper, why isn't incomplete verification a big problem for QMB? Or, if inconclusive verification is acceptable, why isn't inconclusive falsification acceptable? Again, it seems that what was intended to be a critical distinction between models and hypotheses doesn't stand up to close inspection.

Both models and hypotheses are explanatory interpretations that go beyond the data, involve simplifications and generalizations, and are the products of minds that seek to understand. The model advocated by Glass cannot escape whatever functional criticisms the hypothesis receives. Although hypotheses are singled out as sources of bias, QMB gives us no reason think that builders of models do not take the same pride in their creations or become as emotionally attached to them as do builders of hypotheses.

Once we put aside the antiquated definition of a hypothesis as a data-free fictional construct, we find no salient difference between models and hypotheses. We're left with a purely semantic one: if a simplified or idealized statement that summarizes and explains data and makes predictions appears at the end of a paper, it is a "model" by definition; if such a statement appears at the beginning of the paper, it is a "hypothesis," also by definition. While you might argue that it would facilitate scientific communication if there were some fixed, agreed-on difference in meaning between model and hypothesis (I think this viewpoint has merit), I am afraid you'd face an uphill battle to get the scientific community to go along. As Firestein observes, scientists already use the terms interchangeably.

The model is one of the pillars of QMB, and another one is the principle of *inductive power*, which we'll turn to next.

10.C.5 Inductive Power

As I've noted repeatedly, philosophers have not been able to reach consensus about the utility of inductive reasoning (induction) to the quest for truth in science, and many of them concluded that it is of no value whatsoever. Induction continues to cause confusion, though, perhaps partly because the word has different meanings. In one sense, induction is the name of a still-unexplained cognitive process that takes us from the experience of particular instances to the formation of general rules (later, we'll return to the question of whether this is a genuine "process"); I'll call this "cognitive induction." "Induction" is the also name of an abstract philosophical method that is purported to enhance the truth value (or probable truth value) of scientific statements; this is "philosophical induction." Problems can arise if we get the two ideas mixed up, and induction appears in both roles in QMB. There is no controversy about whether or not our brains continuously detect, report, and act on anticipated regularities in the world. We are always engaged in cognitive induction.

What about philosophical induction? Glass reviews David Hume's criticism that the validity of induction rests on the unproven assumption that Nature is uniform. He does not consider it a kiss of death. Instead, while agreeing that induction cannot guarantee future results, he feels that a model's ability to make successful predictions confers "inductive power" on it and that an accumulation of inductive power justifies confidence in the model's future performance. What exactly is inductive power, though?

In essence, inductive power is a subjective estimate of the odds that a model accurately represents the key elements of the data. We can imagine inductive power as an indication of how much to risk on particular outcomes—the best place to build a new road or the amount of life insurance to buy. Saying that a model has a lot of inductive power is equivalent to saying that somebody thinks it is right. This is not meant to trivialize the principle: there may be good reasons to think that a given model is right; experts may have weighed in with opinions based on high-quality evidence and rigorous statistics. Bayesian statistics (Chapter 6), for example, are well-suited to generating quantitative predictions that reflect the strength of hypotheses or models. Ultimately though, inductive power is still a temporary, probabilistic judgment that cannot satisfy the aspiration of science to seek the Truth about nature.

And, no matter how great its value, inductive power confers no protection against the unprecedented event, the rare or unpredictable anomaly, a "black swan"[18] in the sense used by Nassim Taleb, an event that is totally unaccounted for by current models. If Taleb is right, history and science are more decisively determined by black swan events than by the regular, predictable occurrences that support inductively generated models, no matter how much inductive power they have.

10.D Curiosity-Driven Science, QMB, and the Hypothesis

Despite numerous differences, there are also similarities between Curiosity-Driven Science and QMB, especially their attitudes toward the hypothesis. Both Curiosity-Driven and QMB programs mischaracterize it at times, and both admire Newton's famous put-down of it. Moreover, both Firestein and Glass worry about scientists' fecklessness; a lack of spine that leaves scientists "in the grip of a hypothesis" or "forced" by their hypotheses to behave badly; for them, scientists are gullible patsies, intellectually timid and readily swayed by other people's hypotheses. It is a sorry picture of their colleagues, but is there any truth to it? Scientists are social animals and the forging of scientific consensus is a genuine, yet barely understood factor in determining what science "knows."[26] Still, neither Firestein nor Glass acknowledge the fact that science is very competitive or that scientists can be a contentious, contrarian lot who rarely take anyone's word for it, especially when their reputations and egos are on the line. No matter what climate change deniers claim, scientists are not simpletons marching to the beat of the loudest drum. It would be interesting if we could compare how many scientists buy into a popular hypothesis just because "everybody accepts it" to the number that disparage a popular hypothesis for the same reason.

Although the Reproducibility Crisis is the focus of valid concerns in Curiosity-Driven Science and QMB, their tendency to blame hypothesis-based thinking for the problems is hard to understand, even self-defeating on statistical grounds. As we saw in Chapter 8, open-ended search modes, such as the Curiosity-Driven Science and QMB programs, are more prone to generate less-reproducible results than is hypothesis-based research.

Firestein and Glass find fault in the NIH grant review process that they also blame on hypothesis-based research without making it clear how QMB and Curiosity-Driven projects will alleviate the problems. Why should we expect reviewers to be wiser, more consistent, objective, and reasonable, when evaluating projects driven by curiosity or open-ended questioning than when evaluating those based on hypotheses? I suspect that, along with anything else, the NIH's lack of unambiguous positions on different kinds of research causes some of the problems (Chapter 13). Applicants who don't fully appreciate what a hypothesis is or how to write a grant to test one will be at a major disadvantage in any case. An even bigger roadblock may be the absence of consistent, agency-wide standards for judging research. Regardless of the specific issues surrounding grant applications, there is no reason to oust the hypothesis from science even if it is a challenge to write and review a solid, hypothesis-based grant application.

I'll end this section by considering the implication of both Firestein's and Glass's arguments that the hypothesis is particularly likely to foster cognitive bias. I'll take up the topic of bias in general in Chapters 11 and 12, and only want to note here that neither author makes a persuasive case that his preferred alternative is immune from generating bias. Both authors suggest that you can avoid bias by being open-minded, but this is like saying that nations can avoid war by not fighting. If we could ensure good behavior, we wouldn't have to worry about bad behavior. Our minds automatically interpret the world, prefer certain solutions over others, leap to conclusions, and indulge in a host of questionable cognitive processes. No doubt hypothesis users are biased at times, as are Curiosity-Driven scientists and followers of QMB. The problem is with the mind, not the method, and we have to learn to keep an eye on what our minds are up to. In fact, one clear danger is that, in pinning blame on the method—the hypothesis—we'll risk missing out on the advantages it offers.

One thing that Curiosity-Driven Science, QMB, and hypothesis-based science do have in common is a frame of reference: they all accept that empirical observations are the stuff of science. The last anti-hypothesis point of view that I will consider rejects empiricism. It is not the work of a mystic or metaphysician, but the controversial position of a theoretical physicist, David Deutsch. His plan for reconceiving the methods of science and the hypothesis is next.

10.E Conjecture and Criticism (David Deutsch)

While it is true that Deutsch rejects the hypothesis, it is more accurate to say that his program subsumes the hypothesis. He thinks that nature is far weirder than most of us can conceive. For him, our universe is only the faintest wisp in an indescribably intricate structure called the *multiverse*. The multiverse is something like a network of all possible universes, except that network doesn't begin to capture its complexity. I'll return to the multiverse a bit later, but to get a crude feel for it, imagine that you set off in a car from Times Square in New York City headed for Los Angeles. You soon come to an intersection where the road splits, going right and left. In lieu of picking one, you, your car, and all of its contents are duplicated, and one copy of you goes right and the other left, into separate universes, each copy continuing along blithely unaware of the other. And this keeps happening. At every intersection, each copy of you and your car splits, budding off new copies in a fantastic amoeboid way. Imagine the unbelievable complexity by the time you get to the George Washington Bridge! To St. Louis! Yet the roads aren't becoming clogged by the trillions of your doppelgangers because, at every fork in the road your copies cruise off into different universes. The real multiverse is much more subtle and complicated than this, but you can see that Deutsch's thinking takes getting used to. I will skip most of his aggressive program for reorganizing our perceptions of science, philosophy, art, politics, etc., to home in on his vision of the roles of hypotheses, predictions, and testing in science. As you'll see, these concepts need serious revision or reevaluation if the multiverse is going to fit into science.

In a nutshell, for Deutsch, scientific knowledge consists in explanations of natural phenomena that come from an iterative process of *Conjecture and Criticism*. Now, since the procedure of falsification testing is a kind of formalized criticism, you might think that Deutsch's program is actually Karl Popper's *Conjectures and Refutations* in disguise. It's not. Deutsch has bigger ambitions than restating Popper for the twenty-first century; he wants to rescue science, beginning with theoretical physics, from what he sees as its impending identity crisis and set it on a new footing. Whereas Popper believes that the search for scientific Truth proceeds by proposing and testing empirical hypotheses, Deutsch wants to rewrite the rules and dispense with the hypothesis as we know it, and he adds a new element to the search for Truth. As empiricists, we are accustomed to thinking about Truth as representing agreement between our statements about the world and how the world really is. Deutsch plans to wean us away from our dependence on empiricism. He believes that *good explanations* can take its place.

10.E.1 Good Explanations

Why do we have annual seasons? For hundreds of thousands of generations, humans made no progress in answering this or almost any significant question about nature. With a few transient exceptions (e.g., the "Golden Ages" of Greece and Florence), ignorance was the dominant state of affairs in Europe until the 1500s, when the explosion of knowledge called the Scientific Revolution overlapped the equally dramatic societal change known as the Enlightenment. Philosophers (remember this was before science emerged as a separate branch of knowledge) began to rebel against the authority of the Church and the iron legacy of the Greek philosophers which held that logical deductions from accepted facts, *dogma*, would lead to new knowledge. In the breakaway, philosophers began to accept observation and experiment as the way to settle questions of scientific truth—the doctrine of *empiricism* was born.

In David Deutsch's retelling, empiricism was a good start but, by itself, didn't enable us to make significant advances in understanding nature; we needed to develop a culture of criticism and to seek *good explanations*. A good explanation is "hard to vary," which means that every detail of the explanation is irreplaceable; if you change one detail, the explanation falls apart and no longer explains what it was supposed to. *Bad explanations* are "easy to vary," and changing their details has little effect on their explanatory power.

Deutsch's favorite example invokes the accounts of the annual earthly seasons given by ancient mythologies and the one given by modern astronomy. In the myths, some god might cause winter because she's sad and, as her emotions change, the other seasons follow. Because the details of the myths have nothing really to do with the seasons, mythological explanations like this are easy to vary. You can substitute one god for another, alter his or her circumstances, motivations, powers, objectives, etc., and generate equally good explanations for the seasons. Myths are bad explanations. Modern astrophysical theory attributes the seasons to the degree of tilt of the earth's axis of rotation with respect to its orbital plane around the sun. Winter happens in the hemisphere tilted away from the sun, while summer occurs in the other hemisphere. Change any detail of this explanation, say remove the tilt in the axis, and it no longer accounts for the seasons. The explanation is *hard to vary*. You might object that the modern astronomer has a lot more data to work with than the Greek mythologist did, and this is true. In fact, another way of thinking about the hard-to-vary criterion is that a good explanation is more closely connected to observational data, more tightly constrained by data, than are bad explanations.

The parable of the seasons illustrates the concept of hard to vary, but sheds no light on why Deutsch believes that science is outgrowing its dependence on empiricism and the hypothesis, which we'll take up next.

10.E.2 Rejection of Empiricism and Induction

Empiricists believe that we get our information about the world through sensory experience and that experience is the standard for assessing the validity of scientific truths. Deutsch says that empiricism is a dead end, it is "the misconception that we 'derive' all our knowledge from sensory experience." According to his way of thinking, observations cannot provide the logically certain basis required for valid deduction; we can't reason from raw data to deeper truths about nature. Our sensory evidence that the sun comes up every 24 hours is plainly wrong: the sun does not rise; the earth rotates and makes it seem that the sun rises. We are continually deceived by what our senses tell us about the world, and, therefore, reasoning from naïve experience inevitably generates error, not truth.

What about induction? Inductivists claim that we get from observations to explanations through inductive reasoning, but Deutsch declares that inductivism is not merely wrong, it is another misconception: a nonexistent fictional process. He concurs with Hume's take-down of induction—(philosophical) induction is useless because its foundational assumption, the Uniformity of Nature principle, is unprovable—but Deutsch argues further that Hume let the inductivists off the hook too easily. A deeper problem is that induction is unavoidably grounded in empiricism; our sense impressions provide the purported regularities that inductivists reason from, and, since sensory experience is unreliable, inductive conclusions are not only logically weak, they lack any basis at all.

Most importantly, says Deutsch, we *create* the explanations that we seek; they do not arise spontaneously from observations. Instead, science makes good predictions because it develops good explanations, not because it has raw sensory data. Astrophysicists know a tremendous amount about stars and galaxies;

they understand why stars are bright, and why, every so often, one of them blows up in a spectacular burst of energy called a supernova, but no one has direct experience of a star. Astrophysicists know about stars because they have good explanations for them and their properties.

Deutsch concurs with Popper in dismissing inductivists who do not admit that, before they can induce anything from phenomena, they first have to define the phenomena (i.e., they have to conjecture about what to count as phenomena in the first place). Observations, measurements, experience, are all "theory-laden." We "tacitly [rely] on explanatory theories to tell us which combinations of variables . . . we should interpret as being 'repeated.'" Deutsch quotes the Greek philosopher Heraclitus to make the point: "No man steps into the same river twice, because it is not the same river, and he is not the same man." You need theories of "river" and "man" before you can study rivers and men. Therefore, descriptions and categories do not precede theory: they follow theory (and may then, of course, lead to new theory).

And induction comes up woefully short when science needs to make novel predictions. Our limited earth-bound experience of 24-hour sun cycles does not predict when an astronaut in a satellite orbiting the earth will witness sunrise; on the other hand, the astrophysical explanations of the properties of the solar system and of earth-orbiting satellites correctly predict that the astronaut will see a sunrise every 90 minutes.

A key concept for Deutsch is *reach*: the ability of a theory to account for observations other than those that it was constructed to explain. Reach is related to "non-obvious predictions (Chapter 1)," but for Deutsch reach is an expansive and fundamental concept. Reach figures prominently in his rejection of induction—induction goes from particulars to generalizations of the same kind; it is confined by such cases and cannot go beyond them. Reasoning from properties of apples might generalize to properties of oranges, but the reach of real scientific theories takes them beyond the obvious applications. Newton's theory of gravity was constructed to explain the rotation of the moon around the earth. Its reach allows it to account for the paths of man-made, earth-orbiting satellites, as well as earthly ocean tides. Induction can't do that.

10.E.3 Demotion of Testability, Prediction, and the Hypothesis

There is great overlap between Deutsch's program and Popper's *Conjectures and Refutations*, and Deutsch appears to be a sincere admirer of Popper. Like Popper, Deutsch accepts that all our discoveries are ultimately uncertain and that it is unnecessary—in fact, it is a mistake—to try to justify our conjectures about the world. On the other hand, Popper believes that the testability of hypotheses is indispensable for making scientific progress, and Deutsch emphatically disagrees; after all, bad explanations may be testable. He points out that the mythological hypotheses for seasons, though scientifically nonsensical, were testable in principle. Consider. The myths tacitly predicted that all parts of the earth would experience the same season at the same time: winter in Greece would mean winter everywhere. The Greeks, however, had limited information about the shape and extent of the earth. Had they realized that it was spherical, they could have tested the universal season prediction by sailing south of the equator, at which point they would have discovered that their explanation was incorrect. Mythology could devise an endless series of testable stories without hitting on a good explanation. Deutsch concludes that testability alone is not a sufficiently rigorous standard for science.

What about predictions? Deutsch's sees prediction as weaker than testability. The mythological framework made centuries of correct, but vacuous, predictions of seasonal sequences. Predictions are a dime a dozen; you make a prediction when you bet money that your favorite lottery number will be a winner. Lottery betting and the like are examples of "explanation-less" exercises; they are scientifically worthless because they are not part of attempts to explain nature. Many purportedly scientific predictions are little better. *Instrumentalism*, an offshoot of empiricism that emphasizes the predictive capabilities of science and regards successful prediction as a substitute for good explanation (as indeed QMB does; Section 10.B.) is an example of a failed approach. In sum, Deutsch thinks that testability and predictability are inadequate benchmarks for scientific progress.

In the same vein, Deutsch demotes the hypothesis to secondary status because hypothesis testing relies on empiricism, and, since empiricism for him is passé, the testable hypothesis cannot be the apex of scientific standards either. Experience and empirical testing are not irrelevant to science; they can help separate good explanations from bad ones at low levels of scientific analysis. What is of paramount importance though is that we can invent explanations that exist in domains of imagination far removed from anything we can conceivably experience, and that's where David Deutsch wants to go.

10.E.4 Conjecture and Criticism

"Conjecture" is a catch-all term, encompassing guess, explanation, hypothesis, theory, model, and law. For Deutsch, as for Popper, conjectures are products of the creative mind. Their origins don't matter; they are important only for what they say about the world. In Popper's program, we test conjectures and reject the falsified ones. In Deutsch's program we criticize conjectures and reject the bad ones. Testing is an empirical process and criticism is an intellectual one.

We need to evaluate explanations that are not amenable to experimental testing and to distinguish between better and worse conjectures based intellectual considerations alone.

For these reasons and more, Deutsch is relentlessly upbeat about progress. There will be no end of problems for science to tackle, but at bottom, all problems are attributable to lack of knowledge, so they will eventually be solvable. There are, literally, no boundaries to the growth of knowledge except those established by the laws of physics. His vision extends far beyond the scope of experimental science. Why shouldn't we, he asks, be able to take on notions such as "beauty" or "culture" and discover good explanations for them via Conjectures and Criticism?

So far, probably nothing about Deutsch's thinking strikes you as being excessively radical. Unusual, perhaps, but not outlandish, and you might be wondering why he believes that our most basic conceptions of science need major revisions. The answer is that he wants to take science into new dimensions.

10.E.5 Is Theoretical Quantum Mechanics Still Science?

In trying to solve certain otherwise intractable problems, some theoretical physicists have devised such fantastically exotic hypotheses that they are beyond the possibility of empirical testing.[27] This development has become a source of anxiety in physics. In the words of two physicists,[28] "As we see it, theoretical physics risks becoming a no-man's land between mathematics, physics, and philosophy that does not truly meet the requirements of any." What's going on is a "battle for the heart and soul of physics." It may be disconcerting to imagine physics, the "queen of sciences" as leaving (or being banished from?) science, but concerns along this line have been growing for years. Deutsch rises to the challenge of rescuing theoretical physics with his new approach for assessing scientific Truth. Weighty issues are at stake, so it's worthwhile for nonphysicists to peek into Deutsch's world to get a sense of what he's trying to do. Don't worry if you find the following sections somewhat confusing; they are meant as a sketch of scientific problems that have not yielded to conventional problem-solving strategies. The details are not critical.

We'll start by returning to the concept of the *multiverse*. There are many versions of multiverse theory and Deutsch favors a "many worlds" interpretation; that is, the universe that we perceive is only one of a large, potentially infinite number of universes diverging from ours in degrees ranging from imperceptible to enormous. The differences among universes are not arbitrary though. There are rules: the laws of physics hold everywhere, and the universes cannot communicate with each other; each is an isolated entity. A conceptual hurdle for many of us is that we're not supposed to imagine the universes as stretching out contiguously in space. Rather the universes "overlap" and occupy the "same" space. I'll have more to say about this later, but it helps to keep in mind the recommendation of the poet Samuel Taylor Coleridge[29] who said that readers should adopt a "willing suspension of disbelief" in order to grasp the "semblance of truth" that a writer (in this case, me) is trying to express by conjuring up weird and insubstantial images.

Why would anyone waste time on such a crazy idea as the multiverse? As strange as it is, the multiverse conjecture might help rid physics of snags that lurk within the conventional theories, such as randomness and the dual "wave-particle" nature of tiny, fundamental fragments of physical reality. For instance, the physicist Werner Heisenberg theorized that we can never in principle determine the precise position of an electron in an atomic nucleus; it is strictly indeterminate. At best the electron has a random probability of being in one place or another. Though it is integral to the standard interpretation of quantum mechanics, the existence of randomness in physical law has long bothered physicists. Einstein rejected it with the comment that "God does not play dice with the world."[30] Resorting to abstract probabilistic calculations to describe the electron's location represented a cop-out, he implied; if we truly knew what we were talking about, we wouldn't need such calculations. David Deutsch agrees; we shouldn't surrender to randomness.

Physicists like Einstein and Deutsch believe that physical laws are rigorously deterministic, and, at least for Deutsch, the multiverse solves problems associated with randomness and indeterminacy. For a crude physical picture of how this could work, suppose there were a huge national lottery where hundreds of millions of numbered tickets were sold and a scrupulously fair drawing held. The winning number, we would normally say, was picked "at random." But this would be an illusion of our cramped point of view. What we experience as a chance event, when seen from a perspective that takes in the multiverse, is an absolute certainty. The winning number is the one that *had* to win in our universe; its selection was unalterably fixed by the myriad details of its history, from the raw materials and manufacturing of the little numbered balls bouncing around in the lottery machines, to the conditions and precise timing of the drawing. Everything that could possibly have influenced which balls the machines spit out was the result of a definite physical event; no randomness allowed. Variance in any of the details would have resulted in another number's popping up. The most bizarre corollary of multiverse reasoning, however, is that every number must be a winner in some universe (even your number, somewhere, but probably not here).

Back to physics: each tiny physical entity, such as an electron, is like a lottery number and has a defined position, speed, direction—actually, ranges of them—and each parameter is realized in a different universe. There is no shortage of universes. And it's mind-boggling scale is not the most intellectually taxing aspect of the multiverse. In Deutsch's hands, the concept of the multiverse is the tool that rationalizes physics by eliminating the eerie principle that a bit of matter such as an electron is *both* a wave and particle simultaneously, that it has *wave-particle duality.*

Here's the description of wave-particle duality for us lay people: if you send a beam of light (a stream of photons) through a single narrow open slit in an otherwise opaque screen, the pattern on the wall behind the screen where the beam hits is a single fuzzy bar of light; it's what you'd expect if photons were minute solid particles. If you repeatedly fired a BB gun at a screen with an analogous (bigger) slit, the BBs that got through would hit the wall behind in a similar fuzzy-bar pattern. If you shot BBs at a screen with two slits, you'd see two parallel fuzzy bars on the other side, just as you'd expect.

On the other hand, if you send a beam of photons through two identical narrow slits parallel to each other on the same screen, then you don't see a pattern of two fuzzy bars on the other side. Instead, you see a regular array of alternating light and dark bars; an *interference pattern.* This is eerie because we just said that when you have one slit, a photon beam behaves like a stream of miniscule BBs, and now we're saying that when you have two slits, it doesn't.

The interference pattern that you get with photons is like the interference pattern of waves that you see when a single ocean wave passes through two narrow openings in a stone jetty at a beach. Instead of two simple waves emerging on the other side, there is a complex pattern of wavelets as the wave goes through both openings. The wavelets coming out interact by alternately cancelling and reinforcing each other (i.e., interfering with each other). The big deal is that waves create interference patterns, particles do not. To recap: when a photon goes through one slit, it behaves as a particle; when a photon goes through two slits, it behaves as a wave. So which is it? This is the infamous *double-slit* experiment, and it embodies the paradox of wave-particle duality. (There are many excellent videos depicting wave-particle duality and the double-slit experiment online—here's a good one for non-physicists: https://www.youtube.com/watch?v=fwXQjRBLwsQ.)

Everyone agrees on the observations; the question is, what do they mean? One fundamental difficulty is logical as well as physical: a wave is an extended thing—it has volume, it occupies space. An electron has virtually no volume and occupies virtually no space. When waves collide, they interfere with each other, meaning they partially cancel and partially reinforce each other; when particles collide, they bounce off each other. It appears that wave-particle duality is logically self-contradictory, and hence, accepting it means giving up on logic or ignoring the self-contradiction. So much the worse for logic, says the physics Old Guard: don't worry about "understanding" the duality, just "shut up and calculate."[31] If the calculations predict future events reliably, that's all you can hope for.

A paradox like wave-particle duality is anathema to Deutsch. It's self-contradictory nature is a signal that we don't properly understand nature. The very name, "wave-particle duality," is merely a description, it doesn't *explain* anything. The way out of the maze is through the multiverse, where wave-particle duality does not exist. There, photons are always and only particles; however, what seems to be a single photon is actually many simultaneous, overlapping "instances" of photons that can diverge and merge under various circumstances. Each instance of a photon is in a different universe that "overlaps" with ours (i.e., the only one that we're aware of). The many overlapping instances of a given photon are interchangeable and indistinguishable. They are "fungible" in the way that dollars in your bank account are; neither an individual photon nor a dollar in your account have unique identities. No matter how pitifully few dollars might be in your account, the notion of an individual dollar is meaningless; the ones from your paycheck are inextricably intermingled with the ones that your grandmother gave you on your birthday. You couldn't tell them apart if you wanted to. On the other hand, if you withdrew some bills, you could note their serial numbers, draw a mustache on George Washington, etc.—they would become individuals under certain circumstances. Then, if you redeposited them into the bank, they would return to fungibility and lose their identity again.

Deutsch thinks that photons are sort of like that. They may go off and lead differentiated, individualized existences in different universes or remain in a fungible state where universes intersect and repeatedly pop out in one, or in many more than one, universe. Or, like dollars that you hid inside your mattress and forgot about, some photons can remain individualized and never return to fungibility. Permanently individualized photons will have experienced subtly different histories and have acquired different properties; they may be traveling in an array of slightly different directions, for example.

Interference is what we perceive in our universe when two or more separate instances of the same photon cancel each other and return to the fungible state. In the double-slit experiment, some non-individualized photons encounter self-instances and engage in mutual cancellation, some will take slightly different paths through the two slits in the screen and remain individualized. The sum total of all of those cancellations and path deviations produces the net interference pattern that we observe. Apparent wave-particle duality, like randomness, is an illusion created by our limited field of experience; a photon remains a particle, it is never a wave. The wave-like interference pattern is a population phenomenon that indirectly reveals the multiverse-level behavior of photons, and Deutsch believes that the multiverse is in fact "the only possible explanation" for quantum interference.

Although superficial, this overview of reasoning from the standpoint of the multiverse should make it obvious why Deutsch must dispense with empiricist standards and the hypothesis: the multiverse theory cannot be directly tested; it is not falsifiable. We're faced with a choice: either concede that some aspects of nature are forever beyond the scope of rational investigation or redefine rational inquiry to include theories that are hard to vary, though unfalsifiable.

Doesn't this redefining of scientific reasoning throw open the doors to limitless fanciful rubbish? No, says Deutsch, because strict rules remain in place. The multiverse theory is a hard-to-vary explanation for the occurrence of significant physical phenomena. It is severely constrained by observation and other good explanations: the laws of physics are obeyed in all universes, interuniverse communication is not allowed, randomness is not allowed, photons are purely particles, and so on. The theory is specific and can be subjected to criticism, error-detection, and correction. Deutsch admits that the multiverse is a powerfully strange picture of reality but he argues that we can make scientific progress by substituting the goal of explanatory understanding through Conjecture and Criticism for experiential testability.

10.E.6 Summary and Critique

Deutsch's revision of the scientific enterprise makes critics uneasy, although they find his novel ideas thought-provoking.[32] Some conclude that Deutsch is a hard-core Popperian[3x] however, as I've tried to show, Deutsch's program extends well beyond Popper's. The two are in sync when it comes to fallibilism, rejection of induction, and insistence on creativity as the source of hypotheses, and in their vision of science as a means of approaching Truth about nature. I doubt very much that Popper would have considered Deutsch a Popperian, though it is worth recalling that Popper did not believe that unfalsifiable ideas were necessarily worthless[33]

A key unanswered question for Deutsch is how we should undertake the search for good explanations in the context of everyday experimental science. The hard-to-vary criterion may be hard to apply when trying to choose between better and worse explanations except in extreme cases, such as when comparing mythology with modern astronomy.

Maybe the rule should be, "If you can test a hypothesis experimentally with the empirical falsifiability standard, then do it." Deutsch might agree with this rule. "Experience is indeed essential to science," he says, "its main use is to choose between theories that have already been guessed." This process is, of course, central to Popper's philosophy. But if we carry out conventional testing whenever possible and reject explanations that fail the tests as bad explanations, then does the hard-to-vary framework add anything? How does it enhance the usual process of seeking scientific Truth through observation and experiment? Perhaps, except in special cases (e.g., advanced theoretical physics), science doesn't need to abandon its customary practices of hypothesis construction and testing.

What about intellectual areas apart from science? I have skipped over the application of his philosophy to nonscientific fields including esthetics, political science, and cultural studies, but he is confident that Conjecture and Criticism, unencumbered by the demands of empiricism and testability, can make strong contributions to these fields as well. Indeed, *The Beginning of Infinity* can be read as a book-length defense of the proposition that Conjecture and Criticism is the only possible explanation for progress of all kinds. Still, Deutsch is a committed fallibilist who believes that knowledge is always incomplete and that our best hypotheses are perpetually subject to change without notice. Could there be a better explanation for progress than Conjecture and Criticism? He must be keeping an open mind as to the possibility.

10.F Coda

In this chapter, I've reviewed three alternative programs for conducting science that, for the most part, abandon the hypothesis and even the Scientific Method.

Both Curiosity-Driven and QMB programs do make key points that are independent of the hypothesis; however, I think their vigorous rejection of the hypothesis is uncalled for. Their antagonism stems largely from obsolete notions of what a hypothesis is, and both programs reject the hypothesis in part for extraneous reasons that are based on highly dubious assumptions—that it causes bias, that their programs are immune from bias, and others. I believe that, in forswearing the hypothesis, they weaken their own positions, which in many respects are actually quite compatible with the hypothesis. Scientists are not really faced with the all-or-none choices that their arguments take for granted.

In the third part of the chapter, I explored why science might be forced to leave the hypothesis behind as it ventures into territory beyond the boundaries of empiricism. Yet the relationship between Conjectures and Criticism and the hypothesis in most experimental science is not settled, and, at the moment, there seems to be no compelling reason for the average scientist to let go of the hypothesis. Even in the ethereal reaches of theoretical physics, the acceptability of the hard-to-vary criterion as a substitute for more traditional methods is uncertain. When all is said and done, the most enduring contribution of Conjectures and Criticism may be to show how to extend the practice of critical thinking to fields other than science.

As this chapter suggests, many of the objections to the hypothesis raised here—biases, the origin of hypotheses, induction, etc.—are related to matters better dealt with by cognitive science than by the philosophy of science. We turn to these topics in the next two chapters.

Notes

1. Stuart Firestein, *Ignorance: How It Drives Science* (New York: Oxford University Press; 2012).
2. Stuart Firestein, *Failure: Why Science Is So Successful* (New York: Oxford University Press; 2016).
3. David J. Glass, *Experimental Design for Biologists*, 2nd ed. (Cold Spring Harbor, NY: Cold Spring Harbor Laboratory Press; 2011). See also D. Glass and N. Hall, "A Brief History of the Hypothesis," *Cell* 134:378–381, 2008.
4. David Deutsch, *The Beginnings of Infinity: Explanations that Transform the World* (New York: Penguin Books; 2012).
5. The phrase "curiosity-driven" has been used as a modifier of "basic research" (see, e.g., A. Amon, "A Case for More Curiosity-Driven Basic Research," ACSB Award Essay, November 2015, www.molbiolc.org). Amon uses it to differentiate basic science research from applied science research, rather than to differentiate curiosity-driven research from hypothesis-based research, which is what Firestein wants to do. In her essay, Amon advocates using the hypothesis in science, and, except for the term "curiosity-driven," my views are compatible with hers. Nevertheless, I use "curiosity-driven" throughout this book in the sense that Firestein intends it.
6. http://www.ted.com/talks/stuart_firestein_the_pursuit_of_ignorance.html.
7. Numerous readable accounts of the neurobiology of marijuana and the endocannabinoid system can be found online; e.g., https://en.wikipedia.org/wiki/Tetrahydrocannabinol.
8. Institutes of Theology from the Posthumous Works of Rev. Thomas Chalmers (1849) by Chalmers and his son-in-law biographer William Hanna.
9. Attributed to Albert Einstein.
10. Francis Bacon, *The New Organon (Novum Organum): On the True Directions Concerning the Interpretation of Nature*, "Aphorisms (Book One)." Originally published in 1620; James Speeding (Translator), Kindle Book, Amazon Digital Services, 2009.
11. ibid, section LXIX
12. ibid, section XLVII
13. ibid. see section LXXXII for a lucid if poetic account of how science should proceed; not by "groping in the dark" taking experience "as it comes," but viewing experience "duly ordered and digested." Not simply collecting data, in other words, but seeking understanding.

14. Richard S. Westfall, *Never at Rest: A Biography of Isaac Newton* (Cambridge: Cambridge University Press; 1980), pp. 410–414.
15. Isaac Newton to Edmund Halley, June 1686 letter collected at the Newton Project, http://www.newtonproject.ox.ac.uk/view/texts/normalized/NATP00032516. Wesfall, ibid, pp. 269–273.
16. Wesfall, ibid, pp. 269–273.
17. Frank Wilczek, *A Beautiful Question: Finding Nature's Deep Design* (New York: Penguin Books; 2015), pp. 81–83.
18. Thomas Kuhn, *The Structure of Scientific Revolutions*, 2nd ed. (Chicago: University of Chicago Press; 1970).
19. L. G. Brock, J. S. Coombs, and J. C. Eccles, "The Recording of Potentials from Motoneurones with an Intracellular Electrode," *Journal of Physiology* 117:431–460, 1952.
20. Martin Rees, "Tracking Subatomic Physicists," *Science* 343:1434–1435, 2012. Note that Rees seems to be using "curiosity-driven" in the sense mentioned in Note 5, but this highlights the potential confusion associated with the term.
21. Lisa Randall, *Higgs Discovery: The Power of Empty Space* (New York: HarperCollins; 2 012).
22. Bradley E. Alger, "Retrograde Signaling in the Regulation of Synaptic Transmission: Focus on Endocannabinoids," *Progress in Neurobiology* 68:247–286, 2002.
23. https://www.nobelprize.org/nobel_prizes/physics/laureates/1907/press.html.

    In the Nobel Prize Award Ceremony Speech outlining the achievements of Michelson that warranted his receiving the Prize, Michelson is cited for his invention of a very accurate interferometer and some of the uses to which it had been put, including especially the precise measurement of the official standard meter rod. "Your interferometer has rendered it possible to obtain a non-material standard of length, possessed of a degree of accuracy never hitherto attained. By its means we are enabled to ensure that the prototype of the metre has remained unaltered in length, and to restore it with absolute infallibility, supposing it were to get lost."
24. Except as indicated, all quotations in Section 10.C are from the sources mentioned in Note 3.
25. Nicholas N. Taleb, *The Black Swan: The Impact of the Highly Improbable*, 2nd ed. (New York: Random House, 2010), new section: "On Robustness and Fragility" (Incerto).
26. Naomi Oreskes and Erik M. Conway, *Merchants of Doubt: How a Handful of Scientists Obscured the Truth on Issues from Tobacco Smoke to Global Warming* (New York: Bloomsbury Press; 2010).
27. D. Castelvecchi, "Feuding Physicists Turn to Philosophy," *Nature* 528:444–445, 2015; D. H. Bailey and J. M. Borwein, "Data Versus Theory: The Mathematical Battle for the Soul of Physics," https://www.huffingtonpost.com/david-h-bailey/data-vs-theory-the-mathem_b_8886292.html, 2015; A. Frank and M. Gleiser, "Crisis at the Edge of Physics: Opinion," *The New York Times* https://www.nytimes.com/2015/06/07/opinion/a-crisis-at-the-edge-of-physics.html?_r=0.

28. G. Ellis and J. Silk, "Comment: Scientific Method: Defend the Integrity of Physics," *Nature* 516:321–323, 2014.
29. https://en.wikipedia.org/wiki/Suspension_of_disbelief. Quotation is from Samuel Taylor Coleridge from Coleridge, *Biographia Literaria*, 1817, Chapter XIV.

    "It was agreed, that my endeavours should be directed to persons and characters supernatural, or at least romantic, yet so as to transfer from our inward nature a human interest and a semblance of truth sufficient to procure for these shadows of imagination that willing suspension of disbelief for the moment, which constitutes poetic faith."
30. Albert Einstein, quoted in Ronald W. Clark, *Einstein: The Life and Times* (New York: Avon Books; 1971).
31. "Shut up and calculate." Although the quote is widely attributed to Richard Feynman, the physicist N. David Mermin looked into the issue and discovered, evidently to his surprise, that he, Mermin, had originated it. See http://gnm.cl/emenendez/uploads/Cursos/callate-y-calcula.pdf.
32. D. Albert, "Explaining It All: How We Became the Center of the Universe," *New York Times*, August 12, 2011; Review of David Deutsch, *Beginnings of Infinity*, https://www.nytimes.com/2011/08/14/books/review/the-beginning-of-infinity-by-david-deutsch-book-review.html.
33. Bryan Magee, *Philosophy of the Real World: An Introduction to Karl Popper* (La Salle, IL: Open Court; 1985), p. 47.

11
Automatic Thinking
Hypotheses, Biases, and Inductive Reasoning

11.A Introduction

What do hypotheses, biases, and inductive reasoning have in common? The question is at the heart of this chapter. One answer is that they can all arise from what, for lack of a better term, we can call "automatic thinking." I don't have a precise or sophisticated definition for automatic thinking and will have to hope that everybody can agree that thought-producing mental activity is almost continually going on while we're awake. That's what I am talking about. Automatic thinking might be either conscious or unconscious (i.e., we are not always aware of it); regardless, at any moment, we can become aware when a thought emerges into our consciousness, whatever that means. Automatic thinking is fundamentally different from other kinds of automatic neural activity, such as that which controls our heartbeat and that we can't become conscious of. For the most part, I intend to stay out of the debate about consciousness itself, although the topic is impossible to avoid altogether.

Automatic thinking occasionally appears in the spotlight when people are talking about scientific bias; however, it is frequently glossed over or ignored when the topics of hypotheses and inductive reasoning come up, which is too bad because this kind of thinking is key to understanding them as well. If you take automatic thinking into account, you may become more skeptical about advice that tells you to avoid the hypothesis when you're carrying out experimental science and be less inclined to see inductive reasoning as a special aptitude of the human mind.

We don't usually pay much attention to our mental complexity and, instead, work with a common-sense impression of how the mind functions, what the philosopher Daniel Dennett labels "folk psychology."[1] Folk psychology is intuitive, compelling, and often mistaken. We find inductive reasoning persuasive because we rely on folk psychology. In general, folk psychology makes us overconfident in our ability to think freely and independently and ignores evidence that there is much going on in our brains that we don't know about, let alone control.

It goes without saying that scientists have the same built-in cognitive habits that everyone else has: scientists jump to conclusions, use unexamined mental short-cuts, fall prey to biases, go along with the crowd, hop on bandwagons, and engage in other unsound cognitive behaviors. What allows scientists to get beyond the limitations of folk psychology are skepticism, a drive for getting to the truth, and stiff competition with their peers. As the critics that we reviewed in Chapter 10 remind us, a faith in folk psychology continues to confuse discussions of science, and lessons from cognitive science provide a useful perspective. To give it a name, this is a *meta-cognitive* approach: we need to think about how we think, and, to do that, we'll need to ask what sensory illusions, heuristics, fads, and cognitive biases can tell us about science and the hypothesis. Inborn tendencies account for the pervasiveness, indeed, the inevitability, of the hypothesis in scientific thinking. The best we can do is to comprehend and work within our limits. Resistance is futile.

11.B Hypotheses: Always Under Construction

In earlier chapters, we talked about factors that can diminish the influence of the hypothesis—misinterpretation, overt opposition, and neglect, to name a few—but our unacknowledged tendency to generate hypotheses constantly contributes to the problem. To see how, imagine the following scene: when the exams were handed back in class, you were pleasantly surprised to learn that you really nailed it—the highest grade in the class. You did nothing out of the ordinary to prepare for the exam and were expecting to get the respectable but unremarkable score that you always did. "What happened?" you wonder.

You might have asked another question: Why did you feel surprise? It is obvious why you were happy, but your surprise is different. You had expected something that didn't happen, and your expectation is essentially an implicit hypothesis that explained your previous academic performance: namely, that you were a somewhat above average student. This predicted an unremarkable exam grade, and the exam results had just falsified the prediction. Evidently, your hypothesis was wrong, and, in fact, the bolder hypothesis that you are a superior student may already be taking shape in your mind (although you sense that it will be wise to test the original one more rigorously before writing home about the new one).

Hypotheses like this one are trivial and unconsciousness. We generate and test them all the time and hardly realize that we're doing it until a reaction such as surprise suddenly brings them to light. In general, we don't merely register what's going on around us: we constantly try to understand it by making conjectures and comparing their predictions with our experiences. We'll need to take this persistent drive into account if we want to understand the influence of the hypothesis in scientific thinking. No doubt some readers will balk at the thought that an automatic, sense-making mental process should be dignified by calling it "hypothesis generation," and it's true that these are low-level, proto-scientific constructs. However, the constructs look just like genuine hypotheses: they provisionally explain an aspect of the world, they make testable predictions, and they can be rejected if they turn out to be wrong. Furthermore, there is long historical precedent for the conclusion that we do not passively receive and process information from the environment.

The philosopher Frances Bacon, though most famous for championing the causes of inductive reasoning and scientific experimentation, knew that our thoughts are not always trustworthy. He identified four classes of "idols of the mind"[2] that, he said, cause us to commit mental mistakes, and two of them are directly relevant here: the *Idols of the Tribe* encourage us to believe that our perceptions represent the world truly, whereas, in reality, the mind "is like a false mirror" that "distorts and discolors the nature of things by mingling its own nature with [them]." The Idols of the Tribe are rooted in our human nature; nowadays we would say that they are genetically programmed and so are universal human traits. In contrast, the *Idols of the Cave* are those idiosyncratic misperceptions that we acquire individually; they result from the unique collections of preconceptions and blind spots that we pick up through our social and familial interactions, our education, etc. The Idols of the Mind and Idols of the Cave cause us to misinterpret the world unless we know about them and guard against their influences. Bacon wanted to teach his readers to avoid them by adopting the Scientific Method.

In 1738, David Hume[3] also called attention to psychological factors in philosophy, but he was more specific than Bacon. Hume, you'll recall, stripped away the aura of logical invincibility surrounding inductive reasoning, arguing that we believe in the validity of inductive reasoning for psychological reasons. We infer a cause–effect relationship between events from the indirect evidence of *contiguity, priority (temporal order)*, and *constant conjunction*. Putting a finger near (spatial contiguity) a candle flame causes us to feel heat, we never feel the heat before we put the finger near the flame (temporal order), and, finally, every time we do put the finger near the flame we feel the heat (constant conjunction). However, the causal relationship is not logically mandated; we never experience cause directly and couldn't have deduced the relationship if we didn't know anything about flames and heat. Rather, we believe that they're causally linked because of the automatic, unconscious operation of our "imagination," as Hume calls that part of our minds where we join ideas together.

While Hume's analysis of the mind is valid as far as it goes, it doesn't go far into the active mental processes of innate hypothesis generation. The Nobel Prize-winning neuroscientist Eric Kandel[4] credits the nineteenth-century physicist and physiologist Hermann Helmholtz and the Gestalt psychology movement of the twentieth century with the insight that we are constantly making and testing hypotheses about the world. Helmholtz realized that our sensory systems are so crude that, if we didn't actively refine the raw information that we get from the environment, we wouldn't be able to function. "In fact, if the brain relied solely on the information it received from the eyes, vision would be impossible," says Kandel, noting that Helmholtz "concluded that perception must also be based on a process of guessing and hypothesis testing." We rely on inspired guesswork to navigate our environment, though sometimes our guesses are inaccurate. The Gestalt psychologists added that we do not sense single stimuli in isolation, but as parts of integrated wholes; we do not hear a succession of musical notes, we hear a melody. We do not merely react to stimulation, we interpret and try to understand it and we do this automatically and unconsciously. Our perceptions therefore depend on extensive lower level "bottom-up" and higher level "top-down" types of processing.

11.B.1 The Brain Is an Organ for Making the World Intelligible

For many cognitive psychologists, prediction is "an overarching principle of brain function . . . in the service of promoting adaptive interactions with one's environment."[5] Much of the baseline electrical activity of the brain, observed with electroencephalographic (EEG) or functional magnetic resonance imaging (fMRI) techniques when we are physically at rest, the brain's so-called *default mode*, is dedicated to this activity.[6]

Probably the most dramatic illustration of the urge to interpret comes from the classical "split brain" studies[7] on patients who have had major brain surgery to treat their uncontrollable epileptic seizures. Seizures are storms of electrical activity that travel in a bundle of nerve cell fibers (the *corpus callosum*) between the two sides (*hemispheres*) of the brain. Surgically cutting the bundle keeps seizures from spreading and reduces their severity. The operation, however, essentially leaves the patient with two independent mini-brains,[8] and each one processes both sensory and *motor* (muscle movements) information separately from the other. For instance, each eye gets light from both right and left sides of the world (we won't worry about the binocular overlap for the moment) and sends the left-side information to the right side of the brain, and vice versa. Normally, information from both hemispheres funnels into a single *interpreter*[9] region that puts it all together to create a seamless visual experience. Difficulties arose because the interpreter is housed in only one hemisphere, so cutting the bundle of fibers meant that split-brain patients couldn't integrate all the visual information coming into both eyes; their hemispheres saw slightly different images of the world. Similarly, one hemisphere could control only one arm, one leg, etc., and not help coordinate the actions of both sides as it normally did. By studying how these patients with their two mini-brains coped with certain experimental challenges, the experimenters learned how the intact brain ordinarily processed integrated information.

To study the extent of the deficits that the patients suffered, the experimenters devised clever optical devices that let them independently control the visual information that got to each hemisphere. Since one hand was controlled by one hemisphere, the experimenters could find out what visual information each hemisphere received by asking the patient to point with that hand to a picture of the object that was transited to the same hemisphere .

What happened when both hemispheres saw the same picture if the patient's hands were not in sight? Even when her hands were hidden beneath the top of the table at which she was sitting, they both pointed to the correct picture, as expected. What happened if the hemispheres saw different images at the same time? This is where things became really interesting. If the patient's hands were beneath the table top, then she correctly pointed with each hand to the appropriate picture but, as she couldn't see her hands, she was unaware (i.e., her interpreter was unaware) that they were pointing at different pictures. Now, what happened if both eyes saw different images and the patient *could see both hands*? Again, each hand pointed to the appropriate picture (which was different for each hand). And, because both hands were visible, the patient became aware that they were pointing to different pictures. This was a major conundrum because the patient's interpreter had no direct visual information from the other hemisphere about what image *it* saw. Remember, the interpreter is located in one hemisphere and is only getting direct visual input from that hemisphere. Therefore the interpreter had no idea why her hands were pointing at different pictures. What would the patient (specifically, the patient's interpreter function) say if she were asked to explain why her hands were behaving differently?

An interpreter's job is to interpret, as best it can, the information it gets. In the textbook example, one hemisphere was shown a snow shovel and the other one a chicken's claw. When asked why her hands were pointing at those pictures, the patient's interpreter concocted a story that put them together. The shovel, she said, was used to scoop the manure out of the chicken coop. It was a creative, plausible explanation, and, most importantly, it revealed the mind's absolute determination to create meaning, even from fragmentary and contradictory information. Interestingly, the patients seldom offered the obvious, simple, and true alternative response, "I don't know," which illustrates how important it is for us to be confident in our experiences.

While the split-brain experiments provide dramatic illustrations of our urge to make sense of our environment, there are many less dramatic ones.[10] Questions remain, though: Do we ordinarily generate hypotheses consciously or unconsciously, and does it matter?

11.B.2 Thinking About Science Consciously and Unconsciously

We don't understand our unconscious minds, but then we don't understand our conscious ones either. While scientists are beginning to make progress in understanding the objective properties of conscious experiences, the eeriest aspect of consciousness, self-conscious *sentience*, is so difficult that it is designated "the Hard Problem,"[11] and put aside to await future developments. Just because we can't *explain* consciousness doesn't mean that we can't use what we do know about it, however.

11.B.2.a Much of Consciousness May Be Unconscious

Much of our consciousness reflects information processing that takes place before we know it. Consciousness is an effect, not a cause.[12] This is not a novel idea; the philosopher Thomas Huxley said exactly that in 1874. He described humans as "conscious automata,"[13] and explained—essentially—that "we do not run from the bear *because we are afraid of the bear*; we are *afraid of the bear because we are running* from it [emphasis added]." Christof Koch reports that the virtually imperceptible awareness of a rattlesnake near your ankles is exceptionally effective in triggering evasive maneuvers before you know what's there.[14] It makes evolutionary sense to respond to danger first and to think about it consciously second.

Examples of reflexive responses to danger do not prove that all consciousness is a secondary cognitive effect, of course. The nervous system juggles a lot of tasks and, while (somehow) generating consciousness is its most spectacular one, consciousness may be overrated as a causal agent. Innumerable examples, from the fact that we fall for unnoticed advertising gimmick, to our notorious difficulties in making rational judgments, as we'll see shortly, confirm that our unconscious is often in charge. We don't always know what makes us tick.

The official scientific demotion of consciousness from its position as CEO of brain function began more than 30 years ago with experiments done by Benjamin Libet and colleagues[15] who were trying to study consciousness by measuring brain activity associated with it. The reasoning behind their basic experiment was straightforward: the hypothesis that consciousness controls your mind and body predicts that, when you decide to act, the neural activity ("brainwaves") measured by an EEG that signals your conscious intent to act *must precede* the neural commands to the muscles that carry out the action. In effect, if consciousness is the boss, we should be able to overhear the boss giving the orders.

To test this prediction, Libet and colleagues asked their subjects to move a finger when they felt like it and to make a mental note of the exact instant that they decided to move by watching a clock-like device nearby. From past work, they knew that a fraction of a second *before the subject begins to move* her finger, a blip in the EEG will indicate the brain activity being sent to move the finger. The experimenters recorded the finger motion and the subjects' EEG activity. After each trial they could compare the EEG activity with the time the movement decision was made. If her conscious mind *triggered* the decision to move, the brain activity associated with her decision ("I'm going to move my finger . . . now!") must come *before* the finger movement blip. That's the prediction.

Amazingly, Libet's group found that was not what happened. Subjects became consciously aware of their decision to move only *after* the blip signaling finger movement. The brain activity causing the finger to move had already been under way for about half a second by the time they "decided" to move it! A recent fMRI study arrived at the same conclusions as did Libet's.[16] Evidently the conscious command did not start things off. Our sense of conscious control may be an illusion, or simply a parallel sign that, in fact, you're starting to move your finger; in any case, it appears that the CEO is not in charge.

At first, all of this appears to be intensely counterintuitive, but is it? You say that an idea "occurred to me" or that one "popped into my head." Really? Where did it pop in from? Your ideas must happen in your brain yet remain unknowable until they "become" conscious. Why one idea becomes conscious and another doesn't is a mystery, but conscious and unconscious ideas themselves may not differ qualitatively[17]; for instance, a gatekeeper downstream of their origin might determine which is which. While the details of Libet's experiments have been minutely scrutinized and criticized, the experiments did bring the question of conscious influence to the attention of many scientists who had previously remained aloof when consciousness was seen as a subject for philosophers only. Nowadays, "most neuroscientists . . . believe that conscious experiences are consequences of brain activity, rather than causes,"[18] and this conclusion is relevant to the discussion of the hypothesis.

As we've seen (Chapters 2 and 10) some critics of the scientific hypothesis consider it a serious objection that we can't specify how we arrive at a hypothesis or that we don't know what sorts of mental process are responsible for it. The reality is that we're able to manage quite well even if we don't know much about where our ideas come from. I'm not promoting any particular model of consciousness, only suggesting that, since we know so little about either conscious or unconscious thought processes, we shouldn't let preconceptions about one or the other get in the way when trying to analyze our own scientific thinking.

11.B.2.b Unconscious Ideas as a Source of Hypotheses

Would the cognitive source of hypotheses matter even if we knew what it was? The history of science has famous examples of hypotheses that were formed during dreams, the archetypally unconscious state. In the early 1860s, the chemist August Kekule was trying to work out the atomic structure of benzene. At one moment, while "dozing,"[19] vaguely picturing rows of atoms dancing before his eyes, he visualized a snake biting its own tail and realized that the atoms in benzene could be arranged in the shape of a ring; he "spent the rest of the night working out the consequences of this hypothesis." In other words, once he saw it as a possible analogy for benzene's chemical structure, the ring became a hypothesis with specific predictions about benzene's other properties. Kekule's subsequent experimental tests were consistent with the ring model, which was accepted as the best one for many years until it was replaced by Linus Pauling's theory in the 1920s.

In the early twentieth century, pharmacologist Otto Loewi hypothesized that a chemical released by the nerves controlled the heart, but he was stumped about how to test it until a vision of a critical experiment came to him in a dream one night when he was sound asleep.[20] He awoke enough to scribble a note about the experiment to no avail: he couldn't read his scrawled handwriting in the morning. Fortunately, the same dream came to him the next night, and he immediately got up and went to the laboratory, where he took the fluid solution bathing a beating frog heart and applied it to a second, quiescent heart. When the quiescent heart began to beat, as his dream had predicted, the chemical hypothesis of synaptic communication had passed its first critical test and Loewi was on the road to the Nobel Prize in Physiology or Medicine, which he received in 1936.

The moral is that Kekule's and Loewi's hypotheses were products of their unconscious minds and that their origins didn't matter; what the scientists did with their hypotheses did matter, and what they did was to follow the Scientific Method in testing them. You may not find these anecdotes striking enough (though they are) to be convincing evidence that the mind is generating hypothesis-generating activity during ordinary awake behavior. Let's look at some less dramatic evidence.

11.B.2.c Complex Automatic Thinking: Counterfactual Thinking and Memory

The concept "if" is complex[21]; it invites us to consider possibilities that may not have happened. Evaluating "if" statements is so crucial that we can do it automatically. You know that "If I put my hand on a hot stove top, I'll get burned," is true, although you've probably never tested it—deliberately, at least. What does information processing like this mean for understanding hypothesis-based scientific reasoning? It appears that remembering the past is closely related to imagining the future.

The science of memory has been evolving; we know that our personal memories are not *recalled* intact in a preexisting state. Instead, they are actively *reconstructed*. We assemble them as needed from bits of stored information within "a common neural system [that] supports our recollection of times past, imagination, and our attempts to predict the future."[22] The memory system is not like a warehouse filled with data; it is more like a factory for producing certain kinds of thoughts. And it is capable of carrying out the information processing functions that are collectively known as *counterfactual thinking*.[23]

Counterfactual thinking is what you do when you wonder "what if" you hadn't started seeing that person in high school that your parents disapproved of, which led to your hanging out with a new crowd, going to that college, etc. Counterfactual thinking allows you to reconfigure previous experience, to break it down, rearrange, and reorder its parts to create real or potential alternative outcomes. Similarly, it allows you to consider possible future courses of action, such as what would happen if you quit your present job and tried to make a living writing poetry.

Counterfactual thinking is not restricted to personal, *episodic* forms of memory and thinking, but is also pressed into service whenever you "simulate internal models of events and then compare these internally generated models with external reality."[24] According to psychologist Ruth Byrne, the tasks of counterfactual thinking range from idle musing about possible consequences "if only . . . " to supporting "logical, mathematical, and scientific reason, and they underpin complex deductions."[25] Counterfactual thinking is what you do when you hypothesize explanations for phenomena.

We know that the system responsible for counterfactual thinking is hardwired into our brains, and we know something about its architecture. It is partly housed in our *temporal lobes*, the lower sides of our brains. Patients who are severely amnestic because of damage to their temporal lobes lose their ability to recall past events. Perhaps surprisingly, they also suffer deficiencies in their abilities to make future plans.[26]

Nerve cells in the *prefrontal cortex* collaborate with those in the temporal lobes to conduct counterfactual thinking, and damage to the prefrontal cortex also causes distinctive cognitive handicaps. Characteristically, prefrontal lobe patients don't fully consider the consequences of their actions and have trouble learning from their mistakes or making plans, which suggests that the prefrontal lobe normally helps carry out these functions. And, indeed, when healthy subjects relive thoughts of their past actions (e.g., what might have been if they hadn't taken gambles that they took and lost), fMRI studies show that their prefrontal cortex is very active.

Evidently, our ability to create and play with mental images of alternate futures was so important to our primate ancestors that we evolved an extensive, built-in brain network to do the job. This network also makes our lives easier by automatically processing past information and comparing it with our current experience. When our circumstances are safe, secure, and predictable as we expect them to be, all is well and we don't waste a lot of energy taking note of them. Unfortunately, this system, while it simplifies routine tasks, can also make objective scientific thinking more difficult.

11.B.2.d Automatic Hypotheses and Predictions

You get an inkling about your ongoing hypothesis-generating operations when you experience surprises, such as getting unexpectedly good exam results. Surprise indicates that there is a mismatch between what your ongoing unconscious thinking processes predicted and what your conscious mind experienced. Mismatches between unconscious and conscious thought also characterize the phenomenon of *illusions*, both sensory and cognitive, which also suggests that we might learn something about cognitive illusions from studying the more straightforward sensory ones. (Some theoreticians deny the existence of cognitive illusions, but before we can appreciate their arguments we need to understand what it is they're denying.) Let's start with the familiar sensory ones.

11.B.2.e Sensory Illusions

When the information that you perceive about the world conflicts with information that you get from other sources, say, measuring instruments, you say you've experienced an *illusion*, although there is nothing spooky or mysterious about it. The experience of an illusion provides clues about how the nervous system automatically processes information, and it's worth looking into sensory illusions because there is an analogy between how our sensory systems—vision, hearing, touch, etc.—work and how our cognitive systems work.

We've all seen the classic *Müller-Lyer* illusion[27] which shows two medium-length, parallel lines tipped with shorter lines oriented either inward, arrowhead-fashion, or splayed outward. Although the parallel lines are exactly the same length, our unshakable impression when seeing them is that one is longer than the other. The Müller-Lyer illusion is visual, but illusions are a feature of all sensory modalities. An especially convincing nonvisual one is the *thermal illusion* you get after you've immersed one hand in very cool water and the other in very warm water for about 30 seconds and then put both hands at once into a bucket of lukewarm water. Knowing that the final water temperature must be the same for both hands does not help you to feel it that way; it feels quite different to each hand.

Note—and this is extremely important for the later discussion of cognitive illusions that we'll encounter in Chapter 12—that your experience of this illusion does not mean that there is anything wrong with your sensory perception. It does not signal an error in the system or a failure in your instinctual perceptions of touch and temperature. The illusion simply means that there is a discrepancy between what your nervous system is telling you and what an external standard, in this case, measuring instruments, tell you about some aspect of the world. The fact that you experience the water temperature differently with each hand is a sign that your sensory systems are fine; they are working just as they were designed to work.

The thermal illusion is caused by sensory adaptation; the two hands "got used to" different initial temperatures and so, by comparison, the final water bath seems warmer to the hand adapted to cool water and cooler to the hand adapted to warm water. Sensory adaptation is one way that our nervous systems deals with the flood of sensory information pouring in on us every instant. Adaptation filters out information that seems harmless and inessential, and, again, the fact that you experience a sensory illusion means that your system is working the way it should, not that it is malfunctioning. This conclusion seems innocuous in the sensory realm but is far more controversial in the cognitive realm.

11.C Is the Hypothesis Especially Likely to Elicit Biases and Spawn Fads?

The topic of "bias" is multifaceted, so I've broken it down. In this section I focus on the broad general notion of bias and put off most of the complexities for the next chapter, where we'll get into a few of the specific biases that are currently the source of concern regarding the Reproducibility Crisis. The main reason to split up the discussion is that there are two opposed schools of thought regarding bias, and, to understand them, we'll have to explore the concept of *rationality*, and rationality itself is complex. We'll put that subject off until Chapter 12. At the moment, we'll take bias to be an intellectual leaning that may or may not be justified and look at it in relationship to the hypothesis. Critics going back to Frances Bacon have attacked the hypothesis for being unusually prone to evoking bias in the minds of hypothesis-holders. But what does that mean? Does hypothesis-based thinking inherently induce bias? If people, including scientists, form biases, does that mean that they shouldn't develop hypotheses? Is biased thinking always bad practice?

11.C.1 The Hypothesis and Bias

When critics refer to the problem of bias in science, they often mean something like "unjustified" preference for one's own ideas and the neglect or disparagement of alternatives. The term "unjustified" is slippery, though. If a scientist defends her conceptual model that just *seems* right to her or is aesthetically appealing[28,29] and is not in conflict with data, is she wrong? Richard Feynman was renowned for vigorously defending his unique ideas even when he had trouble communicating them to other physicists.[30] The two-time Nobel Prize winner in physics, John Bardeen, was once stymied during an oral examination when he couldn't describe how he'd come up with the correct answer to a challenging question he'd been asked.[31] Neither Feynman nor Bardeen could always justify their conclusions, but how can we understand their strong devotion to their ideas except as a kind of bias? Biases may arise from our unconscious minds and cause us to do or think things that to others might appear indefensible although the defense may be merely inexpressible.

The decisions scientists make as to which experimental avenues to follow can represent biases as well. In their study of synaptic transmission (Chapter 2), Bernard Katz and Paul Fatt chose to focus on one phenomenon in their data and to turn a blind eye to another one. It proved to be a brilliant choice, but they couldn't have fully justified it based on what they knew at the time. In countless cases, scientific "intuition"—many times a polite synonym for "bias"—has led to crucial discoveries, which means that a priori justifiability cannot be a prerequisite for good science. On the contrary, competitive bias that prompts you to defend your hypothesis may benefit science, just as competition improves commercial businesses. Good ideas are not made worse because someone vigorously defends them. Neither all biases nor all competitions are good, but it's an error not to distinguish the good from the bad.

Is it true, as some members of the anti-hypothesis contingent imply, that if you're biased in favor of your hypothesis that you'll somehow profit unfairly? It seems unlikely that this is a major problem. There is no doubt that a dramatic new hypothesis can attract a lot of attention and short-term benefits; however, the long-term advantages should not be exaggerated. Despite the vaunted status of priority of discovery in science, it is rightly said that the major credit goes to her who does it best, not to her who does it first. Moreover, if your alluring, though biased, hypothesis is falsified, then the falsifier reaps more reward than you do. In other words, you're unlikely to enjoy lasting benefits from putting forward a hypothesis that is founded mainly on bias. Science aims to get things right; it claims to be correctible, not infallible. One more point here: even when scientific bad behavior does show up as bias in favor of a hypothesis, as it surely can at times, blaming hypothesis-based science per se makes no sense. Correctly deployed hypotheses do not generate bad science; scientists generate bad science.

Perhaps a misunderstanding of the rewards of science amplifies the perceived dangers of bias. You don't earn respect in science for the same traits that would win you respect in other areas of civil life. We've all heard the politician who brags about his resistance to changing his mind, "if I had it to do all over again,

I'd do it exactly the same way," etc. In the words of one prominent former government official, "Admitting doubt or mistakes is career suicide."[32] Yet essential requirements of the job description of a scientist are a willingness and an ability to change your mind when the facts demand it.

And scientists are admired for different accomplishments than certain other professionals. A good criminal lawyer who gets a guilty client off scot-free is respected, not looked down on. Professional sports teams do not forfeit victories just because a referee missed a call. Refereeing, flaws and all, is considered part of the game. The linemen in professional American football are said commit illegal "holding" infractions on essentially every play and are lionized for their caginess. There are occasional bright ethical lines in sports: if an athlete's cheating threatens the sport's fundamental integrity (say he gets caught using illegal performance-enhancing drugs), then he can be sanctioned. (Lance Armstrong,[33] the disgraced professional bicycle rider, is the poster boy for high-level sports cheating.) Nevertheless, much bad behavior in sports is praised or overlooked, but not punished.

Scientific fraud represents the *n*th degree of bias toward one's ideas. If being biased paid dividends, then Jan Henrik Schoen[34] would be a fantastic success story. Schoen managed to get many fake physics papers into top journals before he was caught, publicly humiliated, and fired. He was not rewarded. Inflexible and unjustified bias in science is a recipe for disaster, or at least for being thought stupid. Scientists who don't change their minds when the evidence goes against them eventually suffer the consequences.

"Bias" is a catch-all term for a multitude of behaviors—good, neutral, and bad—which is why no simple fix will make it go away. Once your hypothesis is public, any biases you might have are open for discussion and correction. Besides being careful and self-critical, applying the lessons of hypothesis-based thinking is the best way to avoid trouble.

11.C.2 The Hypothesis and Scientific Fads

Are hypotheses especially likely to create "bubbles of interest and attention"[35] (which to save space I'll call "fads")? Is there something about a hypothesis that causes otherwise mature, level-headed scientists to behave like the frantic, greedy investors who got caught up in the white-hot housing market of the early 2000s and went bust when the bubble began to burst in 2007? Like biases, fads are often the objects of derision that may not be entirely deserved, and we can ask the question we asked of biases: Are hypotheses bad because they attract followers? It's a reasonable question because science can, no doubt, be faddish. At times, if you're not using the latest technique or into the most fashionable avenues of research, you can feel sidelined. Still, fads, like bias, have their positive and negative sides. Bright young scientists may be attracted to a branch of science because it is flashy—and in rapidly expanding areas of research the funding is good and scientists do need to make names for themselves. A wave of interest in a scientifically and socially important problem can propel needed research. Autism, schizophrenia, Alzheimer's disease, and drug addiction are currently hot topics in neuroscience, and progress in solving any one of them would be a huge boon for society. Or, take the case of Ebola. Over a 20-year span (1990–2010), a total of 783 scientific articles with "Ebola" in their titles came up in an online search of PubMed (i.e., about 39 a year). As the epidemic surged in parts of Africa, so did the number of studies; in 2014 alone, 830 studies of Ebola were published, which probably contributed to its eventual taming (although social and political factors played a key role[36]). In the past, concern about HIV/AIDS triggered an avalanche of research that resulted in HIV being commuted from a death sentence to a treatable disease. If societal interest and scientific incentives create fads, then scientists will follow fads, and that's not necessarily a bad thing.

Furthermore, the dramatic announcement of a new finding can speed up the rate at which it is replicated or falsified, which might ordinarily take many years. In 1989, a report[37] that clean nuclear energy could be generated ("cold fusion") in a desktop device using common laboratory materials and equipment got a bandwagon rolling. If true, the finding would have revolutionized physics and probably altered the course of civilization, with its promise of cheap, limitless energy to everyone on the planet. The report briefly rocked physics and precipitated concerted attempts to test the hypothesis, all of which hastened the virtual extinction of the field when it was not corroborated. A similar flare-up in biology accompanied the "discovery" of bacteria that, supposedly, could live on arsenic instead of phosphorous.[38] If arsenic could be incorporated into DNA and RNA, then phosphorus would not be the mandatory molecular building block of life that everyone thought it was, and the textbooks would have needed rewriting.[39] This finding, too, soon turned out to be a dud. The hypotheses of cold fusion and arsenic-loving bacteria created fads that, because of the intense interest they generated, fizzled, which allowed the scientific record to be quickly corrected.

Naturally, scientific herd behavior is not always praiseworthy. It can deflect young scientists from more productive research paths, delay new and important discoveries, siphon off research funds to narrow specialties, and leave solid, but less-nimble, laboratories in the lurch when scientific winds suddenly shift. The sociology of science, including a tendency to follow fads, is too complex and multidimensional to allow for simplistic generalizations.

The pressing question here is, "What do fads have to do with the hypothesis?" Is there evidence that, for example, Curiosity-Driven Science, or questioning and model building (Chapter 10) are less likely to trigger fads? Has anyone made the case that opening of a new line of research by a Discovery Science project is immune from inspiring groups of imitators to follow suit (i.e. from giving rise to a fad)? If not, then why blame the hypothesis for fads? Conversely, if researchers were drawn to a particular area of hypothesis-based science because they could readily understand it, appreciate its significance, and see how to test it, should we fault the hypothesis? Or credit it for fostering good scientific conduct?

11.E Inductive Reasoning as a Default Cognitive Function

In trying to understand the irrational, as well as the rational, sides to scientific thinking, we've talked about biases and fads, and its time to revisit inductive reasoning in this context because it shares similarities with them. One of the enduring mysteries (for me, anyway) about inductive reasoning is why anyone still believes that it is a special intellectual skill. We've seen (Chapter 1) and dismissed such profundities as "induction is how you justify the expectation that your computer will not suddenly blow up," or that it "is the reasoning by which one determines the simple expectation that the sun will rise and then set at a set time each day. . . . It is a particular predictive model of reality that is verified on a daily basis." And we know that a rationalist would counter that our scientific confidence in the sun's rising is grounded on well-tested physical theories that explain why it does not suddenly explode or go wandering away from Earth. We have no reason to think that these exhaustively corroborated theories are not true, and inductive inferences from having observed countless sunrises doesn't add anything to what we know. On the contrary, at some point in the distant future inductive reasoning will fail—our star will have entered the final stages of solar evolution and will not rise despite having done so roughly a trillion times. Long before then, however, science will have predicted its end (and, we hope, devised a back-up plan). For the most part, we need reasons to think that natural things will *not* continue the way they are, rather than inductive reasons to think that they will.

But, if induction is not a special inferential reasoning process that leads human beings to expect regularity, why do we expect it? Or, to put it another way, does anticipating regularity in the environment qualify as "reasoning" of any kind? It is a fair question. Expecting regularity does not, for one thing, require conscious thought; most of us never cogitate about sunrise at all: we wake up and start our days and there's the sun. Observing and responding to repetitive events in the environment is built into the human cognitive toolkit that was shaped by evolutionary changes that took place millions of years ago when our primate ancestors were eking out a living on the African veldt. Genes that made it easier to find food, shelter, and chances to procreate; genes that allowed us to notice and keep away from danger, etc., were passed on. The culmination of all this evolution is that we have a genetic predisposition to detect meaningful regularities in the environment and to act on them. This is the heart of "inductive reasoning," and it explains why we find its conclusions so irresistible.

Of course, if expecting environmental regularities is wired into our genome, there is no reason to suspect that human beings would be solely blessed with it. All animals would surely benefit by having such an ability, and indeed, they all do. It's why your dog, Snowball, runs into the kitchen when he hears you opening the dog food; it's why the ducks in the pond paddle over to you when you start opening your lunch bag, etc.

In point of fact, I may be guilty of a politically incorrect "taxonism" (an unfair preference for one phylogenetic taxon over another). It appears that bacteria and plants have claims to being inductive reasoners, as well. Bacteriologists have found that the omnipresent gut bacterium *Escherichia coli* (*E. coli*), can learn to evolve in response to regular changes in environmental conditions.[40] The experimenters systematically changed the temperature in which the bacterial colonies lived from 25°C to 37°C and simultaneously lowered the oxygen content in the atmosphere from 20% to zero. After a few weeks (hundreds of generations) "the bacteria had 'learned' to anticipate the drop in oxygen by altering their metabolism just after the temperature change."

Not to be outdone, botanists got into the act by showing that the shoots of garden peas will grow into the arm of a Y-maze where light is "predicted" to be found.[41] Monica Gagliano and colleagues used a Pavlovian-training protocol in which a neutral cue, a breeze caused by a fan, was either paired or not paired with light during a training period. Later, during the test phase with no light, the plants grew into the arm in which breeze had been associated with light during training. The authors concluded that "associative learning represents a universal adaptive mechanism shared by both animals and plants." We naturally take it for granted that the human mental processes that mediate induction are vastly superior to whatever is going on in bacteria and plants, yet at a descriptive behavioral level, it is not obvious that our attitude is warranted.

Actually, you might argue that the Darwinian prevalence of inductive-like behaviors is sufficiently widespread that it turns the philosophical Problem of Induction inside out. We needn't justify our belief that nature is uniform in order to justify induction, as philosophers have always assumed. Instead, we must recognize that the actual uniformity of nature is responsible for the tendency to depend on induction in the first place. That is, the genetic advantage (i.e., continued existence) conferred on life forms having the capacity to sense and act on regularities in nature, demonstrates that nature must be fairly regular on the time spans required for evolution. In a totally chaotic world, expecting environmental regularities would be fatal because there would be none; you'd rarely find food today where you found it yesterday; the tiger would always be popping up at the water hole at unexpected times, and so forth.

Eons of correctly predicting regularity does not mean that it is always a great idea to assume nature is regular, of course. Dinosaur-extinguishing meteors or environment-reshaping human beings can forge immense changes on shorter time scales than it takes most species can adapt. Conversely, you can't make the argument that, because man-made, global catastrophes have not yet happened, they cannot happen. These illustrations show that flawed automatic inductive reasoning can play havoc with our ability to reason scientifically.

11.F Coda

In this chapter, we began to investigate how automatic thinking affects scientific reasoning. Our brain is an organ that is designed in large part to help us make sense of the world by constantly generating hypotheses about what is going on in and around us. The hypothesis therefore is one of the most basic thinking tools, and, rather than fantasizing about how scientists can do away with it, we should concentrate on understanding and mastering our hypothesis-generating tendencies so we can make sure they work to our advantage.

Because hypothesis generation is automatic and mostly unconscious we are not always able to "justify" (i.e., to explain and account for) them, and another word for a tendency to act in ways that we can't consciously and explicitly justify is "bias." Bias is therefore unavoidable, and it may have positive, negative, or neutral consequences. While bad bias can affect scientists, as it does everyone else, there is no reason to associate its ill-effects with the hypothesis in particular. Once we realize and accept the prevalence of bias, we can focus on the more important tasks of ferreting it out and keeping it from adversely influencing our thinking. The danger is not in the bias or the hypothesis, per se, but in the unrecognized and unexamined bias that is associated with tacit and unacknowledged hypotheses.

We also examined the phenomenon of "fads" in science and, again, we noted that the propensity of scientists to follow flashy trends has its good and bad sides. We shouldn't reflexively look down on scientific fads, but we should analyze the hot topic itself and decide whether concerted effort on it is called for: Is it necessary or superfluous? And we should keep in mind that, as was the case with bias, there is no reason to conclude that hypothesis-based research is uniquely prone to attracting crowds of imitators.

Finally, inductive thinking is deeply ingrained in our nervous systems because it represents a primitive survival strategy that allows all animals, and perhaps all cellular life forms, to detect and respond adaptively to regularities in their environments. As this trait seems to be the core of what is most often meant by "induction" (Chapter 1), then there is little cause to consider it a special intellectual property of the human mind. Appreciating the true nature of inductive thinking should make us far more reluctant to want to substitute it for hypothesis-based thinking.

Notes

1. Daniel Dennett, *Consciousness Explained* (New York: Back Bay Books; 1991), p. 314.
2. Francis Bacon, *Novum Organum: Or the True Directions Concerning the Interpretation of Nature.* Originally published in 1620; James Spedding, translator (Amazon Digital Services, Kindle edition; 2009), sections XXXVIII to XLIV.
3. David Hume, *A Treatise of Human Nature: Being an Attempt to Introduce the Experimental Method Into Moral Subjects; Book 1: Of the Understanding.* Originally published in London in 1738; reprinted and edited with an introduction by D. G. C. Macnabb (Cleveland: Meridian Books; 1969), p. 134.
4. Eric Kandel, *The Quest to Understand the Unconscious in Art, Mind, and Brain, From Vienna 1900 to the Present* (New York: Random House; 2012), p. 203.
5. Karl K. Szpunar and Endel Tulving, "Varieties of Future Experience," in M. Bar (Ed.), *Predictions in the Brain: Using Our Past to Generate a Future* (New York: Oxford University Press; 2011), p. 1.
6. Lawrence W. Bursalou, "Simulations, Situated Conceptualizations, and Predictions," in M. Bar (Ed.), *Predictions*, pp. 27–39. Bursalou's situated conceptualizations are patterns somewhat like implicit hypotheses. When a component of a previously stored pattern is experienced, pattern completion in memory gives rise to predictions about what will happen next.
7. Michael S. Gazzaniga, *Who's in Charge: Free Will and the Science of the Brain* (New York: HarperCollins; 2011), p. 82.
8. Y. Pinto, E. H. F. de Haan, and V. A. F. Lamme, "The Split-Brain Phenomenon Revisited: A Single Conscious Agent with Split Perception," *Trends in Cognitive Science* 21:835–851, 2017.
9. Gazzaniga, *Who's in Charge*, pp. 75–104.
10. Daniel Kahneman, *Thinking Fast and Slow* (New York: Farrar, Straus and Giroux; 2011), chapter 7.
11. Christof Koch, *Consciousness: Confessions of a Romantic Reductionist* (Cambridge, MA: MIT Press; 2012) pp. 2–3); Koch cites the philosopher David Chalmers as the originator of the term.
12. Daniel M. Wegner, *The Illusion of Free Will* (Cambridge, MA: MIT Press; 2002); Koch, *Confessions*, chapter 7.
13. Thomas Huxley, as cited in William James, *Principles of Psychology, Volume 1* (Originally published by Henry Holt & Co.; reprinted New York: Dover; 1964), chapter V.

14. Koch, *Confessions*, p. 78.
15. For review, see Patrick Haggard, "Conscious Intention and Motor Cognition," *Trends in Cognitive Science* 9:290–295, 2005. Benjamin Libet's experiments on consciousness have widely been interpreted as having something to do with "free will," a major philosophical thicket that we will stay out of. Suggested readings include the sources mentioned in Notes. 1, 7, and 11; see also Peter Ulric Tse, *The Neural Basis of Free Will* (Cambridge, MA: MIT Press; 2013).
16. C. S. Soon, M. Brass, H.-J. Heinze, and J.-D. Haynes, "Unconscious Determinants of Free Decisions in the Human Brain," *Nature Neuroscience* 11:543–545, 2008. An update of Libet's experiment. Experimenters scanned numerous brain regions simultaneously while the subject chose a button to press. By correlating the activity in each brain region with the final decision, Lau et al. could deduce not only which regions were involved in the decision, but also predict which finger, right or left, was going to move. The brain activity preceding the movement began as much 10 seconds before the subjects reported making their decisions; see H. C. Lau, R. D. Rogers, P. Haggard, and R. E. Passingham, "Attention to Intention," *Science* 303:1208–1210, 2004.
17. Critics dispute the interpretation of the Libet-type experiments, arguing that the preceding activity does not necessarily imply that the path from unconscious processes to conscious thoughts to action is a serial (linear) process, but that the prior unconscious activity causes two downstream consequences that occur in parallel, one leading to the finger movement and the other to the conscious thought. Another possibility is that consciousness represents an emergent process that arises spontaneously from a kind of competition among a number of cortical modules for awareness. Consciousness may not be a process simply caused by some other preceding activity, but an emergent principle that takes place on its own time scale which is always appropriate to that time (i.e., it is not "too late" to be involved in action), as Libet had argued.
18. Haggard, "Conscious Intention."
19. August Kekule's dream; see https://en.wikipedia.org/wiki/August_Kekulé.
20. Otto Loewi's dream; see https://en.wikipedia.org/wiki/Otto_Loewi.
21. Raymond S. Nickerson, *Conditional Reasoning: The Unruly Syntactics, Semantics, Thematics, and Pragmatics of "If"* (New York: Oxford University Press; 2015).
22. Sinéad L. Mullally and Eleanor A. Maguire, "Prediction, Imagination, and Memory," in M. S. Gazzaniga and G. A. Mangun (Eds.), *The Cognitive Neurosciences*, 5th ed. (Cambridge, MA: MIT Press; 2014), p. 605.
23. R. M. J. Byrne, "Counterfactual Thought," *Annual Review of Psychology* 67:135–157, 2016.
24. S. L. Mullally and E. A. Maguire, "Counterfactual Thinking in Patients with Amnesia," *Hippocampus* 24:1261–1266, 2014.
25. Byrne, "Counterfactual Thought," ibid.
26. Ibid.
27. Müller-Lyer illusion of unequal line lengths; see https://en.wikipedia.org/wiki/Müller-Lyer_illusion.

28. Steven Weinberg, *Dreams of a Final Theory: The Scientist's Search for the Ultimate Laws of Nature* (New York: Vintage Books; 1993).
29. Frank Wilczek, *A Beautiful Question: Finding Nature's Deep Design* (New York: Penguin Press; 2015).
30. Richard Feynman, as quoted in James Gleick, *Genius: Richard Feynman and Modern Physics* (New York: Little, Brown and Company; 1992). Feynman tended not to dumb down his ideas to get them across to anyone except physicists. When asked to explain quantum mechanics to a layman, "in terms that I would understand," Feynman replied that "In terms that you understand, I don't understand it."
31. Vicki Daitch and Lillian Hoddeson, *True Genius: The Life and Science of John Bardeen: The Only Winner of Two Nobel Prizes in Physics* (Washington, DC: Joseph Henry Press; 2002), p. 57, quoting J. Bardeen.
32. James Comey, *A Higher Loyalty: Truth, Lies, and Leadership* (New York: Flatiron Books; 2018), p. 105.
33. The disgrace of Lance Armstrong; see https://en.wikipedia.org/wiki/Lance_Armstrong.
34. The Jan Henrik Schön scandal; see https://en.wikipedia.org/wiki/Schön_scandal.
35. Firestein, *Ignorance*, p. 78.
36. Political and social influences in spread of Ebola epidemic; see https://en.wikipedia.org/wiki/Ebola_virus_epidemic_in_Sierra_Leone.
37. The entire saga of Stanley Pons's and Martin Fleischman's ill-fated report of a room-temperature energy-generating fusion process is well told here:

    But see http://coldfusionnow.org/peter-hagelstein-on-the-fleischmann-pons-experiment/ for a contrary view that Pons and Fleischman were, in a sense, right. The fusion process works, but the reaction requires more energy to be supplied than it produces, making it potentially useful for some purposes- but hardly a cure for the global energy crisis. Also see https://blogs.scientificamerican.com/guest-blog/its-not-cold-fusion-but-its-something/.
38. F. Wolfe-Simon, J. Switzer Blum, T. R. Kulp, G. W. Gordon, S. F. Hoeft, J. Pett-Ridge, J. F. Stolz, et al., "A Bacterium That Can Grow by Using Arsenic Instead of Phosphorus," *Science* 332:1163–1166, 2011.
39. Wolfe-Simon et al.'s paper (cited in Note 37), generated great controversy and technical commentary when it was posted online. The end results were that her paper had not demonstrated that arsenic could substitute for phosphorus as a building block of life, but it did generate a number of other interesting insights and questions. The gist of the critical commentary is here: http://www.sciencemag.org/news/2011/05/science-publishes-multiple-critiques-arsenic-bacterium-paper.
40. I. Tagkopoulos, Y.-C. Liu, and S. Tavazoie, "Predictive Behavior Within Microbial Genetic Networks," *Science* 320:1313–1317, 2008.

    Bacteria can learn; see http://www.nature.com/news/2008/080508/full/news.2007.360.html.
41. M. Gagliano, V. V. Vyazovskiy, A. A. Borbély, M. Grimonprez, and M. Depczynski, "Learning by Association in Plants," *Scientific Reports* 6:1–9, 2016.

12
Thinking Rationally About Heuristics and Biases

12.A Introduction

Can you be biased and still be rational? This may sound like a trick question, and it is. In the preceding chapter, I noted that you can assess whether or not a statement was biased by its *justifiability*; if you couldn't justify your position according to the known facts, then it would be, by definition, biased. From this point of view, we all hold countless biased positions that are inoffensive, even sensible, because many factors, including lack of conscious awareness, prevent us from being able to justify them.

We all recognize that from another point of view, biased positions are *irrational*; holding them flies in the face of reason. You can't justify them because, well, they're simply wrong. This is the position we're going to examine in this chapter, and you can see right away that we'll have to confront the issue of *rationality*. This issue is what makes the opening question a trick: whether or not bias is rational depends on how you define rationality. This chapter will cover only a few biases rather than a whole catalog of them in order to consider the broader issues of why we have biases and how they affect scientific thinking.

We ordinarily think of rationality as being associated with concepts like "consistent with norms" and "in accordance with reason," as well as with "logical" and "tending to produce true beliefs." Indeed, technically speaking, "rationality" has many definitions. Wikipedia notes that "rational" has specialized meanings in philosophy, economics, sociology, psychology, evolutionary biology, game theory, and political science." (And mathematics should have been included.) In this chapter, we'll go into two definitions of rationality in some detail because they directly affect how we think about the scientific hypothesis and the reliability of scientific thinking. One definition is based on *logic*, in the formal sense of relating to deductive validity that we discussed in earlier chapters, and the other on *adaptability to the environment*, essentially a pragmatic approach that is related to what works and has worked in the past. An action that is justifiable pragmatically may or may not be formally logical.

12.B Heuristics, Biases, and the Hypothesis

Scientific thinking is often held up as the standard—not exactly a golden one, but an accepted standard—for rational, objective thinking and reasoning.[1] The insights we gain from studying scientific thinking are supposed to find their way into daily life as aids to practical thinking. Nevertheless, scientists are hardly immune from biased thinking. The physicist, Henri Poincare[2] warns, "It is often said that experiments should be made without preconceived ideas. That is impossible. . . . Every man has his own conception of the world, and this he cannot easily lay aside." Poincare thinks that dangerous biases come from unconscious, preconceived ideas and recommends something like multiple, consciously created hypotheses to deal with them. If we have several ideas about a given subject, he believes that although they "will generally disagree," this "will force us to look at things differently." His comments are instructive but don't say much about bias itself: Is cognitive bias rational or irrational, for example? Why do we have biases in the first place?

Bias has become a big concern because of the Reproducibility Crisis; however, it has been at the center of a heated debate in cognitive psychology for a long time. This debate matters to those of us interested in the scientific hypothesis because, as we've already seen in Chapter 10, some critics detect an unsavory connection between the hypothesis and bad forms of bias. We begin with a look into what it means to be biased.

Bias is a complex topic; psychologists distinguish well over 100 different forms of *cognitive bias.*[3] We'll group the great majority into a few broad categories before zeroing in on a few. There is a lot of angst about biases that we can call *social biases* that lead to malicious behavior toward others; that isolate, denigrate, or discriminate against them; or that cause scientific misconduct. Such biases are morally and ethically abhorrent, and they're not what we're talking about here. And I'll distinguish between cognitive bias and biased behavior that doesn't reflect peoples' true beliefs. You might deliberately act in a biased way for a worthy cause—say, advocating for your little brother to be allowed to join the team despite his clumsiness. You think he deserves a chance—not that he is really any good. If you were cognitively biased, you'd be convinced that, contrary to the evidence, he is a good player.

Most of the cognitive biases that we'll be concerned with are universal traits of human thinking that frequently result from using mental shortcuts called *heuristics*. Heuristics are normally useful and readily-available tools for simplifying complex tasks; they come to us so naturally that they're easy to overlook until they create problems. Heuristics and biases arise automatically and unconsciously; we'll go over a few examples shortly.

There are two main schools of thought regarding heuristics, and they are distinguished principally by how heavily they weigh the useful versus the harmful, the rational versus the irrational sides of heuristics. One school is founded on insights derived from evolutionary psychology, the discipline that holds that the foundational structures and processes of our minds are adapted to succeed in the kinds of environment that our ancient apelike ancestors encountered. This school is associated with Gerd Gigerenzer,[4] and it focuses on the advantageous nature of heuristics and biases—their *ecological rationality*. Gigerenzer sees heuristics as "efficient cognitive processes that ignore information"[5] so that we can act quickly and effectively. This school is particularly impressed by cases in which heuristics outperform purely logical problem-solving strategies. Heuristics make sense because they help us navigate the world safely and productively and were, therefore, selected for by evolution.

The other school of thought regarding heuristics, personified by Nobel Prize-winning psychologist Daniel Kahneman,[6] thinks they are "simple procedures[s]" that help "find adequate, though often imperfect, answers to difficult questions."[7] Despite acknowledging its potential benefits, Kahneman's school devotes its attention to the disadvantages of heuristic thinking; to the cognitive errors we commit and the biases we fall prey to when using heuristics.

These opposing points of view have important, though widely divergent, implications for scientific thinking, and, in order to compare them, we need to expand on their interpretations of rationality. We'll begin with Gigerenzer's ideas.

12.B.1 The Fast and Frugal Program: Heuristics as Useful Mental Tools

For Gigerenzer, the epitome of a perfectly designed heuristic is the *gaze heuristic* that makes it possible for a baseball outfielder to catch a fly ball or a dog to catch a Frisbee. While you could, in principle, calculate the flights of baseballs and Frisbees, even if you knew all of the relevant variables—the object's initial speed and launch angle, wind direction and velocity, etc.—and you never do, it wouldn't do you any good. Before you finished the calculations, the thing you're trying to catch would already have hit the ground.

Yet you and your dog can perform flawlessly by following a simple heuristic: if the object is already high in the air, look right at it, and run toward it just fast enough that your eyes remain fixed on it at the same gaze angle. Eventually your path will intersect the object's and, if you don't fumble, you'll make the catch. Modify the heuristic slightly, and it will work while the object is still rising. The gaze heuristic represents an ideal of the *fast-and-frugal* solutions that, for countless millennia, simplified the otherwise overwhelming complexity of the world for us, the species that Gigerenzer calls "homo heuristicus."[8]

Not all heuristics have been programmed by evolution. An example of a culturally advanced and financially rewarding heuristic is to invest your money equally in a number (N) of alternatives (the so-called 1/N heuristic), and leave it alone, rather than darting around buying and selling stock in individual companies. Ingenious people armed with sophisticated data analytic schemes routinely fail to predict the stock market.[9] By sticking with the trivially simple 1/N heuristic, you will miss out on the miniscule chance to strike it rich by investing early in the next colossal winner, like Apple, Inc., but you will avoid going bust by betting on the far more numerous losers.

Though few heuristics are as elegant as the gaze heuristic or as narrowly focused as the 1/N heuristic, most help us do better, most of the time, than we would without them. Our reliance on heuristics is, therefore, *rational*, provided that we keep in mind the words of Nobel Prize-winner Herbert Simon, that we have only "bounded rationality"[10] to begin with. Our mental capacities are limited and, hence, perfect, logical rationality is rarely even an option.

Gigerenzer's program also relies heavily on another of Simon's principles: "satisficing," a Scottish coinage that captures the spirit of "satisfying" and "sufficing." Satisficing is the judicious middle ground. When you are satisficing in making a choice, you are not trying to maximize an outcome; instead, you are weighing your alternatives with a minimal standard, an "aspiration level" that you're trying to achieve, and, as soon as you discover one, you take it without further ado or analysis. *Take-the-best*[11] is an example of a "one-reason" heuristic that we resort to on a daily basis. More elaborate fast-and-frugal heuristics are constructed from basic principles like these, and I'll have more to say about them later. Both bounded rationality and satisficing are on display in the resolution to the *bias-variance dilemma* that accompanies many scientific conundrums.

12.B.1.a The Bias-Variance Dilemma

One of Gigerenzer's pivotal insights is that "less is more," referring to the capability of stripped-down heuristic solutions to surpass more complicated ones in predictive accuracy. He accounts for this by pointing to the bias-variance dilemma, also sometimes called the *bias-variance tradeoff*, which he uses to illustrate how bias resulting from the use of heuristics can be a good thing.[12]

The dilemma comes from the fact that the true relationship between what you observe and the aspect of reality that you want to know about is obscured by unknown amounts and kinds of *noise*—uncontrolled variability in your data. (The variability has its origins in two sources: true individual differences, which causes *sampling error*, and *irreducible error*, which is the random error in our measuring instruments, the environment, etc. Irreducible error affects all measurements, so we rarely give it special attention.) When you're trying to understand the world in terms of hypotheses or models, you have to decide how to deal with variability and still portray reality in meaningful ways.

12.B.1.b The Bias-Variance Dilemma: Example

Say you're interested in predicting young children's heights as they age between 4 and 10 years old. With permission, you go to your local public school and measure the heights of 100 preschool kids. When you plot their heights in inches against their ages in months, the general trend is obvious—they get taller as they get older—and there is much variability; at any given age, some children are shorter and some taller than others. What's the best way to capture the overall relationship between height and age?

One way would be to connect all the dots in the graph, beginning with the height of the youngest child, drawing a line to the height of the next oldest, to the next, etc., finishing at the height of the oldest. The jagged line would accurately represent the growth rate of your group but it would have no predictive value; if you measured a different group of kids of the same ages, the same jagged line would not connect the dots in the second group. As with the model of the coin toss that we talked about earlier, the age–height plot would be entirely *unbiased* and maximally variable.

Alternatively, you could mathematically fit a straight line to your data and use the line to encapsulate your conclusions about children's growth. The fitted line would depict the drift of the growth and simplify the plot enormously, although it would miss all of the variable information about their individual heights. The straight line would be both a crude summary of existing data and an implicit hypothesis. You'd probably feel that, if you measured the heights of other kids in the same age range, the straight line would be a fair approximation of their growth rate as well. Most importantly, because you decided to fit a line, rather than another function, the line would be a form of *bias*. With it, you imposed your opinion on the dots; there was no line in the data. Bias, in this general sense, means weighting information unequally—diminishing some parts of it while paying a great deal of attention to other parts. Bias may be good or bad; reasonable or unreasonable. Bias reveals an inclination toward a particular outcome; it is not a detailed defense of the outcome. This is the tradeoff: increased robustness and predictive cogency against the loss of detail. When we use heuristics, we introduce bias.

Automatic, implicit hypothesis generation represents a type of cognitive heuristic thinking that is usually adaptively accurate, like the generally trustworthy guidance that we get from our sensory systems. Because of them, we confidently act as if the external world is stable, predictable, and benign. Our countless implicit hypotheses about it are rarely falsified even though they're tested all the time; that's why we get away with casually assuming that they're accurate and are taken aback from time to time when we find that they're not.

Still, we shouldn't forget that biases can also throw us off course. Like sensory physiologists who study visual illusions to find out how our visual systems normally operate, cognitive scientists study biases to discover how our cognitive systems normally operate.

12.B.2 Heuristics as the Source of Bad Cognitive Bias

In their *Prospect Theory* of human decision-making under conditions of uncertainty, Kahneman and Amos Tversky[13] detailed what they saw as the "irrationality" of typical human thought, where rationality for them refers to the way in which an omniscient, infallibly logical observer would think. They concluded that, when we use heuristics, the biases we adopt are *errors* in thinking: they should be stamped out, although Kahneman is pessimistic that psychology can teach us to do that.

In *Thinking, Fast and Slow*,[14] Kahneman puts forward a two-system interpretation of heuristics and biases. His vision is that each of us has two fundamentally different cognitive systems—System 1 and System 2—that don't follow the same rules. Roughly speaking, System 1 is at work all of the time processing information automatically in a swift and effortless, though often careless, manner. System 1 gives rise to our implicit hypotheses, heuristics, and biases. System 2 is plodding, logical, and careful, albeit often lazy and hard to engage. When we rigorously examine and test a hypothesis, we're using System 2. Although they can be associated to some extent with separate brain regions, the Systems are mainly convenient labels that designate markedly dissimilar forms of cognition.

Kahneman accounts for many biases as the result of *cognitive illusions*. What are cognitive illusions? They're a bit like the sensory illusions that we discussed in Chapter 11. You experience an illusion when your brain, in automatic-processing mode, presents you with a discrepancy (e.g., you *see* two lines of unequal lengths, but your ruler *measures* two identical lines). Analogous discrepancies occur between our gut-level, instinctive responses to mental challenges and the solutions that we get from dispassionate, methodical analysis. Cognitive illusions can be as seductive as the sensory kind. Here is a quick example: If a ball and a bat together cost $5.50, and the bat costs $5.00 more than the ball, how much does the ball cost? If you reflexively answered, "$0.50," you were the victim of a cognitive illusion. Think about it.[15]

Kahneman and colleagues see cognitive illusions as evidence of *irrational* behavior. This is the point where we need an index of rationality, a standard like a ruler, so we can judge it objectively. If you're accustomed to thinking about rationality as a fuzzy concept with no place outside of philosophy class, you may be surprised to learn that a real-world standard exists; it is called *Expected Utility*

Theory (or just *Utility Theory*).[16] (If you're unfamiliar with the theory, take a minute to go over Box 12.1.)

12.B.2.a What Does Rational Decision-Making Look Like?

Economists have long been obsessed with how people make financial decisions, and lessons about decision-making characteristically begin with examples involving money. Money problems are ubiquitous because everybody can readily relate to them, and, furthermore, these decisions have quantifiable outcomes so we can identify good and bad decisions unambiguously. Later we'll see how to extend Utility Theory beyond the wallet.

Expected Utility Theory says that the psychological value of money is determined solely by its *utility*—the amount of personal satisfaction, measured in goods and services, that you can buy with it. The utility of $1,000 is determined by how much desirable stuff you can get for $1,000. *Rational* decision-makers base their decisions on how much utility they expect to have as a result of their decisions, and they always try to *maximize the absolute amount* of their expected utility (e.g., to have the most money). Nothing else matters. And rational decision-makers always act *selfishly* to maximize their utility; they pay no attention to the needs or wants of others, the benefits to society, etc. (Rational decision-makers don't have many friends.)

As I mentioned, it is standard practice to discuss Utility Theory in terms of money and fiscally related "utilities," however the theory "serves as a model for various psychological processes, including motivation, moral sense, attitudes, and decision-making."[17] In other words, you can evaluate the rationality of all sorts of behavior by looking at whether or not it maximizes the psychological "utility" that's involved and otherwise adheres to the precepts of Utility Theory. That's more or less what you're doing when you weigh the "pros" and "cons" of spending time with a friend who, on the one hand, is a lot of fun to be with yet who, on the other hand, can be a real jerk. Ultimately, you opt for the choice that you expect to make you happiest, or you should if you want to be rational by the lights of Utility Theory.

Illusions, sensory or cognitive, are compelling because they depend on low-level, reflexive information processing; they affect us before we know it. When you put your cold-adapted and warm-adapted hands into the same bucket of lukewarm water, you felt the water temperature differently with each hand and knowing the truth—that the temperature had to be the same for both—didn't change how it felt. Breaking the grip of a cognitive illusion like the bat-and-ball problem can be almost equally hard to do.

Cognitive illusions resemble sensory illusions in ways besides their almost irresistible nature, however. Kahneman and Tversky postulate that we evaluate gains and losses with respect to a given *reference point*, not in absolute terms.

Box 12.1 Expected Utility Theory: A Brief Overview

For the basics of the theory, we need to start with the related notion of *expected value*, which is a statistical term. You can think of expected value as the average value of something that you don't have (or of something that hasn't happened), and it depends on the probability of getting that thing (or having it happen). To calculate expected value, you multiply probability times value. For instance, if someone offered you a lottery ticket that had a 50% chance of winning $100, you'd multiply the probability of the win, 0.5, times its value, $100, and find that the expected value of the ticket is $50.

Utility is a term from economics that rates how you, personally, value something; utility is the amount of *satisfaction* that you get from a particular good or service. Utility is, therefore, *subjective* value. It is correlated with the value of money, but is not equivalent to dollar value because two people might get different amounts of satisfaction from a certain good or service. If you're a vegan, the utility of a meal at an upscale vegan restaurant could be high even if the meal were a bit expensive; if you're a committed carnivore, even a cheap price would probably not increase the utility of a vegan meal for you. The concept of utility even applies to money itself: if you're broke and your rent is due today, you might find a "payday lender's" exorbitant interest rate acceptable, despite the fact that it will cost you more money in the long run. Like expected value, expected utility is an average. Expected utility depends on what something would cost, how much you want it, and the chance that you'll get it.

Modern expected utility theory, often referred to as *rational economic theory*, was formulated by John Von Neuman and Oskar Morgenstern, (see Note 16) and includes a set of logical axioms that define rational decision-making. For example, if you prefer option A over option B, and option B over option C, then, according to the theory you *must* prefer option A over option C. If you chose apples over pears, and pears over bananas, then you must prefer apples over bananas. If your behavior conforms to the axioms, you are acting rationally, otherwise not.

The only thing that matters for economically rational thinkers is the end point; their happiness depends only on the total utility of what they have, not whether they gained or lost in getting that amount. Expected utility theory does recognize that the utility of money does not increase linearly with how much you have; someone whose financial worth is only $1,000 would value a windfall of $100 more than would a millionaire. The relationship, according to Daniel Bernoulli, one of the historical forbears of the modern theory, is logarithmic: the millionaire should value $100,000 roughly as much as the less-well-off person would value $100.

Imagine two people, Joe and Jane. initially Joe has $1,000 and Jane has $3,000. One day, Joe inherits $1,000 from a rich uncle and Jane loses $1,000 in the stock market. Now they both have the same amount of money, $2,000 and, by the rational standard of Expected Utility Theory, both should be equally happy. Clearly, this is ridiculous; everybody knows that Joe will be ecstatic, and Jane will be distraught. Real people care not just about where they are financially, but how they got there. Like the hands in the lukewarm water, Joe and Jane experienced opposite emotions when they were exposed to their new common level of wealth. They were either happy or sad depending on how their fortunes had changed from their individual reference levels, not what their final absolute level was. And this, in turn, suggests that they had "adapted" to their preexisting reference levels of wealth. Rational economic agents don't experience any of these emotions.

And rational agents respond to losses and gains in a symmetrical manner: the prospect of losing $100 is just as unpleasant as the prospect of gaining $100 is pleasant. Do you agree? Would you take a bet on the toss of a fair coin if you were to win $100 if it came up heads, but lose $100 if it came up tails? (If you would normally don't take bets of any kind, please relax for a moment and consider this one.) No? How about if you were offered $150 for the win against $100 for the loss? Still no? Expected Utility Theory says that you definitely should take the bet. The expected value of a gain of $150 is 0.5 × $150 = $75, and the expected value of a loss is 0.5 × –$100 = –$50. The overall *expected value of the bet* is the sum of the two possible outcomes: $75 – $50 = +$25 (i.e., a net gain, meaning that on average you'll come out $25 dollars ahead if you take bets like this). Most people don't take it. Theory be damned, we want a chance to win at least $200 before we're willing to risk losing $100. In our account books, losses "count" for much more than gains. In Prospect Theory, Kahneman and Tversky called this asymmetry *loss aversion*. According to the theory, loss aversion is a powerful psychological bias that comes into a large variety of decision-making processes.

12.C Different Meanings of Rationality; Different Solutions to Problems

At this point we have two conflicting accounts of heuristics and biases: the fast-and-frugal program of the Gigerenzer school and the Prospect Theory of Kahneman and Tversky. How do their competing worldviews affect our understanding of hypothesis-based scientific thinking and practice?

Most of us readily concede that we're not fully economically rational and, considering the personality traits of purely rational economic agents, we're glad that we're not. Still, what does any of this have to do with scientific thinking?

I mentioned that Gigerenzer and Kahneman are often at odds about how to account for basic phenomena. Again, Prospect Theory accepts that traditional logic, the rules of probability, and Expected Utility Theory define *rational* behavior: if your decisions deviate from their rigid framework, then you're being *irrational*. Naturally, no human being is consistently rational, but these are the yardsticks for measuring how irrational you are.

Gigerenzer argues that the pragmatic, achievable standard of *ecological rationality* is the appropriate benchmark for human behavior; ecological rationality is what works, what produces beneficial results in the real world. Its effectiveness follows from its evolutionary origins. Ecologically rational heuristics have worked for millions of years, and, when viewed in this light, they makes perfect sense even when they are at odds with the textbook standard. Ecological rationality is close to what in everyday terms would be considered "reasonable" and is not necessarily rigorously "logical."

I'm going to argue that scientists should have a clear idea of when to apply which standard of rationality, especially when it comes to the topics of hypothesis-based thinking and biases. This is a focal point of the discussion, and, to try to avoid any ambiguity I'll review a few classic cognitive problems in the next section and compare how the two schools approach them. If you're confident that you understand what's going on, you can probably skip ahead to Section 12.C.5. I'll start with the famous "Linda Problem" that I mentioned briefly in Chapter 8.

12.C.1 The Linda Problem

You remember that we had a description of Linda, 31, single, outspoken, very bright, a college philosophy major deeply concerned with issues of discrimination and social justice who participated in anti-nuclear demonstrations. We were asked whether it was more probable (or "more likely") that Linda was (A) a bank teller or (B) a bank teller who was active in the feminist movement. Most people choose (B).

As Kahneman and Tversky point out that, logically, this answer is incorrect. The reason is that *all feminist bank tellers* are, first and foremost, *bank tellers*. It must be the case that Linda, or any random person, is mathematically more likely to be a bank teller—period—than a bank teller with any kind of qualifying property. Think of a Venn diagram with two overlapping circles: one labeled "bank tellers" and one labeled "feminists." Each circle indicates the probability of being either a bank teller or a feminist.

The area where the circles overlap represents "feminist bank tellers," and it *must* be smaller than either of the two main circles.[18] In other words, the probability of being a feminist bank teller must be smaller than being just a bank teller or just a feminist. Indeed, if you think about the large number of female bank tellers in the United States[19] and the complexity of human nature, you'll probably agree that there must be some bank tellers with all of Linda's characteristics except for being feminists, and furthermore, that the total group of bank tellers also includes males. By the rules of probability logic, the answer favored by Prospect Theory has to be right and, as a corollary, selecting answer (B) has to be irrational.

Gigerenzer says the Prospect Theory argument is irrelevant. When it comes to real people's reasoning in this case, the conventional logical rules don't apply. It's true that we are not typically in sync with the dictates of pure logic, but, he says, it doesn't follow that we're irrational; we are rational under the appropriate definition of rationality. Formal logic is an artificial system of rules that are foreign to most people; it is "content-free," meaning that it deals with abstract concepts rather than tangible ones. We considered abstract reasoning like this in Chapter 5, where we saw that statistical hypothesis testing could decide which of the nonexistent species of space aliens was probably scalier than the other. In formal logic the meanings of the words of a statement, its *semantics*, often do not matter (that's why logicians explain the syllogism in terms of antecedents and consequents, P, Q, and R; see Chapter 2). Yet content-free logic may be useless when it comes to social communications because then semantics do matter.

In addition, as pointed out by the philosopher H. P. Grice, there are *conversational rules* that govern our interpersonal communications.[20] For instance, we assume that people are giving us *relevant* and *important* information when they communicate with us (and we occasionally get annoyed—"What's your point?"—when they don't seem to be). In fact, we are adept at picking up the unspoken message (*implicature*) of what we hear (e.g., we know that the question "Do you *really* want to go?" sometimes means, "*I* really don't want to go"). Gigerenzer and colleagues don't accept the interpretation that we fall for cognitive illusions; rather, the semantic content of a statement puts us into a social dimension where perceptive discourse requires sensitivity to underlying intentions along with (or instead of) robotic transmission of data. We are not incapable of being strictly logical; it is just that when obvious word meanings and their context command our attention, then we go with them. Most subjects who are given the Linda Problem do not see it as a puzzle for their abstract reasoning abilities; they understand it as a request for their best guess about Linda's character. Why else, they wonder, would they be getting all the nuanced information about her background? We commonly and often rightly "read into" and respond to aspects of a question that are not, objectively speaking, there. This may not be logical, but it is rational.

Furthermore, the meanings of words are rarely as clear-cut as they'd have to be to support a valid critique of subjects' solutions to the Linda Problem as being logically deficient. In contrast to the crisp denotation of "more probable" that Kahneman and Tversky took for granted, many investigators point out the plethora of synonyms for "probable" that you'll find in the *Oxford English Dictionary*, which include the more flexible "to be expected, anticipated, foreseeable, potential, credible, quite possible" and informal "a good bet, a reasonable bet." Once again, someone working from an imprecise definition could answer the Linda Problem and be both socially rational and inconsistent with the rules of probability logic. Hertwig and Gigerenzer[21] blame the wording of the Linda Problem for the troubles that people have with it and demonstrate that if you rewrite the problem in a way that encourages subjects to see it as a test of logic then "erroneous," socially sensitive answers decrease significantly (though not all researchers find as much support for this "linguistic ambiguity hypothesis"[22]). Evidence like this confirms that people can be analytical when they understand that the context calls for it. This isn't what you'd expect based on Prospect Theory's pessimistic view that people are either rational or, more often, irrational; that our ability to reason abstractly doesn't hinge on the specific words used. This raises the question: Do words matter as much in the overtly quantitative realm, or are we simply hopeless when it comes to numbers?

12.C.2 Probability Versus Frequency

Consider this problem: you know that 10% of the people in a particular town always lie and that 80% of the liars have a red nose; 10% of the people who don't lie also have a red nose. What is the probability that a red-nosed person that you happen to meet is a liar? Most people have a hard time getting the right answer. When it was given to school children or adults, almost none of the children and less than half (47%) of the adults got it right. Kahneman's would posit that our System 1 tries to answer quickly but is just not adept at quantitative problems and that our System 2, which is adept, is hard to arouse and unmotivated.

Now, recast the problem like this: eliminate the intimidating and multifaceted word "probability" and replace it with the evolutionarily friendly and numerically evocative concept of "frequency." You get something like this: in a certain town, 10 out of every 100 people that you meet always lie, and 8 of the 10 people who always lie have red noses. Of the 90 people who don't lie, 9 also have red noses. Now the question: If you meet a group of people with red noses, how many will be liars? Right away you can see that a total of 8 + 9 = 17 of every 100 people have red noses. Therefore, if you meet a group of red-nosed people, you can expect that 8 out of every 17 of them will be liars. In these terms, with actual numbers rather than probabilities, many children and most adults (~76%) get the solution.

Kahneman believes that many of us flail around when confronted with a mystery such as the red nose problem because we suffer from an intellectual malady called *base-rate neglect*. Essentially, we don't know how to incorporate the crucial information about the prevalence of red noses in the town population as a whole into our calculation, and so we pay no attention to it. In contrast, Gigerenzer believes that the real barrier to success is the problem's abstract terms; we need the context of the *natural frequencies* that we normally deal with. Natural frequencies, countable things, exist in the world, "probabilities" do not. Hence, the form in which the problem was presented was at fault, not the subjects' minds. Whereas Kahneman tends to be fatalistic when contemplating our inherent intellectual limitations, Gigerenzer is quite optimistic about the potential for improving our performance through education and policy changes in how information is presented to us. We'll turn to the societal implications of their viewpoints in Chapter 13.

12.C.3 The Wason Selection Task

The notorious Wason Selection Task, invented by psychologist Peter Wason,[23] highlights another distinction between logical and ecological rationality. In this task there are four cards in front of you, marked D, F, 3, and 7, you're told that each card has a number on one side and a letter on the other. You're supposed to test the "hypothesis" that "all cards with D on one side have 3 on the reverse side." The question is: What cards *must* you turn over to test the hypothesis? If you've never seen this one, try it before going on.

Ready? The minimal correct answer is that you must turn over the D and 7 cards.[24] How'd you do? The great majority of test takers (~75%) don't get it right. The reasoning goes like this: obviously, you have to turn over the D card, but turning over the 3 is pointless; the instructions we're given don't state that *only* D's are associated with 3's, so you learn nothing if the 3 card doesn't have a D on the back. And even if there is a D on the back, the proposed hypothesis could still be false. The acid test is the 7 card; if it has a D on the other side, the hypothesis is toast, yet if there is a different letter on the other side, then you have no reason to reject it. Again, Kahneman's conclusion is that people who don't do well on the task are irrational.

This is all well and good, but Gigerenzer and colleagues point out that if you change the labels on the cards, you can get a problem that is formally identical to the original but that almost everybody gets right. For example, imagine that the cocktail lounge that you own is going to lose its liquor license if anyone underage is caught drinking alcohol in it, and you resolve to check the customers carefully. You go to a booth where there are four customers with cards in front of them, and one side of each card identifies the beverage the customer is drinking while the other side lists his or her age. The cards are marked: Beer, Coke, 25, and 16. Which ones *must* you turn over to find out whether or not you're in trouble?

Probably, like most (~75%) subjects, you got this one right—you flipped over the cards marked Beer and 16. What happened? The problem has gone from being from an abstract, mechanical reasoning puzzle to one that is meaningful to human social circumstance. It's obvious that you don't care about how old the Coke-drinker is or what the 25-year-old is having. The evolutionary psychologists Leda Cosmides and John Tooby[25] see the improved performance on this form of the Selection Task as evidence of a built-in "cheating detector" that humans have evolved to ensure that members of a society obey its rules. Although not all psychologists are comfortable with that interpretation[26] many do agree that how you perform on the Wason Selection task depends on whether it is presented in terms that you care about; basically, its ecological rationality. Since our thinking abilities are adapted to our ancient ancestors' needs, it appears that instilling the ability to solve abstract, *content-free* problems was not high up on evolution's to-do list.

12.C.4 Framing Effects or Efficient Choosing

Imagine the following, admittedly pretty unrealistic, scenarios:

Scenario 1: You're given $1,000 by a well-funded (!) researcher; now choose between:

A. getting an additional $500 for sure.
B. taking a 50% chance to win an additional $1,000, otherwise getting nothing more.

Scenario 2: You're given $2,000 by the same well-funded researcher; now choose between:

C. losing $500 for sure.
D. taking a 50% chance to lose $1,000, otherwise losing nothing at all.

In Scenario 1 most of us choose the sure thing, choice A, get $500 for sure, and, in Scenario 2, the gamble, choice D, a 50% chance of losing $1,000.

The punch-line is that our choice pattern violates the rules of Expected Utility Theory, which only looks at the overall value (utility) of what we have. In contrast, we humans pay attention to the way in which the choices are expressed; that is, how they are *framed* by the wording. Choices A and Care sure things;

they guarantee that we'll walk away with $1500. Choices B and D are risky bets; they both offer the chance of having either $1,000 or $2,000 at the end. If we were logically consistent, we'd pick either choices A *and* C, or B *and* D, but most of us mix things up. Our behavior conforms to a general rule, however: we are *risk-averse* when outcomes are framed as potential gains, as in Scenario 1; that is, we play it safe when we stand to gain and prefer the guaranteed money over the chance of having more money ("a bird in the hand. . ."). When outcomes are framed as potential losses, as in Scenario 2, we become *risk-seeking*; we prefer to risk losing even more money rather than to voluntarily give up a smaller amount. This pattern of choices counts as irrational behavior in the world of Prospect Theory.

The fast-and-frugal heuristics program accounts for the pattern of choices that most of us made in Scenarios 1 and 2 in terms of adaptive strategies—quantified as mathematical *process models*—that predict when people will choose to stand pat and when they will take a chance. The fast-and-frugal advocates think it is absurd to imagine that we ordinarily go through an involved, Utility Theory–compliant, quantitative calculation when facing such choices. Instead, we can use a *priority heuristic*[27] that makes a few simple comparisons to judge the relative benefits of a choice. We needn't go into the details here, but, in brief, the scheme starts with a *priority* rule that looks at the minimum possible gain of each alternative, the chance of minimum gain, and the maximum possible gain, and incorporates a *stopping rule* that says when to quit and decide. The heuristic adheres to the principle of *satisficing* (aiming for an outcome that is *good enough* without trying to maximize returns). The net result is that, using the priority heuristic, you can, in principle, reproduce the apparently anomalous pattern—risk-avoiding for gains, risk-seeking for losses—in a plausible way that does not imply that you're behaving irrationally.

Which account of human behavior—Prospect Theory or the priority heuristic—is the right one? Although a definitive answer is not available, in head-to-head comparisons where computer programs representing the two theories compete to see which one can best predict or postdict some forms of human choice behavior, the priority heuristic often comes out on top. Still, knowing that a heuristic process could account for behavior is not the same as saying that people *do* rely on it when making choices.

It is worth noting that Prospect Theory and fast-and-frugal heuristics have entirely distinct perspectives on the role of *emotion* in shaping decisions. Prospect Theory sees emotions as a dominant force in driving irrational behavior. Indeed, "emotional" and "rational" are often used as antonyms in descriptions of people's behavior. While the fast-and-frugal school doesn't deny that we have emotions, it does downplay their significance. Members of this school want to know not so much what moves us to act, but what we do and why we do it, and their answer is that we act for adaptively sound reasons. Gigerenzer believes that the concept of emotion is patched into Prospect Theory and its successors as a sort of fudge factor to make up for the shortcomings in the classical economic Utility Theory. That is, since Prospect Theory is predicated on the same standard of rationality that underpins Expected Utility Theory, factoring in emotion is just a kludgy way to "repair" the classical theory.[28]

12.C.5 Are Scientists Naturally Rational?

Do scientists, perhaps because of their nature or training, think differently, more logically than nonscientists? Maybe the insights regarding cognitive quirks simply don't apply to scientists? Data from two independent sources suggest otherwise. In one experiment, I gave 100 neuroscience students, postdocs, and faculty a modified version of a task that has been used to assess the effects of hidden influences on our thinking.[29] One group, roughly 50% of the subjects, randomly selected, received one form of a question and the rest of the subjects, another form. The process was anonymous, and I didn't know who answered, let alone how they answered. The task was a test of an occult "priming effect" in which simply seeing a completely meaningless number can influence people's responses to other. The first question was:

> "Are there more than (x number) of windows in the US Presidential White House?" where "x" was "242" for one group and "56" for the other. The subjects checked "yes" or "no."

The second question was identical for both groups:

> "How many windows do you think there are in the US Presidential White House? Write your best guess here__________." Eighty-one answers came in.

The first question was a decoy—its purpose was to place two different numbers before the subjects. Prospect Theory predicts that although the numbers in the first question were incorrect (the real number is 147) and entirely irrelevant to the second question, they would "prime" the guess in the second part. That is, people's answers to the second question would be influenced by the number that they saw in the first question. Indeed, the group that saw 56 guessed 115 ± 23 ($n = 39$) while the group that saw 242 guessed, 243 ± 21 ($n = 42$).

If the number in the first question did not affect the subjects, you'd expect that their answers to the second question would be random and indistinguishable between the groups; as it was, their answers diverged significantly and systematically in the direction predicted by the theory. If you saw a low number in question one, you guessed a relatively low number for question two, and vice versa. The "anchoring index" (difference in group means/difference in group "anchors") was (243 – 115)/(242 – 56) = 69%, typical for this kind of question.[30] It appears that these highly educated, presumably analytical thinkers were every bit as susceptible to this subtle cognitive influence as everybody else.

In an independent study, neuroscientist Raymond Dingledine posed four questions derived from Kahneman and Tversky's work to members of basic science departments at Emory University School of Medicine[31] to test the hypothesis "that a scientifically literate population would respond more objectively to the survey questions originally posed in the 1960s and 1970s" (i.e., by Kahneman and Tversky). Dingledine's conclusion was succinct: "[The hypothesis] was wrong." That is, modern biomedical scientists were no more objective in answering the questions than were the college students tested 50 years ago. In particular, Dingledine identified believing in "the law of small numbers," "intuitive pattern seeking," and ignoring the "base rate" as cognitive errors that affected the scientists he tested.

Kahneman argues that the priming and anchoring effects show that we have a not always rational urge to attain "associative coherence" in our thinking. This is not the only possible explanation for these effects, and an alternative favors the ecological psychology interpretation: the anchoring-type problems are artificial and nonrepresentative traps that we fall into because we normally and legitimately anticipate that people who ask us questions are not out to trick us. If they tell us some number, we expect it to be relevant. Under this interpretation, the priming and anchoring effects would have their own adaptive rationale.

The debate about how to interpret thinking that does not conform to the strict rules of logic goes on, but however you look at the debate, it seems that the hypothesis that scientists are inherently more rational (when rationality is defined by abstract systems of logic and Expected Utility Theory) than nonscientists has been falsified.

12.D What We Do and What We Should Do: The Hypothesis and Rationality

I took this detour into the nature of rationality because we're going to evaluate scientific bias in the context of rationality, and it is important to know about the two standards. Both of them come into play in scientific thinking, though not in the same ways. And when it comes to the hypothesis, we have to choose. Probably the most valuable lesson that we learn is not that "humans are irrational" but that "scientific thinking does not come naturally to most people, including scientists."

Earlier in this chapter, I noted that scientific thinking has been held up as a model for rational thinking in general, and yet everyone agrees that "bias," as a form of mistaken thinking, creates problems for science. We've considered two viewpoints on how rationality and biases are related. One take on the conflict between Gigerenzer's and Kahneman's theories is that they do not genuinely differ on substantial matters,[32] "core claims," but are at odds in the arena of "rhetorical flourishes," those flamboyant beliefs that each side holds but which it cannot fully support. Indeed, it often appears that they are arguing past each other about two separate questions: How *do* humans think, and how *should* they think? This is a variant of the split between "descriptive" (what actually happens) and "normative" (what ought to happen) analyses that cognitive scientists, philosophers of science, and others often make. It implies that the teachings of both schools of thought are complementary and equally applicable, and, in one sense, this implication seems entirely reasonable.

In another sense, the teachings are incompatible and, since we're interested in rationality and science, we need to know whether the descriptive or the normative side is more important. We have to choose: Which is the most appropriate for scientific thinking? The answer is: it depends on what you're trying to accomplish.

12.D.1 Critical Scientific Thinking Requires By-the-Book Logic

When it comes to critical scientific thinking, there can't be any real debate: we must be narrowly focused, logical, and rational in the traditional sense. As the hypothesis (and its cousins: theories and laws) is the premier logical structure in science, we are obliged to apply the psychological principles that take formal logic as the measure of rationality when we're evaluating hypothesis-based scientific thinking. If a hypothesis makes a prediction that is logically inconsistent with evidence, the hypothesis is false; if a prediction does not logically follow from a hypothesis, testing the prediction does not truly test the hypothesis, etc. There are right and wrong answers. Kahneman's standards for *rational* and *irrational* thinking are also ours when we design and test hypotheses. Conventional logic is what holds science together.

What constitutes a logical argument is not always unambiguous and scientists can disagree about whether a given test does or does not adequately test a hypothesis. Nevertheless, scientists believe that True answers do exist and that, guided by empirical evidence, we're capable of finding them (despite remaining unsure if we've been successful). The logical ideal is easy to state and hard to achieve, and one reason is that our automatic thinking generally doesn't conform to the demands of textbook logic. It adheres to the built-in guidelines that we've acquired through the ages.

12.D.2 Biases and Ecological Rationality

As many commentators have noted, scientific thinking is a learned skill, and the primary hurdle that we have to clear in acquiring it is unlearning, or learning to modify, our normal thinking habits. The first step must be to recognize that the problem exists.

The concept of ecological rationality goes a long way toward explaining why scientific thinking is so hard for so many of us. Ecological rationality gives rise to thinking that it is pragmatic and practically useful though not necessarily compatible with scientific reasoning. We default to its heuristic solutions reflexively, and it is persuasive because it is fluid and often correct—or "correct enough"—that we can function efficiently in daily life. The dilemma for scientists is that heuristic thinking goes against the grain of the logical thinking that we need to be expert in.

We should depend on heuristics only when they are appropriate and avoid them when they're not. When operating in their critical analytic mode, scientists gravitate toward the logical thinking standard. For instance, in the course of their staunch defense of ecological rationality, Gigerenzer and his colleagues offer cogent critiques of, say, Prospect Theory, that conform to conventional logical reasoning. They formulate, test, and evaluate their hypotheses in good agreement with the guidelines suggested in Chapter 2.

The first step in dealing with the potential problems caused by automatic, heuristic-based thinking must be to recognize when we need to switch into the rigorous logical mode. It would be helpful if we had an external signal to tell us when the scientific mode is called for. Putting on a white lab coat may help those of us who wear them to shift cognitive gears and get into the science-thinking frame of mind. Along this line, I've wondered (not too seriously) if maybe those of us who don't wear lab coats could put on a special "science-beanie" to prompt us to take a formal cognitive stance and abide by the rules of cold, abstract logic. You'd take off the beanie when you were being imaginative and dreaming up hypotheses or asking questions. Taking off the beanie would also signal to critics gnawing at the old worn tropes about how rigorous scientific thinking "stifles creativity" that all is well; disciplined as well as fancy-free thinking are both compatible with doing good science, each in its time.

To streamline the following discussion, I will adopt Kahneman's terminology and position: deviations from logic are irrational or at least unfortunate when they happen.

12.D.2.a The Currency of Abstract Thought: What Have We Got to Lose?

We all keep mental accounts of gains and losses[33] in many facets of our lives. "I owe you one," we say when someone does us a favor; we incur a "debt of gratitude" and plan to "pay it back" to square the social account. Likewise, guilt, regret, pride, and self-respect are currencies in the self-image accounts that we try to keep in balance. Scientists, too, keep mental accounts, and their account balances can affect how they behave, whether it is their willingness to state their hypotheses or their tendency to act in biased ways. Our mental accounts are so important that innate cognitive alarms go off to warn of imbalances.

We do not like to suffer loss of any psychological goods; indeed, the mere *expectation* of loss can trigger loss aversion; we suffer if we are to lose the *possibility* of a gain that we hadn't received, and a region of our brains is especially attuned to detecting loss.[34] Loss aversion strongly influences how we make risky decisions when we are uncertain, and uncertainty is a dominant motif in science. It is not that people are congenitally allergic to risk, it's just that we're inconsistent in how we confront it. The threat of loss on their mental accounts of self-esteem may affect scientists' behavior.

12.D.2.b Loss Aversion and the Hypothesis

I noted earlier (Chapter 9) that the majority of scientific papers that I reviewed steered away from an explicitly stated hypothesis, and my survey respondents cited a few of their reasons for not stating them. If you think about what we've been saying about self-images and mental account balancing, you might recognize a reason. Let's suppose that scientists assess their self-esteem in units, call them *creds*, and that the better their reputation among their colleagues, the greater their self-esteem, the more creds in their mental accounts. Achieving success, publishing a good paper, getting a grant, being thought reliable and careful by their peers, all add to their accumulation of creds, while failure, being thought sloppy, unreliable, or untrustworthy, etc., deplete it. Putting forward an explicit hypothesis announces their personal investment in an idea. They stake creds on it and therefore stand to gain or lose depending on how it fares.

Imagine that you have made a new discovery and have a novel hypothesis to explain it. As you write up your work for publication, you face a choice: Do you make your hypothesis explicit, or do you put forward a less-structured essay and hope that your readers will get the message anyway? If you don't state your hypothesis, then you might not get credit, or be cited, for it. (*Citations*, also known as "hits," are published references to your work that acknowledge your contribution; citations are coins of the realm in science: they add creds to your account.) If your hypothesis becomes widely accepted, you'll be better known and be invited to visit other places and give seminars about it, which are other ways of amassing creds. Contrariwise, not stating your hypothesis risks not getting all the creds that you deserve, and, what would be worse, if you don't state it explicitly, someone else might and claim the glory (yes, this happens in real life). Such factors strongly favor not beating around the bush and getting your hypothesis out into the public eye.

On the other hand, the very thought of stating your hypothesis directly causes you anxiety; it could be a dicey move. Your competitors could take advantage of the research plan so clearly laid out by your hypothesis and jump ahead of you. After all, its key predictions are dead obvious; you're virtually giving your opponents directions to follow and, maybe, prove you wrong. By putting your hypothesis out there in plain terms for all to see, you will have stuck your neck out; your fear of the loss of creds if it is falsified is almost palpable. Or, what if people think that you're being pretentious and making a big deal of a modest insight ("I hypothesize . . ."). Does anybody really to do that, or will it somehow mark you as a phony? You begin to fret that the costs of stating your hypothesis may outweigh the benefits. You wonder if it would be smarter to remain vague and to keep your head down. Maybe you won't get all the creds you deserve but at least you won't be gambling with the ones that you have. The main thing is not to lose face! In the end, you opt for a safe, somewhat rambling narrative style in your paper, trusting that your friends will know what you mean and telling yourself that the data are really the most important thing anyway.

In this made up but not entirely far-fetched scenario, *loss aversion*, the fear of losing creds, determines how scientists present their results. Although scientists undoubtedly don't consciously go through such deliberations, unconscious apprehension about losing creds may well affect what they do. I'll come back to these concerns and how you can address them in Chapter 14.

12.D.3 Confirmation Bias: Loss Aversion, Cognitive Dissonance, or Ignorance?

Confirmation bias is the tendency to seek and hold onto information that agrees with, or "confirms" your hypotheses and to ignore disconfirming information. Confirmation bias is always bad and is often linked to having a hypothesis. Bizarrely enough, though, some critics imply that the problem lies with the hypothesis, not the bias itself. They say that if we want to reduce bias, we should stop doing hypothesis-based science. On the contrary, if you want to avoid confirmation bias, failing to propose and test hypotheses is emphatically the wrong way to go.

12.D.3.a Confirmation Bias, Loss Aversion, and Cognitive Dissonance

Loss aversion applied to personal creds can explain confirmation bias. We see ourselves as intelligent, thoughtful observers of nature and we expect our beliefs about it to be true. Evidence that flies in the face of our cherished hypotheses threatens a loss of creds, which we try to prevent by amassing confirmatory evidence and evading the disconfirmatory kind.

A psychological stimulus that can bolster a tendency toward confirmation bias is a need to reduce *cognitive dissonance*. Cognitive dissonance is the unpleasant feeling set up by "the state of tension that occurs whenever a person holds two cognitions (idea, attitudes, beliefs, opinions) that are psychologically incompatible."[35] It is an emotion that arises from a threat to our sense of self-consistency and that impels us to reduce the threat. The archetypal case is a cigarette smoker who, knowing that smoking is bad for him, does one of two things: quits smoking or convinces himself that smoking is not so bad, that "everybody's got to go sometime," etc., and keeps on smoking. Both strategies bring him some peace of mind by reducing the dissonance. Cognitive dissonance is not restricted to justifying ("rationalizing") unhealthy personal habits, however. Carol Tavris and Elliott Aronson[36] argue that the urge to reduce cognitive dissonance drives professional psychotherapists to concoct unfalsifiable hypotheses to defend their tenuous diagnoses. Would it be surprising if, for the same reason, a scientist whose ego got caught up in his hypothesis had trouble conceding that it was wrong and therefore searched for more reasons to think it was right?

A desire to circumvent the emotional tolls of loss aversion or cognitive dissonance are possible contributors to confirmation bias, but ignorance is another one, although it is frequently overlooked.

12.D.3.b Confirmation Bias and the Hypothesis

Blaming the hypothesis for confirmation bias is exquisitely ironic because the originator of tests widely used to study confirmation bias, Peter Wason, attributed the problem to a failure of his subjects to seek out potentially disconfirming evidence to test their hypotheses.[37] Here's an example of one of his tests: you are given a set of three numbers—2, 4, and 8—that conforms to a rule, and you have to figure out the rule by gathering information and creating and testing hypotheses about what the rule is. To gather information, you propose a set of three numbers that you think might follow the rule and you're told whether you're right or not. You can continue like this, proposing sets of three numbers and getting feedback, for as long as you like; there is no time limit and no penalty for wrong proposals. After you've gathered enough information and believe that you know what the rule is, you get one chance to state your hypothesis about it. This is the payoff, the whole point of the exercise, so you want to be as sure as possible before committing yourself. In Wason's day (1960s), his subjects used pencil and paper, and he provided the feedback on their preliminary proposals in person. You might be interested in trying an online version of the test before going further; it is easy to do, gives you real insight into the problem of confirmation bias, and is anonymous. Here's the link: https://www.nytimes.com/interactive/2015/07/03/upshot/a-quick-puzzle-to-test-your-problem-solving.html (Spoiler alert: I'm going to analyze the test next, so if you want to get the full effect, you should pause and try it now.)

Wason wanted to know if ordinary people (well, college psychology students, anyway) behaved as scientific reasoners were supposed to. Did they rigorously test their hypotheses before putting them forward? Wason designed his test in such a way that using inductive reasoning to generalize a rule from the sample sequence alone was virtually guaranteed to fail. Nonetheless, he found that most subjects were satisfied if they guessed a couple of correct sequences before they were confident enough to state their hypothesis, which was usually incorrect. The few subjects whose first hypothesis was correct guessed at many more sequences and, in particular, tried many nonconforming sequences before declaring their hypothesis about the true rule, which was simply "a series of three increasing numbers."

Likewise, 77% of the participants in the online version failed to do enough exploration to find at least one wrong sequence of numbers before stating their hypothesis. In both classroom and online, the great majority of people quickly homed in on one pattern—for example "an increasing series of even numbers"—entered a few numerical sequences that conformed to their notion, and then declared their incorrect hypothesis.

A popular psychological interpretation of the Wason test results is that people want to hear that they are right and don't like to be told that they're wrong. Using inductive reasoning, they stick with what works. They don't suggest sets of numbers that might be wrong to avoid what Wason calls "the disenchantment of negative instances." The phenomena that we considered in the previous section, loss aversion and cognitive dissonance, probably factor into the "disenchantment," although Wason does not pursue the matter. In any case, it is possible that subjects in his tests make only suggestions that they expect will earn them "yes" answers for emotional reasons.

No doubt there is some validity to this interpretation; however, Wason thought there must be more to it. Most people trying to solve a puzzle want more than a pat on the head and praise for making a good try: they really want to solve the puzzle. The goal in the Wason test is to find the true rule governing the series, and if you don't rigorously test your hypothesis you'll probably fail to find the rule. Surely, this intellectual failure would sting their egos more than proposing a nonconforming set of numbers would, so why didn't the students work harder to test their hypotheses?

To discover why, Wason asked them to talk through their reasoning out loud, and, furthermore, he had them try again if their first hypothesis was incorrect. Surprisingly, on hearing that they had failed, many subjects stuck with their original strategy or made only trivial changes before declaring another hypothesis. In the extreme cases, they simply restated their initial incorrect rule in slightly different words and without testing any more sets of numbers. Wason called this "magical thinking," as in sorcery, where the success or failure of spells hinges on precisely how the words are spoken.

Wason concluded that the majority of the students were simply unaware of the power of falsification to test their hypotheses. They clung to their superficial inductive solutions, not because they were seeking emotional reassurance, but because they didn't know what else to do. Confirmation bias was the product of ignorance.

One bright note: Wason found that many students, on hearing that their first hypothesis about the proposed rule was wrong, did change their plan of attack and carefully check their next hypothesis with several negative and positive instances before submitting it. (Kahneman might say that failure jolted their System 2s into action.) However that may be, it is clear that critical scientific thinking habits can be strengthened by instruction and feedback. If this interpretation of the Wason test results is correct, then the solution to the problem of confirmation bias is not less hypothesizing, but an increased emphasis on testing hypotheses by trying to falsify them.

Kahneman thinks about confirmation bias itself in different terms.[38] He says that the bias arises because our System 1, in trying to make sense of the world, first looks for reasons to believe that a proposition is true. The concept is that if you can't *believe* a statement is true, you can't very well *judge* if it's true. Only after System 1 has understood a statement by believing it can our analytical System 2 scrutinize it and decide whether it actually is true or not. Unfortunately, System 2 is lazy, and skeptical examination is hard, so without determined effort to engage System 2, we are often stuck with what System 1 comes up with—an apparently "confirmed" proposition that we haven't thoroughly vetted. Improving our scientific thinking skill requires rousing our System 2 and forcibly directing its attention toward the task at hand. (Maybe the beanie would help?)

As I've mentioned, despite rejecting traditional standards of rationality, the fast-and-frugal school of thought does recognize that people at times must act in accordance with conventional logic. We face two major stumbling blocks when we try to solve problems in logic and probability. first, these kinds of problems are typically cast in abstract, unnatural language that we can't process, and second, our educational systems don't adequately prepare us to think logically. Along these lines, Gigerenzer has championed a push to make biomedical communication more understandable to patients and doctors by switching to frequency-based language[39]—think about the red-nosed liars. And he argues that, starting in the early school years, teaching the fundamentals of probability thinking in the tangible terms of *natural frequencies* can improve society's *risk literacy*.[40]

I noted that others[41] have called attention to the areas of convergence of the programs of Kahneman and Gigerenzer on some of the practical issues surrounding rationality. I'd like to draw the optimistic conclusion that we can improve our ability to think scientifically, provided that we start by acknowledging that, for whatever reason, we're not born scientific thinkers and be willing to exert concerted effort.

12.D.4 "It Ain't What You Say, It's the Way That You Say It . . . ": Framing Effects Matter

Those of us of a certain age remember the early days of credit cards when you had to pay an additional amount, a "surcharge," if you wanted to use a card to buy an item in a store. Since then, things have changed, yet, in a way, not much is different. Nowadays, if you fill up your car's tank at your local gas station, you are likely to get a "discount" if you pay with cash. There is never a word about surcharges, though the end result is the same as it was in the early days—you pay more if you pay by card. If the net effect is the same, why did the language change?

Richard Thaler,[42] the Nobel Prize-winning economist, associated this overwhelming—though strictly speaking irrational—preference for the revised wording with a phenomenon he named the "endowment effect"; we really hate to give up what we already have. He explains that it is relatively easy to tolerate not getting a discount and paying the "regular price" by credit because, well, we don't exactly *own* the discount, but we definitely do own our money and do not want to have to "pay extra." Kahneman and Tversky call it *framing* when we let the individual words of a choice, rather than its overall meaning, determine our preference. Framing effects seem to influence our behavior in countless ways.

12.D.5 Framing Effects, Loss Aversion, and Publication Bias

"Negative data" are the Rodney Dangerfield of scientific results: they get no respect.[43] No one wants to publish them, which is a major reason for *publication bias*. The journals are full of positive research findings, while negative ones languish, forgotten in file drawers. The bias skews the literature and prevents scientists from learning about previously discovered dead ends, thus slowing progress and wasting resources, as others rediscover what doesn't work. Setting policies, haranguing people, or offering incentives and rewards for publishing negative data will probably help reduce publication bias, but we could do more if we took into account why scientists make their decisions. Besides disincentives, what factors might make a scientist hesitate to publish negative results, especially ones that are maligned as "failures" because they falsify hypotheses? Framing effects and loss aversion may be involved

The power of negative connotations of *negative* data and *failure* shouldn't be underestimated. After all, if positive data and successes are the shining ideals of science, what good are negative data and failures? At best they are unworthy goals; at worst, they're entirely pointless. Why would anyone spend time and effort pursuing them? Who wants to be the guy who seeks them out; how many creds do you lose if you publish such stuff? I can't prove that scientists entertain these very thoughts but, given the effort that they spend building up their stores of creds, it is likely that framing effects play into the bias against publishing negative data.

If such considerations do influence our mental calculus, it should be relatively cost-free, although not necessarily easy, to counteract them: we can stop degrading the value of negative data by not associating them with failure and loss, and we can start touting their value in identifying fruitless avenues of research and eliminating inadequate explanations. We can counter the concern that negative data will clog the journals by distinguishing between mindless "scientific stuff"—measurements for the sake of measurements—and data that bear on important hypotheses, concepts, and variables. In any case, accurately defining and rewarding all of the activities involved in testing hypotheses will be a major step forward.

12.E Coda

Heuristics are mental shortcuts that we rely on instead of exhaustive, complex reasoning processes. Heuristics usually reflect biases that, in many circumstances, serve as simple hypotheses. Evolution has selected for heuristics that, while generally beneficial, are not infallible. Whether we consider heuristics to be rational or irrational depends on what we mean by rational. The common standard is a formal, explicit logical framework that is appropriate for scientific thinking but can be ill-suited to everyday demands. A more approachable standard is ecological rationality; a pragmatic framework that says, roughly, that what is right is what works. Errors can result when we fall for cognitive illusions that occur when the immediate, direct answers provided by a heuristic disagree with the answers mandated by logical analysis.

There are many kinds of biased thinking. Two that are often identified as causing major problems in science are *confirmation bias* and *publication bias*. Following Peter Wason, we conclude that rigorous hypothesis-based thinking is the antidote to confirmation bias. Publication bias is made more likely by disparaging our results as "negative data" and "failures." In general, we find problems easier to understand and solve if they are framed in terms we readily connect with the concrete circumstances that we are familiar with.

We return to these topics in Chapter 13 and review additional ways of dealing with them.

Notes

1. Numerous books on the general topic of critical thinking make this point: Caleb W. Lack and Jacques Rousseau, *Critical Thinking, Science, and Pseudoscience* (New York: Springer; 2017) and Galen A. Foresman, Peter S. Fosl, and James Carlin Watson, *The Critical Thinking Toolkit* (Malden, MA: John Wiley & Sons; 2017) are good recent sources.
2. Henri Poincare, *Science and Hypothesis* (CreateSpace Independent Publishing Platform; October 11, 2013), preface.
3. Kinds of cognitive bias: https://en.wikipedia.org/wiki/List_of_cognitive_biases.
4. Gerd Gigerenzer, *Rationality for Mortals: How People Cope with Uncertainty* (Oxford: Oxford University Press; 2008), is an accessible entrée into Gigerenzer's oeuvre and you'll find copious experimental details in the collection of research-related papers by him and members of his school of thought in Gerd Gigerenzer, Ralph Hertwig, Thorsten Pachur (Eds.), *Heuristics: The Foundations of Adaptive Behavior* (Oxford: Oxford University Press; 2011).
5. G. Gigerenzer and H. Brighton, "*Homo-Heuristicus*: Why Biased Minds Make Better Decisions," in G. Gigerenzer, R. Hertwig, and T. Pachur (Eds.), *Heuristics: The Foundations of Adaptive Behavior* (Oxford: Oxford University Press; 2011), pp. 2–26.
6. Daniel Kahneman, *Thinking, Fast and Slow* (New York: Farrar, Straus and Giroux; 2011). This book provides a good overview of Kahneman's thought as it has evolved beyond the seminal work he did with Amos Tversky; e.g., Amos Tversky and Daniel Kahneman, "Judgment Under Uncertainty: Heuristics and Biases" in Daniel Kahneman, Paul Slover, Amos Tversky (Eds.), *Judgment Under Uncertainty: Heuristics and Biases* (Cambridge: Cambridge University Press, 1982).
7. Kahneman, *Thinking*, p. 98.
8. G. Gigerenzer and H. Brighton, "*Homo heuristicus*" ibid.
9. Nicholas Nassim Taleb, *Fooled by Randomness: The Hidden Role of Chance in Life and in the Market*, 2nd ed. (New York: Random House; 2005).
10. The term "bounded rationality" was coined by the Nobel Prize-winning political scientist, psychologist, economist, and artificial intelligence pioneer, Herbert Simon; see Bryan D. Jones, "Bounded Rationality," *Annual Review of Political Science* 2:297–321,

1999, for a concise introduction. Simon also introduced the term "satisficing" (i.e., satisfying and sufficing) to the American public.
11. Gigerenzer and Brighton, "*Homo Heuristicus*," pp. 6–9.
12. Ibid., pp. 10–12.
13. Kahneman, *Thinking*, "Prospect Theory," chapter 26, pp. 278–299. The program that Kahneman and Tversky initiated is sometimes called "heuristics and biases" but all opposing programs, including "fast-and-frugal," also deal with heuristics and biases. To avoid pointless confusion, I'll refer to all of Kahneman and Tversky's ideas under the rubric of Prospect Theory and, when necessary, Kahneman's more recent work done since Tversky's death as his "two-system" approach.
14. Kahneman, *Thinking*, "Two Systems," part I, pp. 19–105.
15. If the bat and ball together cost $5.50 and the bat costs *$5.00 more than* the ball, the bat must cost $5.25 and the ball must cost $0.25.
16. Expected Utility Theory (Hypothesis). See https://en.wikipedia.org/wiki/Expected_utility_hypothesis. An accessible discussion of rational economic theory is found Reid Hastie and Robyn M. Dawes, *Rational Choice in an Uncertain World*, 2nd ed. (Los Angeles: SAGE; 2010).
17. Gerd Gigerenzer, Ralph Hertwig, Thorsten Pachur (Eds.), *Heuristics: The Foundations of Adaptive Behavior* (Oxford: Oxford University Press, 2011), p. 153.
18. The area of overlap must be smaller than either of the two main circles, provided, of course, that the circles are not exactly the same size and totally overlap (i.e., I assume that all feminists are not bank tellers and all bank tellers not feminists). In this case, class of feminist bank tellers would be the same size, though not larger, as the class of bank tellers.
19. Number of "tellers"; most, not all, are in banking: 345,709 in the United States in 2015 (https://datausa.io/profile/soc/433071/). Total number of tellers, 502,700 (https://www.bls.gov/ooh/office-and-administrative-support/tellers.htm); 5 of 6 are women (http://www.nelp.org/content/uploads/NELP-Data-Brief-15-Minimum-Wage-for-Bank-Workers.pdf). Putting the two together, approximately 418,000 bank tellers are women.
20. H. P. Grice (who also published as H. Paul Grice and Paul Grice) was a British philosopher who studied word meaning in conversational contexts. He named the concept of "implicature," essentially, unstated word meaning; https://plato.stanford.edu/entries/implicature/.
21. Ralph Hertwig and Gerd Gigerenzer, "The 'Conjunction Fallacy' Revisited: How Intelligent Inferences Look Like Reasoning Errors," *Journal of Behavioral Decision Making* 12:275–305, 1999.
22. C. F. Chick, V. F. Reyna, and J. C. Corbin, "Framing Effects Are Robust to Linguistic Disambiguation: A Critical Test of Contemporary Theory," *Journal of Experimental Psychology: Learning, Memory, and Cognition*. 42:238–256, 2016.
23. C. Badcock, "Making Sense of Wason: Parallel Mentalistic/Mechanistic Cognition Resolves the Controversy," *Psychology Today*, The Imprinted Brain, posted May 5, 2012; https://www.psychologytoday.com/us/blog/the-imprinted-brain/201205/making-sense-wason

24. Some people feel that, strictly speaking, we should turn over the F card as well, since we weren't told that each card had a number and a letter. The main point, however, is that most people choose the 3 card instead. For an overview and critique of the four-card test, see https://www.psychologytoday.com/blog/the-imprinted-brain/201205/making-sense-wason.
25. L. Cosmides and John Tooby, "Cognitive Adaptations for Social Exchange," in Jerome H. Barkow, Leda Cosmides, John Tooby (Eds.), *The Adapted Mind: Evolutional Psychology and the Generation of Culture* (New York: Oxford University Press; 1992), pp. 181–206.
26. Badcock, "Making Sense of Wason."
27. E. Brandstätter, G. Gigerenzer, and R. Hertwig, "The Priority Heuristic: Making Choices Without Trade-Offs," *Psychological Review* 113:409–432, 2006.
28. Ibid., p. 4.
29. Kahneman, *Thinking*, "Anchors," pp. 119–128.
30. Ibid., p. 124.
31. Raymond Dingledine, "Why is it so hard to do good science?" *eNeuro*, September/Octorber, 2018, 5(5), e0188, 1–8.
32. R. Samuels, S. Stich, and M. Bishop, "Ending the Rationality Wars: How to Make Disputes About Human Rationality Disappear," in Renée Elio (Ed.), *Common Sense, Reasoning & Rationality* (New York: Oxford University Press, 2002), pp. 236–268.
33. R. H. Thaler and C. R. Sunstein, *Nudge: Improving Decisions About Health, Wealth, and Happiness* (New York: W. H. Norton; 2015).
34. B. Seymour, N. Daw, P. Dayan, T. Singer, and R. Dolan, "Differential Encoding of Losses and Gains in the Human Striatum," *Journal of Neuroscience* 27:4826–4831, 2007.
35. Carol Tavris and Elliot Aronson, *Mistakes Were Made (but Not by Me): Why We Justify Foolish Beliefs, Bad Decisions, and Hurtful Acts* (New York: Harcourt; 2007).
36. Ibid., chapter 4.
37. P. C. Wason, "On the Failure to Eliminate Hypotheses in a Conceptual Task," *Quarterly Journal of Experimental Psychology* 12:129–140, 1960.
38. Kahneman, *Thinking*, p. 368.
39. Gerd Gigerenzer, *Simply Rational: Decision Making in the Real World* (New York: Oxford University Press; 2015), pp. 1–106. A recent example of health statistics given as frequencies rather than probabilities indicates that Gigerenzer's call for more user-friendly language is having an effect; see https://www.cancer.org/cancer/prostate-cancer/about/key-statistics.html
40. Michael Bond, "Decision-Making: Risk School," *Nature* 463:1189–1192, 2009.
41. Samuels et al., "Ending the Rationality Wars."
42. Richard H. Thaler, *Misbehaving: The Making of Behavioral Economics* (New York: W. H. Norton; 2015), p. 18.
43. Rodney Dangerfield was the stage name of a well-known American comedian whose standard tag-line was "I don't get no respect." See https://en.wikipedia.org/wiki/Rodney_Dangerfield.

PART III

PRESENT POLICIES AND THE FUTURE

13
The Hypothesis in Science Education

13.A Introduction

Why don't scientists state their hypothesis when they have one? This question has been lurking throughout much of the book, and I've touched on several possible answers, including cognitive factors in the previous chapter Our educational background is also important. Survey results (Chapter 9) showed that fewer than 20% of biomedical scientists had received a significant amount of formal instruction in scientific thinking, and 70% of us had almost none. What we do know we acquired informally via the apprenticeship system that defines our graduate and postgraduate education—discussions, trial and error, and exposure (we absorb through a sort of "osmotic" process). Because we learn from mentors, reviewers, and colleagues, all whom have idiosyncrasies and biases of their own, our training varies in quality and quantity. While 89% of scientists surveyed said they felt confident in their knowledge of the hypothesis, the osmotic method isn't wholly satisfactory, as 92% of them also said that formal training in hypothesis-based methods would be "useful or highly useful." Our haphazard training in scientific thinking could help explain why many of us aren't more explicit in putting forth our hypotheses. We can't say much about the informal system, but we can ask "What sorts of formal educational information are available to scientists?" This chapter explores answers to this question.

Science education is not only important for the development of professional scientists. Many more people take science classes in high school and college than become scientists, and these nonscientists form the nucleus of a scientifically aware citizenry whose education is in the interests of modern society. However, polls of thousands of Americans over the past 20 or so years have repeatedly shown that only about 25% of us have even a rudimentary grasp of how science works.[1]

How do we get our information about the hypothesis? To get a rough idea, I sampled sources at three levels of our educational system. Science education standards in public school systems vary widely, so I initially focused on national science education standards and policies. I then shifted up a level and evaluated books on critical thinking that are aimed at university students and the educated lay public. Last, I consulted the policies of two federal agencies that fund scientific research, the National Institutes of Health (NIH) and the National Science

Foundation (NSF). Although an exhaustive investigation was not possible, we can get idea general impression about what information on scientific thinking is available for students and others.

There are good, publicly available sources that I don't cover. For instance, *The Khan Academy*, an online trove of knowledge on all sorts of educational topics, has lectures on the Scientific Method and the hypothesis.[2] Another first-class, publicly accessible website is maintained by the University of California, Berkeley, *Understanding Science: How Science Really Works.*[3] The site defines many terms, discusses misconceptions associated with scientific concepts and practices, and presents a range of age- and grade-appropriate instructional materials supplemented by interactive graphics. *Wikipedia* has a cursory article (as of November 2018) that skims over several meanings of "hypothesis" including logical and statistical as well as scientific hypotheses without providing much detail. While science professionals and students might consult such web resources individually, I decided to limit my focus to materials targeted at specific groups.

The information in the bulk of the materials that I assessed was remarkably inconsistent. There does not seem to be a common pool of principles covering hypothesis-based reasoning, and the hypothesis was often misrepresented when it was not ignored altogether. There are plenty of opportunities for improvement in educational, social, and governmental policies regarding scientific thinking, and, at the end of the chapter, I will make a few suggestions for how to accomplish this.

13.B The Hypothesis in Science Education

In an unsystematic pilot study, I briefly reviewed 13 high school–level science textbooks (7 biology, 4 ecology, 1 chemistry, 1 geology) to see how they covered the Scientific Method and the hypothesis. Twelve of the 13 had a brief (2- to 7-page) section entitled something like "The Nature of Science" and sketched the topics with degrees of quality and depth ranging from good to poor. Their messages were diverse and even if each one made sense in isolation, you couldn't put them together. For example, at one school, I happened to read, back-to-back, two descriptions of the Scientific Method in different textbooks: the first said that the Scientific Method was one of two major principles that all science was based on, the second that the whole notion of a Scientific Method was mistaken because scientists don't follow a method. What should a student make of it all? I got the impression from informal conversations with science teachers that the contradictions wouldn't cause much difficulty because the teachers tended to omit "all of that stuff" from their lessons. And indeed, in 9 of 12 books the word "hypothesis" only appeared in the introductory section; evidently the concept that a hypothesis might play a role in the process of science was not taken too seriously.

Perhaps the omissions are not surprising. Perhaps it is understandable that pre-college textbooks that cover a single scientific area, such as biology, focus on the factual content of that science or how the facts are gathered observationally, instead of being deduced from controlled experiments. Textbook publishers, it seems, may have little incentive to worry too much about broader issues of the philosophy of science. National, nonprofit educational organizations, on the other hand, should be above narrow commercial or social interests; indeed, they could help mold those interests. Perhaps they keep the flame of scientific wisdom alive?

A major source for national science education policy is the National Science Teachers Association (NSTA),[4] which is the largest organization of pre-college science teachers in America. A special focus of NSTA training is on the Next Generation Science Standards (NGSS), which were developed by a consortium of national scientific societies[5] including NSTA. The NGSS is intended to serve as a guide to teaching science from kindergarten through high school (K–12).

13.B.1 Teaching the Hypothesis at Pre-College Levels

Being a grade school science teacher must be one of the hardest teaching jobs. In addition to the usual classroom and curricular hassles, the science teacher has to stay abreast of the blinding pace of change in science and its technical dimensions—the websites, computer apps, and devices that affect access to scientific information—all the while keeping a weather eye on state and local regulations and being alert to societal sensitivities. And science teachers must be prepared for push-back from students who come to class primed to disbelieve their subject—resistance that few, say, arithmetic teachers have to counteract.

When it was rolled out 2013, the NGSS program was the first major overhaul of science education standards since 1996. NGSS is founded on the principle that students must be actively engaged with science problems, discovering and testing their own ideas—"Less Memorizing, More Sense Making!" NGSS identifies core themes in individual science disciplines, frames each theme in age-appropriate contexts, and links themes across disciplines via "cross-cutting concepts." NGSS emphasizes what scientists actually do (e.g., "use and develop models", "get evidence to investigate and test concepts," "construct explanations") rather than on what scientists know. NGSS recommends assessment of student achievement through performance measures, not multiple-choice exams that encourage empty "teaching to the test." And it stresses that scientific knowledge is provisional, that it represents the best information that we have at the moment, and that it is constantly updated as new evidence comes in. As an overarching philosophy, this is outstanding. As of November 2017, NGSS had been adopted in 19 states and the District of Columbia.

13.B.1.a NGSS, NSTA, and the Hypothesis

In NGSS, the concept of scientific thinking is taught through examples and is reinforced throughout all grades. While this attention is much needed, there are some areas of concern. For instance, while NGSS and NSTA correctly attach great importance to scientific words such as *hypothesis, prediction, theory, law,* etc., the essays on these topics lack unity.[6] I'll skip over *theories* and *laws* because[7] they are conceptually similar to hypotheses—conjectural, testable, falsifiable, and provisional generalizations—and because there was no consensus on how to define them. Most importantly for my purposes, theories and laws are either rare or absent from certain sciences (e.g., biology, neuroscience), or they have completely different connotations in, for example, psychology and sociology than they do in physical sciences. In contrast, all sciences recognize hypotheses and predictions as valid and applicable concepts.

Unfortunately, the NGSS/NSTA philosophy doesn't take a definitive stance on what "hypothesis" and "prediction" mean or on the relationships between them. This is a problem because you can't fully comprehend the Scientific Method or what is involved in testing hypotheses without being clear on what these terms mean. Now at first, these look like simple, easily correctible oversights: define the words and get on with more important matters! Unfortunately, because of a fundamental NSTA precept that I'll get to shortly, it may not be easy to fix them.

13.B.1.b Definitions

Although the word "hypothesis" is mentioned several times in the NGSS program description, it is not unambiguously defined. As an example, one statement of high school objectives says that students should know that "A hypothesis is used by scientists as an idea that may contribute important new knowledge for the evaluation of a scientific theory"; another notes that, "Scientists often use hypotheses to develop and test their theoretical explanations." So far, so good, but what *is* a hypothesis? The authors don't say, and the essays on it elsewhere in the program don't always agree.

One essay claims that a hypothesis is a "prediction with an explanation."[8] In addition to its confusing wording, this definition makes it hard to follow advice on how to conduct certain classroom exercises. A section describing a laboratory exercise says that students are supposed to formulate their hypotheses "after the lab [while] doing the data analysis" and another instructs that during data analysis "is not when the hypothesis is to be formed"[9] and requires that students state their hypothesis before the exercise. Several articles explain that you make a prediction after you have a hypothesis, but these articles don't explain how the prediction relates to the hypothesis or clarify what testing a prediction has to do with the hypothesis.

According to one essay "A hypothesis is an educated guess about the cause of a phenomenon," whereas "a prediction is an educated guess about the expected result of a specific experiment."[10] This sounds fine and the authors lucidly discuss the nature of scientific thinking, even mentioning *falsification*. Then, at another moment, we read that, when it comes to hypotheses, "there's no guessing involved."[11] You wonder how readers seeking enlightenment can put it all together.

A key principle in scientific thinking is how predictions are related to hypotheses. One NGSS-compliant program called Hypothesis-Based Learning (HBL)[12] seems to agree, noting that, "remarkably there is much confusion concerning definitions of hypothesis and prediction." HBL says that a "hypothesis is an explanation," that "a prediction makes the hypothesis a subject for rational thought," and, finally, that a prediction is "given by compound 'if-then' logic,"[13] (i.e., if the explanation is true and such and such experiment is done, such and such results will be observed). This is consistent with the answer that I advocate (Chapter 1), although it is incomplete and, regrettably, is the last sentence of the essay. Will it suffice to clear up the "confusion" about the hypothesis? A fuller account of the logical relationship between predictions and hypotheses would have explained why valid conclusions follow from the outcome of an "if-then" test.

Vagueness surrounding foundational principles may contribute to another problem in science education, as a vignette in *Teaching for Conceptual Understanding in Science*[14] related by science educator Page Keeley suggests. Using an anonymous checklist of 14 possible definitions of "hypothesis" (six good ones and eight poor ones), Keeley queried a group of middle school teachers about their understanding of the term. Nearly everyone in the group strongly preferred two of the poorer descriptors: "an educated guess" and a statement "used to prove whether something is true," and they were "adamant" in their choices. Keeley argues that teachers need to be aware that these phrases perpetuate students' misuse of words and, presumably, contribute to their misunderstanding of science. Though admittedly small and informal, the sample results suggest that some science teachers may lack the background to lead students to a nuanced understanding of scientific thinking. The NSTA website should be first and foremost a reliable resource for science teachers.

In short, the genuine value of NSTA/NGSS is diluted to an extent by an abundance of divergent views and an absence of direction as to which ones are most worth listening to.

13.B.1.c Off-the-Shelf Science Education

To help science teachers manage the curricular demands of NGSS, educational companies offer ready-made teaching plans that not only summarize scientific advances and principles and present topics and materials for classroom use, but also lay out guiding philosophies for teaching science. Argument Driven Inquiry[15] (ADI) and Claims-Evidence-Reasoning[16] (CER) are two examples.

Both ADI and CER teach that *empirical evidence* is critical in formulating and defending a scientific conclusion. For example, a teacher may give her students a scientific dataset—say, one made public by NASA[17] relating to cycles of sunspots—and ask them to interpret it. Students learn to support their interpretations—their *claims*—of what the data show and, guided by the teacher, debate their insights and try to reach a consensus. Encouraging students' active engagement with evidence is obviously a great idea.

Given their promising premises, I was surprised to learn that neither the Scientific Method nor the hypothesis are even mentioned, let alone taught, in either program. The topics are not forbidden, as I was assured by an author of one of the program guides; individual teachers can bring them up if they want to, but the concepts have no official place in the curriculum. The written materials create the impression that an optimal outcome is a group of children who have all contributed to the discussion. In itself, getting all students involved is a highly commendable goal. Yet the message that everyone's point of view is as good as everyone else's can be a slippery slope. (It reminds me a little of the time when, in early grade school, my son returned from a sports-oriented field day with an eye-catching ribbon proclaiming that he had won a "Participant!" award.) In science, there are better and worse ideas, and the worst ones are discarded. If we don't acknowledge this, we foster societal debates that end inconclusively with a statement that "there are scientific supporters on both sides," even when there are 97 highly qualified experts on one side and 3 vocal opponents on the other. Sometimes you have to make a decision in favor of one side, and the goals of harmony and participation must give way to the critical thinking that takes place within the Scientific Method.

My concern is that, if neither national norms nor state educational agencies are clamoring for instruction in critical scientific thinking, then educational companies will not want to risk potentially rocking the boat by bringing it up. The short-term pedagogical gains may come at the cost of a scientifically less savvy populace.

13.B.1.d Don't Find "Fault"

Why can't we just tell the students what they need to know about the rigors of hypothesis testing, falsification, rejection, etc.? The problem is that, under the NGSS/NSTA aegis, we can't do that, at least in so many words. A central tenet of the NSTA educational philosophy is that science teachers must strictly avoid "fault words"[18] (e.g., *right, wrong, true, false*). It's as if you weren't allowed to call out "warmer" or "cooler" to the blindfolded child searching for the prize in the kids' party game.

Perhaps to help students get over their anxiety about rejection and to be more willing to offer their own hypotheses, the NSTA/NGSS programs teach that hypotheses are never rejected or falsified, they are either "supported" or "unsupported" by evidence. Obviously, we don't want to traumatize students, and it is true that working scientists often say that evidence supports or is "consistent with "their hypothesis. It is also true that, if enough evidence goes against a hypothesis, science throws in the towel, "unsupported" becomes "falsified," "rejected," or "wrong." Avoiding "fault words" obscures the ultimate purpose of vigorous hypothesis testing and falsification, which is to get to the Truth as best we can.

13.B.1.e Abstract Ideals Have Concrete Significance

Are we supposed to avoid fault words because they allude to ideal states (e.g., Truth) that are not attainable? This, too, seems misguided. "At some point" in a scientific investigation the hypothesis must be rejected. Naturally, rejection won't usually occur at the first sign of trouble. As I've discussed elsewhere in the book, scientific decisions can be complicated. The shining ideals of "Conclusive Falsification" or "Scientific Truth" may be literally unattainable, but the mere fact that we can't attain them shouldn't prevent us from teaching students about them. We don't dismiss the ideals of "Honesty," "Justice," and "Freedom" simply because we can't always achieve them.

Trying to account for scientific progress without calling on the concepts of right and wrong, true and false, etc., subtly endorses the myth that confirmatory evidence is the goal of science. This brings us back to the realm of inductive reasoning and away from the search for scientific Truth.

13.B.1.f Right, Wrong, and the Real World

There are two additional reasons—one practical, one abstract—not to avoid fault words when talking about fundamental concepts. Practically speaking, students who enter the scientific workforce expecting their cherished opinions to be treated with kid gloves will be in for a rude shock, but they, at least, will eventually get straightened out. The greater danger is that we're educating the public to consider all hypotheses with some empirical support as being equally sound. If the standards of truth and falsehood are not in place, even as abstract goals, then the basis for making societal decisions that depend on scientific outcomes is on shaky ground. If we're not taught about better and worse hypotheses, then we not equipped to judge the claims of science that affect society. And we're more likely to believe that the mere existence of conflicting evidence indicates there is a stalemate and that no real-world decisions are warranted until "all" of the data are in.[19] All the data will never be in, and everybody's opinion is not equally valid.

In summary, many pre-college students may not be getting a realistic picture of the hypothesis or its place in scientific thinking. But perhaps this is understandable; perhaps a little sugar-coating is necessary to make science palatable at that stage and is justified by the increased level of participation children show when they're shielded from fault words. Are materials aimed at older students more honest?

13.B.2 Critical Thinking for Advanced Students and Science Consumers

Among the resources relevant to scientific reasoning are the online sites I mentioned earlier and books that deal with "critical thinking," including scientific thinking. written for post–high school students and others. Many of the books on critical thinking advance the view that anyone can learn to think carefully and skeptically. Typically, scientific thinking, especially as it relates to the hypothesis, is somewhat of a tangent to their main thrust, which can range from distinguishing science from pseudo-science,[20,21] to establishing the philosophical underpinnings of critical reasoning,[22] to describing the skills needed for rational decision-making under uncertainty.[23] The hypothesis is one possible approach to critical thinking, and science is one arena of many in which critical thinking might be applied. Therefore, although these books make valuable contributions they aren't part of the science curriculum.

Four books do cover the scientific thinking and the hypothesis in detail. The textbook by David Glass (Chapter 10A) is intended for university science students, and Stuart Firestein's books (Chapter 10B), although not textbooks, are also intended mainly for post–high school science readers. As I've criticized the way these books cover the hypothesis, I won't add anything here, except to state the obvious: they do not provide the reliable instruction regarding the Scientific Method that I, and I believe the respondents to my survey, are looking for. One textbook that does aspire to provide formal instruction in scientific thinking is the subject of the next section.

13.B.2.a Is Astrology a "Marginal Science?": The Program of Giere, Bickle, and Mauldin

In their influential textbook, *Understanding Scientific Reasoning*, philosophers of science Ronald Giere, John Bickle, and Robert Mauldin[24] propose a step-by-step program to help post–high school students and nonscientist readers understand and evaluate scientific findings. The book is loaded with examples of how to apply their program to historical and contemporary scientific problems, and it offers study questions as well as advice for understanding scientists' decision-making process. Principles of experimental design, including randomization, double-blinding, statistical practices, and decision-making under conditions of uncertainty, are covered in depth. Giere et al. are devoted to the proposition that scientific thinking is a skill that anyone can master with practice, and their book is intended for people wishing to acquire that skill, including college students. Their treatment of the hypothesis is somewhat abstruse and misleading, although that is a minor objection. There are more serious drawback to the program, which we'll get to. First, the program itself.

13.B.2.b Models, Hypotheses, and Claims

For Giere et al., there are three kinds of hypotheses: *theoretical, statistical* (their statistical hypothesis is different from mine), and *causal.* All science depends on *models* which, they add, can be theoretical entities. In fact, "thinking scientifically about anything requires constructing a model." A model depicts a real-world phenomenon, although a model is not the same as a hypothesis. Rather, a *claim* that a model accurately depicts a phenomenon is a *theoretical hypothesis.* In other words, scientists do not try to find out if a model is true; they try to find out if a *claim* that a model is true, is true. Imagine that we want to test our conceptual model that "Grass is green." We would do experiments, not to discover if grass is really green, but to find out whether our claim that "Grass is green" is true. Only a philosopher could love such a distinction, and I suspect that it has never entered the mind of any experimental scientist.[25] In any case, Giere et al. do state that "all general scientific claims are hypotheses." By comparing a model's real or conceptual properties, or its predictions, with the world, we can determine whether the model "fits" the world and, thereby, whether the claim about it is true or false.

13.B.2.c The Program

If you want to understand a scientific finding, according to Giere et al., you apply a six-step algorithm, which is modified slightly for the various kinds of hypotheses and decision processes that they evaluate throughout the book. This is how it works for the theoretical hypothesis:

1. Identify the aspect of the real world that is being studied.
2. Identify the theoretical model whose fit with the real world is at issue.
3. Identify a prediction based on the model to find out what data should be obtained.
4. See if data that bear on the model is available.
5. Ask if the data agree with the prediction. If the data do not fit the predictions, reject the model. If they do fit, go to step 6.

6. Ask if there is another *plausible* hypothesis that could explain the fit between the data and model as well as or better than the present hypothesis. If there is no other hypothesis, accept the present one provisionally. If there is another one, either reject the current hypothesis or decide that the situation is *inconclusive*. (Emphasis added.)

Giere et al. consider the idea that science is distinguished by a Scientific Method to be "doubtful, at best,"[26] and they don't mention either Karl Popper or the principles of demarcation or falsification. They do say that science is meant to "explore how the world works" by doing experiments and making observations designed to "help [scientists] decide which of several possible ways the world *might work* is the most like the way *it really does work.*" They state that science tries to understand the world, though they focus attention on exploration and observation, which for them leads more-or-less directly to the creation and evaluation of models.

This image, while reasonable in some respects, depicts science as a soft, largely passive endeavor that does not engage in aggressive searching for answers to profound questions. Scientists gather up data through experiment or observation and then match the data with model predictions; they never critically *challenge* a model, let alone try to falsify it. And any bit of data seems to be as good as any other; the main thing is whether the data agree or disagree with a model. Implicitly, the more agreement the better. Giere et al. point to a few caveats (agreement counts only when such agreement would have been very unlikely "if the hypothesis were clearly false"[27]), but at heart their program prizes verification and confirmatory evidence as progress. They do not spend much time talking about how you choose between models, although they do bring up Expected Utility Theory, which we'll get to, and prefer a model that *fits* a collection of data better than does another model.

These weaknesses are perhaps debatable and, in any case, are not the real problem with Giere et al.'s philosophy. The inconsistency between tolerance and indecisiveness, on the one hand, and unreachably high standards, on the other, is a tip-off to its more serious drawbacks.

13.B.2.d Concerns About the Giere et al. Program

Giere et al. feel that it is largely a waste of time to try to distinguish systematically between science and nonscience or pseudo-science. Certain human activities—religion, art—are obviously not science, they admit, but other fields are on a legitimate fringe of acceptability. Their willingness to consider scientific claims to be *undecidable*, rather than false, leads the authors to accept Freudian psychology, astrology, extraterrestrial visitation, reincarnation, and extrasensory perception (including clairvoyance) as *marginal sciences*. For Giere et al., any field that uses "models to represent the world" and that makes an "appeal to empirical data" to support its "hypotheses" may qualify as marginal science.

If what we know about extraordinary claims is not convincing enough to warrant incorporating them into genuine science, or if is there is another *plausible model* for existing data, then Giere et al. insist that we must withhold final judgment pending more information. Not wrong, these claims exist in limbo awaiting stronger support. Evidently, we are supposed to ignore evidence that a "marginal science" (like astrology) has been rigorously debunked, discredited, or requires violation of the known laws of physics, as several of them do. The authors' reluctance to take a firm stand and declare these belief systems to be scientific nonsense is alarming and not only because they give aid and comfort to devotees of the supernatural.

In striking contrast to their readiness to accept astrology and reincarnation as marginal sciences, the authors place impossibly high demands on legitimate scientific hypotheses. Using "global warming" (anthropogenic global warming; i.e., man-made global warming—increasingly and more appropriately known as "climate change," although we'll stick with their term to avoid confusion) and the "greenhouse effect" caused by atmospheric carbon dioxide (CO_2) as a case study, the authors consider two competing hypotheses to explain recent warm weather trends: (1) global warming and an alternative (2) natural temperature fluctuations, reportedly duplicated by a model that does not incorporate atmospheric CO_2. Giere et al. conclude that the existence of two plausible competing hypotheses means that we can't reasonably draw any conclusion; that we're obliged to hold out for "undeniable evidence" before making a judgment.

The net result of having the two widely divergent standards—one too lax and one too strict—is to expand the zone of undecidable "marginal" science. In effect, Giere et al. sanction placing newspaper column astrology and rigorous climate science into the same category.

And that is not the only area in which their program flirts with intellectual irresponsibility. They do not mention the mass of evidence that is consistent with the man-made global warming hypothesis or that the overwhelming majority of climate scientists accept that this change is caused by human activity.[28] They do not discuss the different standards of validity that science adopts depending on the purposes of the information—basic or applied science—it gathers. They do not discuss the role that consensus plays in deciding scientific questions. However, the most pernicious fallacy in their argument is that the truths of science are determined by the criterion of *undeniability*. Remember, this is the standard that they set for man-made global warming, and it flatly contradicts their statement that all scientific conclusions are provisional. If a statement is provisional, it is "deniable" at some level.

Much of *Understanding Scientific Reasoning* addresses scientists' and policy makers' obligations to act despite being faced with uncertain facts. The authors advocate using *Utility Theory* (i.e., Expected Utility Theory; see Chapter 12) and again they use man-made global warming as an illustration. To apply the theory, you have to assign values to variables, such as the cost of taking action (e.g., trying to stop global warming vs. doing nothing) and of different outcomes (e.g., global warming occurs or it doesn't) and the probabilities of each outcome happening. You plug these values in a simple matrix of costs and probabilities, and choose the option having the greatest expected value[29]. However, Expected Utility Theory calculations are meaningless unless they're based on realistic assumed values. The expected utility of doing something depends on both the cost of the outcomes and the chance that the outcomes occur. You multiply the apparent cost times the probability to determine the expected utility of an action. Hence, even an outcome with only a tiny probability of happening could be worth avoiding if the possible costs of inaction are high enough. You may believe that it is extremely unlikely that you'll die at a young age (the probability of your early death is very small), but if your spouse and children would suffer greatly without your financial support (the potential cost of your dying is very high) you may decide to buy life insurance even though you're young and healthy.

When it comes to man-made global warming, Giere et al. cite "considerable controversy among scientists," effectively tabling the question of whether it is even real or at least making the probability of it's happening appear relatively small. The worst costs of man-made global warming that they entertain are associated with "flooding of major cities" and "the destruction of farmland." This sounds as if it could be fairly bad, but the vague and limited assessment of damage seems to diminish the likely costs of global warming happening. Thus we have a dubious and not especially high, probability that global warming is real coupled with costs that are ill-defined, though perhaps manageable; in sum, there is probably nothing to worry too much about.

How would their example change if they took into account actual evidence that "around 95% of climate researchers actively publishing climate papers"[30] agree with the conclusion that global warming is being caused by human activity? This would tend to make the probability of global warming seem significantly higher. How about the total costs of global warming? Giere et al. do not consider, let alone assign values to, potential costs associated with mass human starvation, worldwide strife, or widespread animal extinctions that could result from global warming. Such costs are likely to be extraordinarily high but hard to assess in the abstract. A more tangible assessment was provided by the administration of US President Donald Trump, not known as a bastion of liberal nervous Nellies on global warming, which released a

US government report that estimates that the costs of global warming to the US economy amount to hundreds of billions of dollars, perhaps as much as 10% of the US gross domestic product, by the end of the century.[31] This is a genuine prediction that the American people will have to pay significant out-of-pocket costs.

Rather than stressing the requirement for realistic estimates of probabilities and costs, Giere et al. breezily conclude that "What would be required to make the decision process [regarding man-made global warming] look different to policy makers is *undeniable evidence both that doing nothing will lead to warming and that doing something will prevent it* [emphasis added]." Perhaps the authors were being ironic and meant to imply a distance between what they themselves believe and what they anticipate that "policy makers" think? It is not clear. What is clear is that they do not call attention to the fiction of "undeniable" evidence, and their omission takes on ominous overtones, as they conclude that, without undeniable evidence for both positions, the only rational decision regarding global warming is to do nothing.

I do not know whether policy makers have read or been influenced by Giere et al.'s book, but it has clearly been read by many ordinary citizens. The discussion of man-made global warming is a perfect example of how something as seemingly minor and esoteric as a mistaken view of scientific reasoning could have far-reaching real-world consequences. Collectively, citizen opinion of science has a huge effect on how seriously we take the alarms about global climate change. Citizens who are poorly informed about how science works cannot make the best decisions.

13.B.3 Suggestions for Improving Science Education About the Hypothesis[32]

1. Scientific educational organizations should develop and promulgate consistent core definitions of the key terms and principles of scientific thinking, especially those related to hypothesis, prediction, and testing.
2. Remove "fault words" from the taboo list and instead describe the process of science and explain how the fault words fit into the quest for scientific progress.
3. Stress that it is not wrong to have an idea that turns out to be wrong; that science progresses by replacing worse ideas with better ones.
4. Distinguish between the objectives of applied science and basic research science and emphasize that applied science is guided by the best scientific information available, while basic ("pure") research seeks to achieve the ideal of Truth.

5. State that the goals of pure science to have all of the data and to achieve Truth are ideals that inspire us to understand nature to the greatest degree possible, even though we cannot actually achieve perfect knowledge.
6. Explain that it is a mistake to judge the actions of applied science by the goals of pure science.

13.C The Professional Scientist and the Hypothesis

Professional scientists hardly ever have the chance to get formal instruction regarding the hypothesis and Scientific Method after they've earned their final academic degrees. There are, however, resources available to those individuals that might fill the void. I've mentioned some online websites and books on the topics, but as a group, scientists do not consult these outside sources. There is one source of information that scientists do consult: the agencies that review and fund research. Do these agencies offer the sort of information that could help scientists fill in gaps in their education?

The most important federal agencies that fund research for biomedical scientists are the NIH and the NSF. Since scientific thinking is a major—maybe *the* major—point of a scientific grant application, you could imagine that NIH and NSF would be rich repositories of information about scientific thinking. Are they?

The 27 agencies that make up the NIH focus on particular diseases or disease-related areas, and they support all kinds of science (basic and applied, including clinical, hypothesis-based, Discovery Science, etc.) that are intended to benefit human health. There are NIH-wide policies regarding grant applications, which each agency supplements with its own specific rules. Biomedical scientists pay close attention to what NIH has to say and what information it offers.

Prompted in part by the recent hue and cry concerning reproducibility in science, *rigor* and *transparency*[33] have become major buzzwords at the NIH. The desire to enhance rigor and transparency has given rise to four guidelines meant to increase the reliability of scientific findings. Two of the four are fairly technical and won't concern us. Instead, I'll concentrate on those dealing with "rigor" and the "scientific premise."

13.C.1 Rigor and Transparency

The policy on "rigorous experimental design for robust and unbiased results," says that "Scientific rigor is the strict application of the scientific method." The *scientific premise* is an innovation.[34] A grant application must have a strong premise, and a premise is not the same as a hypothesis. Instead, the premise includes the rationale for the project, which could include a hypothesis, and an evaluation of the strength of the existing evidence that supports the rationale. Despite the importance of rigor and the scientific premise, the site does not explain the Scientific Method or the scientific hypothesis.

Indeed, the word "hypothesis" does not appear anywhere in the policy, in the linked Frequently Asked Questions, or in an associated blog post.[35] Apparently the official policy of these agencies takes it for granted that scientists know about hypotheses. Fair enough, you say. The instructions are intended for professional working scientists who, it is presumed, only need to know how to apply for research grants, not how to think scientifically. Nevertheless, elsewhere on the NIH site, we learn that not all grantees do understand what a hypothesis is. An NIH blog writer[36] gives a dictionary definition of "hypothesis," including a link to a definition suitable for "children," adding that, "I am providing this as a public service. Recently I received a proposal to review from 'scientists' [*sic*] whose hypothesis was actually a list of aims and methods. Perhaps it is wrong to assume that everyone with a science degree knows what this word means?"

Although the blog writer is incredulous, the responses to the "Hypothesis Overdrive" blog[37] that I discussed in the Introduction documents disagreements among scientists about what a hypothesis is. And, as someone who served on numerous NIH review panels, I can attest that a proposal betraying ignorance about the hypothesis was not a rare event. Thus, a fuller explication of the subject would not seem out of place.

Are matters like scientific thinking perhaps covered by the individual NIH divisions, rather than the umbrella website? To find out, I searched "hypothesis" on each of the 28 NIH subcomponent websites (27 institutes plus the Director's Office). The results were amazingly diverse. The number of hits on the various sites ranged from less than 10 to 34,500 (the large numbers mainly represented hits on the abstracts of applications funded by the division in question or papers in the scientific literature tracked through the National Library of Medicine). A few citations linked to NIH pages with helpful hints on writing a successful grant application ("be clear," "use subheadings," "no more than 20 words in a sentence," etc.) without touching on weightier topics, such as the hypothesis.

Several sites gave a *pro forma* nod to the topic (e.g., "committed to new innovative hypothesis-driven research"), however, the nature and quality of the information relevant to the hypothesis was spotty. For example, the primary references on the National Institute of General Medical Sciences (NIGMS) site are to the "Hypothesis-Overdrive" blog with its strongly anti-hypothesis bent. Only three, the National Institutes of: Allergy and Infectious Disease (NIAID); Mental Health (NIMH); and Drug Abuse (NIDA)[38] offered substantive advice on scientific thinking and writing. The most entertaining discussion that I found deals with "Stuffed Animal Science."[39] which is intended to teach school children the basics of scientific thinking and is admirably straightforward.

Advice regarding hypothesis-based applications ("Application Missteps") in grant preparation from NIAID[40] presents helpful examples of more and less "focused" hypotheses. "Unfocused" means that a strong central hypothesis is lacking; the hypothesis can be too broad (e.g., "inflammation is a key etiological component of autoimmune diseases") or too descriptive (e.g., "we will evaluate changes in transcriptional signatures in the involved tissues following infection"). Unfortunately, the second example is not a hypothesis at all. The site implies that multiple hypotheses are desirable, but then notes that "Optimally, your experimental results *should be able to prove or disprove your central hypothesis*," implying that proof is possible and arguably perpetuating a misunderstanding of science.

The NIMH discussion of the hypothesis in grant writing is fairly extensive and is the only one that I found that gives the reader the invaluable information that a good hypothesis-based application will lay out a "win-win" outcome, meaning that, the site explains, even if the central hypothesis of the grant is disconfirmed, the data will constitute an important scientific advance. This is, of course, is the very essence of Karl Popper's program.

A search on "hypothesis" on the NSF website yielded nothing similar to the NIH's instructions for applicants. However, the NSF site is a font of data about the state of US science education, including international comparisons of science student achievements (we do not do well, although there are extenuating circumstances). The site also tracks Americans' attitudes on a variety of topics, including global warming, medical research, and even, for a while, astrology.[41]

In summary, useful information on the NIH and NSF websites regarding hypothesis, prediction, testing, etc. is scattered and inconsistent. This is especially lamentable in the case of the NIH because of its position as the predominant funder of US biomedical research and, therefore, an oracular source of technical information for scientists and bioscientific knowledge for the general public.

13.C.2 How the NIH Can Help Educate Its Profession and Lay Clients About Scientific Thinking

1. Develop and make available acceptable NIH-wide definitions of hypothesis, prediction, test, etc.
2. Resolve the issues raised in the "Hypothesis Overdrive?" blog regarding NIH's position on hypothesis-driven research. Is NIH favorably disposed to hypotheses or not?

3. Provide instructions, for both grant applicants and reviewers, of exemplary hypothesis-based grants, including examples.
4. Elevate the worth of misnamed "negative data" by emphasizing how a well-designed, hypothesis-testing grant advances science by falsifying substantive hypotheses.
5. Establish guidelines that assist grant applicants and especially grant reviewers in distinguishing between hypothesis-based and non-hypothesis-based (e.g., Discovery Science) research and in evaluating them appropriately.

13.D Right Responses to Error: Lessons from the Airline Industry

I began this chapter with the question of why scientific authors often avoid stating their hypothesis when they have one. The discussion so far has centered on the paucity of trustworthy information about the hypothesis throughout the school years and on into the professional career. Apart from formal sources of information, a myriad of other influences affecting scientists' thinking that, collectively, constitute the values of the scientific community. In this last section, I want to consider a cultural force that affects how scientists behave; namely, the way in which the scientific community views falsification of hypotheses and the practice of labeling experiments that falsify hypotheses as "failures."

"Success hinges on how we react to failure." So says Matthew Syed in *Black Box Thinking: Why Most People Never Learn from Their Mistakes—But Some Do.*[42] Making mistakes is inevitable; the larger issue is how we respond to them. Syed contrasts the way that the airline industry responds to failure with that of clinical medicine. Errors in both sectors can be fatal. While it is difficult to make direct comparisons, the stark numbers are staggering: With tens of millions of flights worldwide each year, members of the International Air Transport Association had a fatality record of approximately 1 fatal accident per 8.3 million takeoffs in 2014.[43] In US medical facilities alone, estimates range from 44,000 to more than 440,000 "premature deaths associated with preventable harm" each year, with an average estimate of approximately 210,000[44] The US Centers for Disease Control (CDC) reports an annual total of about 149.5 million hospital visits (in- and outpatient combined) to US hospitals.[45,46] Taking the estimate of 210,000 annual preventable hospital deaths, we get a rate of preventable deaths in hospitals, per 8.3 million visits, that is *more than 13,600 times greater* than the airline fatal accident rate. Of course, people are more complex than airplanes, they come to hospitals because they are sick, and caring for sick patients is much harder than flying planes. Still, the conclusion that there is an unconscionably vast difference between airline safety and hospital safety seems inescapable. Syed argues that the ways in which each industry reacts to error is largely responsible for the gap. Flying errors trigger serious, systematic inquiries into their causes. The personnel involved are encouraged to come forward with information, and the black-box data is analyzed. The overriding purpose of the inquiry is to learn from the mistakes, not assign blame for what went wrong.

In medicine, there is a pervasive fear of failure and the burden of unrealistic expectations. When perfection is expected by patients, lawyers, supervisors, and the doctors themselves, then admitting to a mistake amounts to acknowledging incompetence. Medical errors engender blame, often in the form of expensive and reputation-scarring lawsuits. Doctors develop habits of denial that keep them from recognizing and owning up to error. For Syed, this fosters a culture of "evasion" in medicine that is rooted in a dysfunctional attitude toward failure.

The good news is that change is possible. The airline industry was not always a paragon of virtue when it came to dealing with the nightmare of plane crashes. Years of investigation and analysis led to the conclusion that two factors are essential if large entities are to learn from their mistakes: there must be (1) a system that harnesses error for making progress and (2) a *mindset that enables the system to flourish* [emphasis added].

Falsification of a hypothesis is not a "mistake" or a "failure" in the senses of the mistakes and failures in the airline industry or medicine. But perception can be everything, and many critics—and scientists themselves—view a falsified a hypothesis as an indication that something went wrong, and this, I believe, enforces a reluctance to express hypotheses explicitly. The good news here is that science has a system that, in principle, is set up to profit from its "mistakes." The other factor, the mindset that sees error as opportunity for improvement, is harder to inculcate and gets less attention. The question is, "Is the current mindset of the scientific community open to making changes that can help science move forward?"

You can avoid failure by creating goals that are so vague and loose that nobody, including you, can say whether you've achieved them or not. You can't learn from mistakes if you can't identify them. We may be witnessing the embryogenesis of a culture of evasion in the lack of a stated hypothesis in published research, in the overt opposition to hypothesis-based science voiced by critics, in the disorganized approach to teaching students about scientific thinking, and in the diminished status of the hypothesis in the eyes of granting agencies, such as the NIH. We should do our utmost to stop a culture of evasion from developing, and one big step toward stopping its development is to promote regard for the Scientific Method and its most valuable tool, the scientific hypothesis.

13.D Coda

Educational resources concerning the hypothesis are disappointingly variable in quantity and quality. The apparent lack of formal training in scientific thinking and hypothesis-based reasoning that was evident in my survey results (Chapter 9) probably reflects this educational deficiency to a large extent. Instruction on even the most fundamental concepts is so inconstant in depth and quality that it's impossible to achieve a common understanding of scientific thinking among students and teachers. The disagreement among viewpoints extends from pre-college to post-university levels.

Popular books cover scientific thinking only as one part of the broader category of critical thinking. The value of textbooks targeted to university-level teaching which include good practical advice on the mechanics of scientific procedures is undermined by inadequate or faulty accounts of the Scientific Method and the hypothesis.

The situation confronting professional scientists is no better. The large federal granting agencies, which could be sources for explicit and authoritative information on the basics of scientific thinking do not fulfill this role. When combined with what for most of us is a shaky educational background in scientific thinking, scientists in biomedical and other sciences are left with little in the way of solid, dependable advice about how to reason scientifically. There are opportunities to enhance the extent and validity of available information that would be worth looking into.

Notes

1. The National Science Foundation (NSF) posts the results of polls of 1,500 to 2,200 Americans taken by educational and other institutions over the years. One measure that aggregates responses to several science-related questions indicates that roughly 40% of Americans know something about science, but only about 25% understand how scientists go about their work (i.e., by testing theories, hypotheses, etc. https://www.nsf.gov/statistics/seind14/index.cfm/chapter-7/tt07-09.htm). The numbers show no improvement from 1995 (when the aggregate number was first calculated) until 2014.
2. The Khan Academy; https://www.khanacademy.org/science/biology/intro-to-biology/science-of-biology/a/the-science-of-biology.
3. Understanding Science: How Science Really Works; https://undsci.ber2ley.edu/article/intro_01. Publicly accessible website established by the University of California, Berkeley. One of the best general resources for science students, teachers, and anyone else interested in an easy to understand overview of what science is all about.
4. National Science Teachers Association (NSTA); http://www.nsta.org/.

5. The Next-Generation Science Standards (NGSS) were developed by the National Research Council, the National Science Teachers Association, the American Association for the Advancement of Science, and Achieve, nonprofit educational organization, based on the NRC's *K-12 Framework for Science Education* http://ngss.nsta.org/About.aspx. See also NGSS development https://www.nextgenscience.org/developing-standards/developing-standards.
6. NGSS is closely associated with NSTA, and many essays posted on the NGSS/NSTA website expand on topics that are integral to the NGSS program. The succeeding discussion draws heavily on these essays. See http://common.nsta.org/search/default.aspx?action=browse&text=hypothesis&price=&type=&subject=&topic=0&gradelevel=&sort=1&page=0&dep=&coll=&author=. See also http://ngss.nsta.org/Practices.aspx?id=1&exampleID=318. http://ngss.nsta.org/NSforCC.aspx?id=4&detailID=69.
7. See Note 3 in Chapter 2 regarding theories, laws and hypotheses. As the distinctions among are not precisely defined, I have lumped them together throughout the book.
8. Kimberly J. Davis and Tracy L. Coskie, *Hypothesis Testing: It's Ok to Be Wrong*, http://static.nsta.org/files/sc0902_58.pdf.
9. Students must state their hypothesis before doing the laboratory exercise; see http://ngss.nsta.org/Resource.aspx?ResourceID=54.
10. Louise M. Baxter and Martha J. Kurtz, *When a Hypothesis is NOT an Educated Guess*; http://static.nsta.org/files/sc0104_18.pdf.
11. Page Keeley, *To Hypothesize or Not*, http://static.nsta.org/files/sc1012_24.pdf.
12. Kristy van Dorn, *Science Sampler: Hypothesis-Based Learning*, http://static.nsta.org/files/ss0511_57.pdf.
13. Kristy van Dorn, Mwarumba Mavita, Luis Montes, Bruce J. Ackerson, and Mark Rockely, *Hypothesis-Based Learning*, http://static.nsta.org/files/ss0401_24.pdf.
14. Richard Konicek-Moran and Page Keeley, *Teaching for Conceptual Understanding in Science* (Arlington, VA: NSTA Press; 2015), pp. 10–11.
15. Argument-driven inquiry (ADI), https://www.argumentdriveninquiry.com; http://www.activatelearning.com/#home-section/.
16. Claim, evidence, reasoning (CER); program offered by Activate Learning.
17. Resources for teachers from NASA are available at https://www.windows2universe.org/?page=/teacher_resources/sunspotplot_edu.html; http://static.nsta.org/files/ss0009_40.pdf.
18. See Note 8; "fault words."
19. Naomi Oreskes and Erik M. Conway, *The Merchants of Doubt: How a Handful of Scientists Obscured the Truth on Issues from Tobacco Smoke to Global Warming* (New York: Bloomsbury; 2011), p. 267.
20. Caleb W. Lack and Jacques Rousseau, *Critical Thinking, Science, and Pseudoscience: Why We Can't Trust Our Brains* (New York: Springer; 2016).
21. Massimo Pigluicci, *Nonsense on Stilts: How to Tell Science from Bunk* (Chicago: University of Chicago Press; 2010).
22. Galen A. Foresman, Peter S. Fost, and Jamie Carlin Watson, *The Critical Thinking Toolkit* (Chichester: Wiley Blackwell; 2017).

23. Reid Hastie and Roby Dawes, *Rational Choice in an Uncertain World: The Psychology of Judgment and Decision Making*, 2nd ed. (Thousand Oaks, CA: Sage; 2010).
24. Ronald N. Giere, John Bickle, and Robert F. Mauldin, *Understanding Scientific Reasoning*, 5th ed. (Belmont, NY: Wadworth Cenage Learning, 2006).
25. The distinction between a "model" and a "claim that the model is true" might be related to the *correspondence theory of truth* which I mentioned in Chapter 1; that is, although we cannot directly compare a physical thing, such as a model, with an abstract concept, such as truth, we can directly compare a *statement about a model* with a *statement of a concept*, such as truth. The need to keep these sorts of considerations in mind may be why you decided to go into science and not philosophy.
26. The full context is "The idea that science is distinguished by the scientific method is . . . doubtful, at best, because scientists use methods that are as varied as the subjects they study." This is the same misinterpretation of Scientific Method that I discussed in Chapter 1.
27. "*If hypothesis were clearly false*, which is to say, if some significantly different model provided a better fit to the real world" [emphasis in original]," p. 33.
28. Most scientists accepted the verdict of global warming even in 2005; the fifth edition of the book came out in 2006; see for example, Naomi Oreskes, "The Scientific Consensus on Climate Change," *Science* 306:1686–1688, 2004. See also https://www.skepticalscience.com/global-warming-scientific-consensus-intermediate.htm and an article on the US government report released by the Trump administration on the true costs of climate change: https://www.nytimes.com/reuters/2018/11/23/us/politics/23reuters-climate-change-usa.html.
29. They fill in the matrix like this: the value "0" is assigned to the box representing "doing something and having warming occur anyway"—it is the worst outcome. "Doing something and avoiding warming" gets 0.9; "doing nothing and having warming occur" is assigned 0.1; and "doing nothing and having no warming occur" gets a value of 1.0—it is the best outcome. According to the matrix analysis, "do nothing" is the best choice. Even overlooking the gross simplification behind these numbers, the analysis ignores any mention of realistically complex outcomes—what if the worst cases are near-global famine, widespread animal extinctions, wars, the degradation or destruction of civil society, etc.? How would these outcomes be costed in the analysis? What about the benefits of reducing air pollution, even if warming doesn't occur? The list goes on. No doubt the authors are aware of such complications. Unfortunately, it is likely that the sole exposure to scientific thinking that many students will ever have is the teachings of this book. It is deplorable that the authors have chosen to trivialize such a serious issue without at least alluding to the profound issues at stake.
30. See Note 28.
31. Costs of global warming to U.S. economy; https://www.nytimes.com/2018/11/23/climate/us-climate-report.html
32. William F. McComas, *The Nature of Science in Science Education: Rationales and Strategies* (Dordrecht, Netherlands: Kluwer Academic; 1998).
33. NIH Rigor and Transparency (formerly, Rigor and Responsibility): https://grants.nih.gov/grants/guide/notice-files/NOT-OD-16-011.html; https://www.nih.

gov/research-training/rigor-reproducibility. " Two of the cornerstones of science advancement are rigor in designing and performing scientific research and the ability to reproduce biomedical research findings" (see https://grants.nih.gov/reproducibility/index.htm).

34. https://nexus.od.nih.gov/all/2016/01/28/scientific-premise-in-nih-grant-applications/. https://grants.nih.gov/reproducibility/index.htm-rigor and transparency.
35. Michael Lauer, *Open Mike*; https://nexus.od.nih.gov/all/category/blog/
36. Definition of hypothesis, courtesy of NIH: http://nihgrants.blogspot.com/search?q=hypothesis.
37. Jon Lorsch: https://loop.nigms.nih.gov/2014/03/hypothesis-overdrive/.
38. The most informative material about the hypothesis that I could find at NIH was from the NIAID (https://www.niaid.nih.gov/search/niaidsite/hypothesis); NIMH (https://www.nimh.nih.gov/search.jsp?query=hypothesis); and NIDA (https://search.usa.gov/search?utf8=%E2%9C%93&affiliate=www.drugabuse.gov&query=hypothesis&commit=Search) websites.
39. Stuffed animal science; https://www.drugabuse.gov/stuffed-animal-science.
40. Information r9garding the hypothesis in science from NIAID: https://www.niaid.nih.gov/grants-contracts/application-missteps-unfocused-hypothesis-specific-aims; and https://www.niaid.nih.gov/grants-contracts/prepare-your-application.
41. National Science Foundation reports: https://www.nsf.gov/statistics/2016/nsb20161/#/downloads/report.
42. Matthew Syed, *Black Box Thinking: Why Most People Never Learn from Their Mistakes—But Some Do* (New York: Penguin; 2015). The "black box" in the title refers to the famous "black box" carried on commercial airplanes—officially, it is a "flight recorder," and it is bright orange, not black—to record information about the plane's mechanical performance, pilot actions, voices, and other vital statistics that help investigators discover what went wrong after a crash. Syed is British, and the typical American sense of "black box" as a mysterious machine or source of information is not what he has in mind.
43. Airline travel safety: http://www.telegraph.co.uk/travel/news/2015-was-the-safest-year-in-aviation-history/; http://www.planecrashinfo.com/cause.htm.
44. https://www.propublica.org/article/how-many-die-from-medical-mistakes-in-us-hospitals. John T. James, "A New, Evidence-Based Estimate of Patient Harms Associated with Hospital Care," *Journal of Patient Safety* 9:122–128, 2013.
45. Estimated total visits to US hospitals during this period is the sum of the overnight stays lasting at least one night (23,749,000; data for 2015) plus 125,700,000 ambulatory visits, for a total of 149,449,000 total visits.
46. Statistics for overnight hospital admissions from https://www.cdc.gov/nchs/fastats/hospital.htm; ambulatory visits from https://www.cdc.gov/nchs/data/ahcd/nhamcs_outpatient/2011_opd_web_tables.pdf. (data for 2011).

14
How to Improve Your Own Scientific Thinking

14.A Introduction

"Scientific reasoning requires a certain discipline and we should try to teach this discipline," so said Richard Feynman,[1] though he noted elsewhere that it actually wasn't taught in any course that he knew about. Apparently, according to my survey (Chapter 9), it mostly still isn't.

You might find "discipline" a bit off-putting, all grim self-denial and flavorless meals; however, improving your scientific thinking doesn't have to be aversive. Scientific thinking is orderly, not regimented or formulaic; it is certainly not uncreative or Feynman, an extremely creative scientist, wouldn't have recommended it. The goal is to train your cognitive abilities so that you can learn and practice science productively, even enjoyably. True, it's not always easy to interpret data or use the hypothesis, but you already know that science is a challenging career. In any case, mastering any complex skill—basketball, piano, or scientific thinking—can be fun and stimulating while it increases your capabilities.

And it's not as if stilted engagement in disciplined thinking is mandatory every minute that you're doing science—spontaneous, unconscious mind-wandering is as important as self-abnegation—and, in any case, you have the logic of the hypothesis to guide you. In this chapter, I want to offer a few suggestions for acquiring the skill of thinking with the hypothesis. The first part outlines the exercises of *finding* and *diagramming* hypotheses in published work and your own. The second part suggests ways of thinking about your own thinking that may help you improve its quality.

You won't find these suggestions in conventional science textbooks. Rather, to shake things up, I've collected tips and principles that are not specifically scientific and that may seem unfamiliar, even odd, in the context of the Scientific Method. The hope is that if you come at the subject from a novel point of view, you'll see the discipline that Feynman is talking about in a new light.

14.B Find the Hypothesis

Scientific papers built around an implicit hypothesis are very common (Chapter 9). One problem with implicit hypotheses is that, if authors don't state their hypothesis directly, readers, especially students, can have difficulty extracting the message and, furthermore, may get the impression that most people don't rely on hypotheses. Psychologists (and advertisers and politicians) are well aware that, in general, most of us like to do what others do. In *The Structure of Scientific Revolutions*, Thomas Kuhn[2] argues that social or perceptual influences strongly shape scientists' behavior, including, we assume, how scientists communicate their work. Probably many scientists don't state their hypotheses because other scientists don't state theirs. Making implicit hypotheses explicit could help break this cycle.

A good place to start is to become aware of the hidden hypotheses in papers. For instance, one method, *find the hypothesis*, is to practice identifying and bringing out the hypothesis in scientific reports. You can do this in the privacy of your own room, of course, but a typical venue is the weekly journal club or lab meeting. The presenter begins by laying out the hypothesis of the paper, whether it was implicit or explicit, and traces its logic throughout the presentation. If the hypothesis isn't explicitly stated and you're the presenter, then you have to find it, and you can begin by pinpointing the main theme of the paper; there are generally clues in its title and abstract. You may find that relatively few papers formally assert and follow a hypothesis[3]. More likely, you'll come across indirect statements or hints: the authors will say that a certain subject is "poorly understood," imply that an existing model is "incomplete," or note that some aspect of nature has not been "well-studied." These are clues about where to find the hypothesis; they are not the hypothesis itself. To find it, ask yourself what the authors are trying to explain; whether it is feature of the abstract or buried in the text, the explanation is probably the hypothesis.

You also have to be aware that, because the current usages of "hypothesis" and "prediction" are inconsistent, even if a hypothesis or prediction is stated, you can't always be certain that the words mean the same thing in the paper as they do in this book. I often find "predictions" sprinkled almost whimsically throughout papers: a pleasant if isolated surprise when one pops up. Of course, some published work does genuinely lack a hypothesis (Chapter 4), but even in those cases, it is a good exercise to classify the kind of science involved. Frequently, a paper will start off sounding like a pure Discovery Science project: for example, the pattern of gene expression in a certain condition is not known and the authors want to "characterize" it, yet, as soon as the characterizing is done, they switch into hypothesis-testing mode with little fanfare.

After you've identified the probable hypothesis, look at the experiments and figures and try to work out how they relate to it and to each other. Do they follow a logical sequence ; are they arranged in a way that makes sense to you? Does each experiment test a bona fide prediction of a hypothesis? Could the test falsify a hypothesis? Finally, read the discussion and conclusion carefully. Not infrequently, you'll find that the implicit hypothesis of the paper only becomes evident when you see how the authors summarize the data and their significance. In the end, if you can see how all the pieces fit together, the odds are excellent that you'll have found the hypothesis.

Two things to watch out for are extraneous observations that, while interesting and important on their own, have little to do with the main theme of the paper, and experiments loosely connected to the theme that not test predictions. Ask yourself whether the main conclusion of the paper would have changed if an experiment had been left out. If the answer is no, then it probably falls into one of these categories. (Incidentally, I'm a big fan of off-hand observations because they can enrich a report, and the fact that they're not an integral part of your line of reasoning is no reason not to mention them. Just call them what they are—intriguing side notes—and make sure they are not mixed up in the main flow of your argument or, I assure you, some referee will misunderstand and ding you in the review.)

If you can find the hypothesis in the paper you're going to present and see how, even if unstated, it forms the scaffold for the work, you and your audience will get much more out of your presentation than you'd get by simply reciting the data. And you won't have to resort to that deadliest and emptiest of phrases to get you from one slide or idea to the next; "then they wanted to look at . . .," (Whenever I hear that I think, "Why? Why did they want to look at it? Help us out here!")

In the next section, I'll review an outstanding paper in neuroscience that tests an unstated hypothesis and use it to illustrate how to find and diagram an implicit hypothesis.

14.B.1 Diagram the Hypothesis

When you're trying to understand an abstract concept, it's often very helpful to make a sketch or diagram of its fundamental parts and their relationships, and there are many methods of diagramming.[4] I've used a tree diagram in Figure 14.1A to show how a generic hypothesis-based paper might look. There is nothing unique about this scheme; you might identify different elements or a more elaborate tree in a paper that tested more than one hypothesis or that carried out extensive preliminary studies before getting down to the main job of testing the central hypothesis.

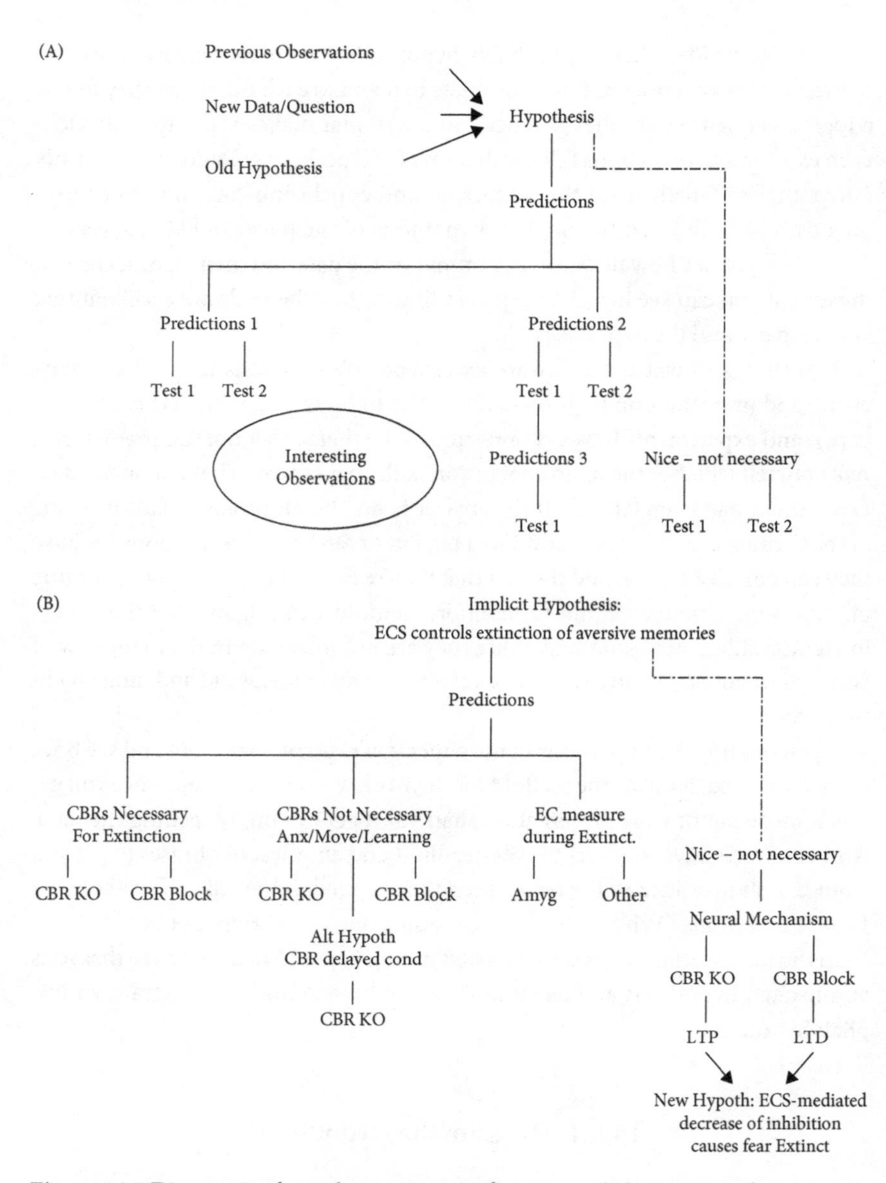

Figure 14.1 Diagraming hypotheses in scientific papers. (A) Diagram of a generic hypothesis-based paper. The hypothesis, which may be explicit or implicit, leads to two kinds of predictions: mandatory predictions that are logically deduced from the hypothesis, and optional—nice, but not necessary—predictions that are not logically demanded by the hypothesis but which may lead to complementary information. Logically required predictions are connected to the hypothesis by solid lines, optional experiments by dashed lines. Each prediction is subjected to one or more tests. The paths of reasoning from the tests back to the hypothesis are illustrated in Figure 2.1. An average paper may also report on "interesting observations" that are only loosely coupled to the main theme of the paper. (B) Diagram of the report by

In their beautiful and widely cited *Nature* paper (>1,000 citations as of May 2018), Giovanni Marsicano and colleagues[5] tested an implicit hypothesis about the role of the endogenous cannabinoid system, which involves the brain's own "marijuana" (Chapter 2). Figure 14.1B shows how you might diagram their paper. The authors do not state their hypothesis or spell out its predictions (they often call their predictions "assumptions") in the body of the paper, and they say only that they want to study the involvement of the endocannabinoid system in memory formation. Nevertheless, their title strongly implies a hypothesis: "The Endogenous Cannabinoid System Controls Extinction of Aversive Memories." They studied a part of the brain, the *amygdala*, that is a center for the control of fear. As the paper unfolds, they test three logically necessary predictions of this hypothesis and one that was "nice, but not necessary."

Some background: neuroscientists are interested in fear (aversive) memories for two reasons. First, all animals, including humans, learn to fear danger in their environment (*fear conditioning*) and, whenever possible, learn to let go of their fear when danger no longer threatens (*fear extinction*). Hence, understanding how fear memories are formed and abolished would tell us about a fundamental neural process. Second, fear memory is obviously relevant to human maladies such as posttraumatic stress syndrome (PTSD). Unlike most people, PTSD patients can't "just get over," or *extinguish*, the memory of their traumatic experience, and its persistence can make their lives miserable as they continue to respond emotionally to nonexistent threats. If we understood the scientific basis of fear extinction, we might be able to help these patients.

Figure 14.1 Continued
Marsicano and colleagues (see text). In this paper the authors tested the implicit hypothesis that the endocannabinoid system (ECS) in the amygdala (Amyg) controlled the extinction of aversive fear memories in mice. Their hypothesis predicted that cannabinoid receptors (CBRs) would have to be available so the endocannabinoids could bind to them. They tested this prediction by either deleting the CBRs (CBR KO) or blocking them pharmacologically (CBR Block). Their hypothesis also predicted that a number of behaviors that resemble fear responding should not be dependent on CBRs, which they tested with CBR KO and CBR Block. Finally, their hypothesis predicted that the level of ECs should increase in the amygdala but not other brain regions. The authors tested the optional prediction that synaptic connections between neurons in the amygdala could be strengthened by learning behavior, and they carried out several tests of this prediction. These tests led them to proposed a refined hypothesis based on a specific cellular mechanism for fear extinction mediated by the ECS.

The authors chose a conventional behavioral test in which a (male) mouse gets a short, mild electrical shock to its feet through the metal wire of its cage floor and responds by remaining motionless ("freezing") for a few seconds. If a brief, audible tone regularly sounds immediately before the shock, mice soon learn to associate the tone with the shock and to freeze as soon as it comes on—a Pavlovian, or classical conditioning, task. When the animal has learned the tone–shock association, he has become *conditioned* to fear the tone. Now if you give him trials on which the tone sounds but no shock follows, he gradually realizes that the tone no longer announces that a shock is coming and stops getting into his freezing position. This phase of the experiment is called *extinction*, which is not the forgetting of an old fact, but the *learning* of a new one: the tone is now harmless.

The (implicit) hypothesis that amygdalar endocannabinoids (ECs in Figure 14.1B) control fear extinction makes three major predictions: (1) cannabinoid receptors (CBRs) must be activated for extinction to occur; (2) activation of cannabinoid receptors affects the fear responses specifically and not associated behaviors (such as the auditory response to the tone, ability to move, to sense the environment, or to feel anxiety) that could be mistaken for fear; and (3) that, when the tone sounds, endocannabinoids will be produced in the amygdala but not in other brain areas. Marsicano and colleagues tested these predictions by either genetically deleting the CBR, "knocking it out" (CBR KO) or blocking it with a drug, and by directly measuring endocannabinoid production (EC measure) in the amygdala. Figure 14.1B diagrams their paper. Solid lines indicate that the three main predictions are logically required by the implicit hypothesis, meaning that, if they are false, the hypothesis is false. (Note that, although the authors don't state their central hypothesis about extinction, they do state and test one alternative hypothesis for memory consolidation.)

Figure 14.1B makes another important point. The authors also report the results of several electrophysiological studies in a brain slice model of possible cellular mechanisms (the phenomenon of long-term potentiation [LTP] and a related phenomenon, long-term depression [LTD], that I mentioned in Chapter 7) whereby endocannabinoids might affect fear conditioning and fear extinction. Notice that, because the cellular details of this extinction process have not been fully worked out, the outcomes of these experiments were not truly predicted by the hypothesis. At best the results would be "nice, not necessary" for the conclusions of the study. And, no matter how they turned out, they could not falsify the main hypothesis. I've indicated this nonobligatory connection by a dashed line. Although I'd read the paper before, I hadn't realized this aspect of it until drawing the diagram. As it happened, the extra electrophysiological results were largely consistent with their main hypothesis and illustrate how such results can enrich and extend a report, in this case by suggesting a new hypothesis for a cellular mechanism of endocannabinoid-mediated fear extinction.

Incidentally, "nice, but not necessary" experiments are the sorts of things that I had in mind when discussing the Reproducibility Crisis (Chapter 7). You remember that the Reproducibility Project teams tried to replicate the last experiment in a paper and, if they couldn't replicate it, they put a check-mark in the "irreproducible" column of their ledger. The entire paper was graded as irreproducible. However, if the last electrophysiological experiments in the Marsicano et al. paper failed a replication trial, their major conclusion would be unaffected.

Finding a hypothesis "in the wild" and diagramming it gets you caught up in the reasoning of the paper and can give you valuable insight into what conclusions it truly allows, as well as an appreciation of its logical architecture. In the next section, I'll go over some suggestions for training yourself how to think in ways that can make the diagrams of your own papers more elegant.

14.B.2 Finding and Diagramming Your Own Hypothesis

If you are like the student I mentioned in the Introduction, you may be wondering about your own hypothesis. Your mentor may have dusted off a stack of old data and said "there's a PhD, thesis in there somewhere." you may have started doing a few follow-up experiments, yet you feel that what comes next is not blatantly obvious. A good way to begin sorting things out is by asking yourself a series of questions:

- What are you "looking at" in the lab?
- Why? What's interesting about it?
- Do you see a problem that you can solve? Something you don't understand?
- What do you think might going on? Can you think of another explanation?
- Can you test your idea?

If the answer to that last question is yes, that is probably your hypothesis. Try diagramming your thoughts to get a sense of what the parts are and how they may fit together.

Always ask yourself: What is the best conceivable outcome of my experimental tests, and will they tell me anything, really? The purpose of doing this is not to bias your outcomes: it is to keep you from wasting time and help you design better experiments. If the best imaginable scenario is uninformative, you might want to rethink why you'd be doing those tests.

Now let's step back and look at a few more general ways of improving your ability to organize your thoughts and communications effectively.

14.C Think About Thinking

Cognitive ease[6] is a state in which people are less careful, less critical, more likely to accept superficially reasonable statements and, therefore, more likely to fall for cognitive illusions or otherwise make mistakes. One message of this chapter is that, to improve your thinking skills, at times you'll need to disrupt this state of ease by treating your thinking as a deliberate act, rather than just going with the drift of your ideas. I want to begin with an affliction called the *curse of knowledge* because it affects not only scientific thinking and communication, but also our own self-awareness.

14.C.1 Beware the Curse of Knowledge

In a *Sense of Style*,[7] Steven Pinker defines the curse as "a difficulty in imagining what it is like for someone else not to know something you know." Now think back on a time when you sat through a lecture in which almost every word, from the initial rationale of the talk to its conclusions, left you feeling baffled. It is likely that the talk was given by someone firmly in the clutches of the curse; someone who didn't realize that, although you were an intelligent, interested listener, you did not have all of the background knowledge that he had. Had he been attuned to what you (and probably 80% of the audience) might not know, everyone, including the speaker himself, would have gotten much more out of the experience. I suspect that many scientific authors do not state their hypotheses explicitly because they suffer from the curse of knowledge.

Obviously, you want to avoid the curse when you are trying to communicate, and this can be hard to do. Pinker points out that completely abolishing the curse would demand not just putting yourself in others' shoes, though that should help, but putting yourself in others' brains, which is impossible. He emphasizes that you can alleviate the effects of the curse by avoiding jargon and too many abbreviations, knowing your audience, and other tips.

While all this is good advice, it doesn't explain why we are susceptible to the curse in the first place. We know full well that others don't really know what we know—that's why we're trying to communicate with them! so why are we so obtuse about the need to get on the same wavelength with them?

Although it is not the primary issue for him, Pinker notes that we are unaware of how abstractly we think about things that we know well. We pack large amounts of information into idiosyncratically organized chunks of knowledge and forget to unpack them for everyone else. In the same vein, we don't explain the logic behind our reasoning. In Chapter 11, we found that our reasoning and our knowledge are often unconscious, hidden from ourselves as well as from others. Without special effort to find out, we often don't know what we know. We take for granted that we fully understand the details of those high-level abstractions in our brains, though we sometimes don't. In other words, we can be victims of our own curse of knowledge.

The first step in overcoming the curse, says Pinker, is to appreciate how "devilish" it is. Likewise, becoming aware of the curse of knowledge in science, making our thoughts and reasoning concrete and explicit, thinking about what we're doing in forming and testing hypotheses, is a first step toward improving scientific thinking.

As a complementary suggestion, I'd like to advocate for a writing style that may help reduce the abstract nature of scientific communications. Along with an increasing number of scientific journal editors, I recommend that we use the "active voice" in writing (see Box 14.1). That is, do not write, "the following experiment was done . . . ;" write "we did the following experiment . . . " (and don't use "we" if you mean, "I"). If you take ownership of your ideas you will probably think about them and present them in more direct and easier-to-follow ways than when you're filtering them through the passive voice.

14.C.2 Take an *Outside* Point of View

Daniel Kahneman tells the story[8] of a committee that he was once on that was charged with writing a new statistics textbook. Kahneman asked members of the team how long they thought the project would take to complete. Everyone estimated about 2–3 years. When Kahneman singled out a member who had contributed to several multiauthored textbooks how long those projects had actually taken to finish, the expert reevaluated his answer and admitted that each one had needed 7–10 years and that about 40% of them were never finished. (In the end, the present team took 8 years to complete their statistics book.) For Kahneman, the moral of the story is that, despite his experience, when the expert was thinking as a team member, he had gone along with everyone else's optimistic projection. He had taken an "inside" perspective. When he stepped back and viewed the current project objectively, as someone "outside" the committee would, his assessment of their likely success became more realistic.

We need to take an *outside* view of our own thinking, trying to see it as an external observer would. This does not conflict with the advice to use the active voice when actually writing, of course. This is an extension of the meta-cognitive view that I mentioned earlier; we need to be thinking about our own thinking in order to find ways to make it better.

Box 14.1 The Passive Voice and the Hypothesis

"The data [evidence, results] suggest . . .," "the data show . . . ," "the data imply. . . ." If you took it literally, this way of talking would make no sense. "The data" are what we get from doing experiments—columns of numbers, graphs, images from high-powered microscopes, and so on; data just sit there, not suggesting, showing, or implying anything. We human beings look at the data; we interpret, infer from, think about them; and we finally communicate our thoughts about what they mean. When we assign "the data" the leading role, we're using what's called the *passive voice* in English. In general, when the entity that is carrying an action is not the subject of a sentence and may be left out of it entirely, the sentence is *passive*. For instance, "the data suggest" probably means "I inferred [something about the data]." Why do scientists tend to talk in such a roundabout way? Custom, convention, habit. Duke University has a nice overview (https://cgi.duke.edu/web/sciwriting/index.php?action=passive_voice) that includes these insights.

The passive voice has adherents."Considered to be objective, impersonal, and well suited to science writing, the passive voice became the standard style for medical and scientific journal publications for decades" (http://www.biomedicaleditor.com/active-voice.html).

"In conclusion, the use of the passive voice encourages precision and probity and when used correctly can generate as much passion and stimulation as the skilled use of the active voice. The active voice encourages carelessness, partisanship, and as used by many of its adherents, does no favors to the English language or science" (http://www.sci.utah.edu/~macleod/writing/passive-letters.html).

On the other hand, there are those who object strongly to the passive voice:

"Most scientists use passive voice either out of habit or to make themselves seem scholarly, objective, or sophisticated. Scientists have not always written in passive voice. First-person pronouns such as I and we began to disappear from scientific writing in the United States in the 1920s when active voice was replaced by today's inflexible, impersonal and often boring style of scientific writing" (Randy Moore, Editor, *The American Biology Teacher*).

Indeed, some journals are trying to swing the pendulum back toward the active voice:

"Nature journals like authors to write in the active voice ('we performed the experiment . . .') as experience has shown that readers find concepts and results to be conveyed more clearly if written directly" (https://www.nature.com/authors/author_resources/how_write.html).

"Choose the active voice more often than you choose the passive, for the passive voice usually requires more words and often obscures the agent of action. Use first person, not third; do not use first person plural ["we"] when singular ["I"] is appropriate" (http://science.sciencemag.org/content/141/3578/305).

The dispute about passive and active voice focuses mainly on science writing style, and style is not my concern. Rather, I worry about how language shapes our thinking. I suspect that a strong bias in favor of the passive voice makes us less inclined to be forthright about our hypotheses. Hypotheses do not spring up spontaneously; scientists invent, draw predictions from, and test them. Perhaps, in order to avoid the disembodied and faintly pompous passive construction, "The hypothesis suggested itself that . . .," we excise the hypothesis from our writing. If there is no hypothesis, then we don't have to account for where one came from. In any case, by omitting the actor—the scientist—from the report, the passive voice obscures the reasoning process involved. In contrast, by adopting the active voice, "I hypothesize . . ." or words to that effect, we make the reasoning plain, and direct responsibility and give credit where it belongs: the scientists themselves.

14.C.3 Do a "Premortem"

Seventy-three seconds after its launch on January 28, 1986, the Space Shuttle *Challenger* blew up in mid-air, killing the seven crew members and virtually halting the US space program for more than 2 years. Following the tragedy, the Rogers Commission conducted a thorough review to determine what caused it and how it could have been prevented. An investigation into the causes of a failure, conducted *after* the failure has occurred, is called a *postmortem*.

Kahneman recommends undertaking a similar, thorough investigation of your plans *before* you act. He calls this a *premortem*.[9] The idea is to assume that you already put a plan into effect and now, a year later, have learned that it failed. That's the starting point of the exercise: it definitely failed. You have work out what went wrong. Ideally, you enlist coworkers and ask them specifically to identify the fault(s) in the plan.

You can apply the premortem technique to your scientific thinking as well. Imagine that you've written a paper or a grant application that went through a review process and was rejected with the comments that it is "hard to follow," "diffuse," "not well-organized," etc. Ask yourself—or better, ask your colleagues—to read it and tell you why it was rejected. Take their advice seriously.

14.C.4 What Can You Explain Now that You Didn't Set Out to Explain?

In his biography, *Genius: Richard Feynman and Modern Physics*, James Gleick reports[10] that Feynman would often ask a young theorist: "What can you explain now that you didn't set out to explain?" Good ideas have considerable "reach" in David Deutsch's terms (Chapter 10.C); they go well beyond the circumstance in which they first arose. Einstein's General Theory of Relativity got a big boost in acceptability when astronomers realized that it could account for a slight anomaly in Mercury's orbit that had troubled them. Einstein had been unaware that his theory had such reach and was pleased to learn that it did.

On a far more humble scale, I mentioned (Chapter 2) an experiment that a colleague and I did where we found that when we stimulated a principal (P) cell, incoming signals from inhibitory (I) cells transiently disappeared because a mysterious signal from the P cell briefly silenced the I cell. Later, we found that there were two separate groups of I cells: one whose signals disappeared, one whose signals didn't. Why the difference? We had no idea. Meanwhile, another group of investigators discovered that cannabinoid receptors were only present on one type of I cell.[11] This was an isolated finding, what looked like a naked fact, until the mysterious signal that the P cell was sending back to the I cells was discovered to be an endocannabinoid.

Suddenly, everything fell into place: our temporarily disappearing I signals must be coming from the I cells bearing cannabinoid receptors. The endocannabinoid hypothesis, which began as an explanation of how P cells could regulate the output of I cells, now suggested that there was a defined network of I cells that would be uniquely affected by cannabinoids because only they had the cannabinoid receptors. This, in turn, could mean that brain activity affected by cannabinoid molecules—the ones in marijuana in addition to the natural ones—would be disproportionately regulated by these particular cells. Researchers now suspect that the endocannabinoid hypothesis reaches far beyond its origins to brain activity involved in highly complex human behaviors, everything from the development of play behavior in young males[12] (rats, anyway) to epileptic seizures,[13] as well as the fear responding that we just reviewed.

The overarching message is that you can hardly answer Feynman's question—"What can you explain that you didn't set out to explain?"—without thinking a lot about your data and your hypothesis.

14.D Cognitive Lessons from Behavioral Economics

In the preceding chapter I made an analogy between the way we evaluate real physical objects, including money, and the way we evaluate ourselves, our self-esteem, measured in fictitious units called *creds*. If we push the analogy a bit more, we get a new perspective on the hypothesis and scientific thinking.

14.D.1 Take a Chance

How much we're willing to expose our egos or how much we need to protect them depends on the balance in our cred accounts. How many creds can we afford to risk in putting forward a hypothesis? If we play it too safe, we short change ourselves, our colleagues, and the taxpayers, donors, and funding agencies who support us. We should be willing to take intellectual chances, as Stuart Firestein[14] recommends. Maybe we can apply that same boldness to our intellectual projects. Granted, encouraging and rewarding risk-taking will mean that the scientific community and its funders must be willing to undertake structural and cultural changes in how science is done. We can take modest, and inexpensive steps in the direction of intellectual boldness by being less timid and more forthright in stating our hypotheses explicitly.

14.D.2 The Greater Good

In a way, science faces a situation like the one described by the behavioral economist and Nobel Prize winner Richard Thaler[15] who once asked a group of high-level money managers at a large financial firm if they would each take on a mildly risky project that had a good chance of turning a big profit for the firm. The odds of success were good, but success was not guaranteed. A large majority, 85%, of the managers reported (anonymously) that, despite the favorable odds, they would not take on such a project because, if it failed, they would look bad; it would reflect poorly on their individual records.

When the firm's CEO was asked what he would want his managers to do, he said he hoped that they would all take on the risk. The CEO's reasoning was straightforward: if each project had a good chance of succeeding, the combined odds for success in the group of them were so good that the firm, as a whole, would be almost guaranteed to come out ahead, even if a few individual projects didn't pan out. For the firm to maximize its gain, all the managers would have act courageously, so the challenge was to merge the perceived self-interests of the individual managers with the overall goals of the firm, to ensure that managers would be rewarded for taking chances.

Likewise, the scientific community stands to profit when individual scientists are rewarded for sharing the fruits of their scientific training, their thinking, especially their hypotheses, more openly than they now do. In fact, science may be in a better position than the business firm studied by Thaler. Whereas losing a monetary bet would cost the firm in real terms, putting forward a legitimate hypothesis that is falsified is a benefit, not a cost, to science. We advance by trial and error, as Popper stresses. Error is an inevitable and valuable part of the process. Remember the engineers at the TESTCo tire company (Chapter 2): you can only make better tires by trying lots of designs, most of which are destined to fail.

One problem is that we're raised from an early age to want to be right. We've always been rewarded for it. How can we turn around and see our falsified hypotheses for the positive contributions that they are? One possibility is that we can learn from the economic rule to *ignore sunk costs.*[16]

14.D.3 Ignore Sunk Costs

At one time or another probably most of us have been confronted with a situation like this: we've spent a significant amount of money to go to a future event, say a basketball game, and then, when game day arrives, we're ill—headache, plugged sinuses, the works. We begin to dread going to sit for hours with hundreds of screaming, semi-inebriated fans while feeling terrible to boot. We go anyway because, if we don't "the money will be wasted."

This decision makes no sense to a rational thinker: the money that's been spent is gone, and you won't get it back no matter what you do—it's a *sunk cost.* "Honoring" sunk costs by allowing them to influence your future behavior is irrational, and if you honor them you are committing the *sunk cost fallacy.* Folk wisdom recognizes the fallacy when it warns you not to "throw good money after bad" and to "cut your losses." The fallacy is probably linked to *loss aversion* (Chapter 11) because the anticipation of loss of money or creds scares us into sticking with a set plan even when sticking with it no longer makes sense. And the sunk cost fallacy goes beyond money matters: countries keep fighting obviously lost wars—"So our soldiers will not have died in vain"—although continuing to fight will cost thousands more lives. You should keep fighting only if you think you have a chance of winning the war, not because of the heroic sacrifices your soldiers have already made. Likewise, you shouldn't go to the game just because you bought the tickets; you should only go if you expect to enjoy going more than you'll enjoy staying home.

The sunk cost fallacy can influence scientific thinking. Our egos get involved. You invest creds when you state a hypothesis, and, in one way, this is a good thing because you'll deserve the rewards that come from advancing science with a new idea. But what happens if your hypothesis is wrong? What do you do? Doggedly defend your hypothesis because you've invested creds in it? Put up objections or carry out ad hoc revisions? In short, do you insist on its correctness simply because you're afraid of "wasting" your creds? If so, you may have fallen for the sunk cost fallacy.

If your hypothesis is truly wrong, let it go. You made a positive contribution by getting others to think about a part of nature in a new light. You may have prompted your competitors to build on your idea. Be proud of that and move on.

Notice that the question here, as it was with the game tickets and the wars, is not whether you should abandon your hypothesis at the drop of a hat; if you have good reason to think that it is correct, then, by all means, defend it. But focus on finding new ways of testing it or challenging the rival hypothesis, not on protecting your ego.

14.D.4 The Sunk Cost Fallacy and Opportunity Cost

The sunk cost fallacy may affect individual scientists well before their results are published. I am pretty sure that I am not the only researcher who stayed with an experimental project for too long, working on it past the time when I began to realize that it wasn't going to lead to the nifty discovery that I had first dreamed that it was. In part, I kept at it to keep from acknowledging to myself that I'd already wasted a lot of time, effort, and money on it. A classic case of falling for the sunk cost fallacy.

Unfortunately, committing the fallacy was not my only blunder. By sticking with the disappointing project, I also lost opportunities to work on more productive ones. Economists, because they are economists, reckon missed opportunities as real monetary costs, called *opportunity costs*. If you spend $5 on the latest designer cup of coffee, you're not only out $5, but you've given up the chance to buy something that might have had more lasting value. Evaluating the opportunity costs when it comes to buying and selling is straightforward; you know how much money is involved. Opportunity costs are not always monetary; if you stay home alone to read a book, you give up the chance to go to a movie with your friends.

People are more willing to abandon an investment if there is not a lot at stake. If you've spent only a little money for those basketball game tickets or got them for free, then you are more likely to stay at home nursing your cold. If you've spent a fortune on them, doing the rational thing is much harder. However, it is not only what you actually paid that is important: how you perceive the cost matters, too, and the influence of a cost wears off with time. People who've paid up-front membership fees to join a health club around New Year's Day are conscientious in going to the club at first, but then taper off, only to experience a new burst of energy when the semi-annual bill comes due or bathing suit season arrives.

In other words, if real or perceived costs to you are low, you should be better at ignoring sunk costs. When it comes to expressing your scientific thoughts, making a habit of explicitly stating your hypothesis would make it seem less like a risky exception and would decrease the costs associated with it. Fewer personal creds would be on the line, and it would easier to walk away from a falsified hypothesis.

The history of science assures us that the overwhelming majority of hypotheses will, sooner or later, be falsified at some level. Ideally, there would be no downside to putting forward your hypothesis explicitly or, contrariwise, no reason to slip it, *sotto voce*, into your papers in an implied state. We invest creds to take part in the intellectual exchange that is science, hoping to gain and accepting the risks of loss. This vision will undoubtedly strike many readers as naïve. This is where the ability to think like a trader comes in handy.

14.D.5 "Think Like a Trader"

How can we overcome our anxiety about losing creds whenever we put forward a hypothesis? Maybe we can learn a lesson from people whose livelihood depends on risking money for profit: the financial traders on Wall Street who buy and sell shares in companies. If their shares go up in price, they make money; if their shares go down, they lose. In order to stay sane, traders take a broad view of their holdings: they don't invest too much ego or hope in any one investment. Instead, they make lots of investments and expect to win some and lose some (winning a few more, of course). If you think like a trader, you are less likely to suffer from loss aversion, to be more willing to let go of a lost cause, or be less likely to fall for the sunk cost fallacy.

What would thinking like a trader mean for a scientist debating about whether to express a hypothesis? You can put forward multiple hypotheses for any one phenomenon, and you can step back and take the long view of your career.

Generating multiple hypotheses is probably the closest scientists come to actually behaving like a trader. After you've thought of one explanation for your phenomenon, you put it aside and think of another. Besides demanding that you think seriously about the phenomenon, generating multiple hypotheses lets you stay a decent emotional distance away from each of them. When thinking like a trader, you know that some of your ideas will be right and some will be wrong. Having multiple hypotheses is like having a diversified financial portfolio with investments in several different stocks—or even better, in different kinds of entities entirely—if you want to be buffered against sweeping changes. Typically, when the prices of stocks go up, the prices of bonds (municipal bonds, government bonds) go down. If you're a creative scientist you can try to come up with hypotheses that are quite different kinds of explanations. In any case, with multiple hypotheses, you don't have all of your eggs in one basket.

Another way to think like a trader is to take the long view and realize that throughout your career you'll be in the position of advancing hypotheses many times. If they are carefully thought out and rigorously tested, there is every reason to think that some will be correct. Your intellectual investments will undoubtedly pay off, but only if you make them in the first place. The way a trader would.

14.E How to Discover Your Own Hypothesis

This book focuses on the hypothesis, yet as I've noted a couple of times, hypotheses are products of our unconscious minds, and we don't know much about how our minds work. There are a few methods, or collections of hints, that may be useful if you're stuck and need a boost in developing a hypothesis, however. For many people, these techniques help train their intuition.

A hypothesis is an explanation. Your problem might be either that you don't have a clear picture of what you need to explain or that, while you know exactly what you want to explain, you're having trouble explaining it. The following sources of inspiration may jostle your mind enough to become unblocked.

14.E.1 Induction

Whatever the process of induction is, there is no denying that the effort of trying to find and generalize from a pattern in your data is beneficial; it may call attention to a gap in your understanding that needs to be filled and perhaps how to fill it. In other words, looking for regularities in your data is a good way to generate a hypothesis.

14.E.2 How to Solve It

A (maybe *the*) classic text for those who seek to enhance their problem-solving ability is *How to Solve It: A New Aspect of Mathematical Method*, second edition,[17] by George Polya. If you're a math-o-phobe, don't be put off by the reference to math in the title. Polya's suggestions are good for solving nonmathematical problems as well. You don't even have to work through the relatively simple geometric problems he does use; just read the words and get the message. There is a lot of excellent advice about how to develop your own "common sense" and a convenient "Short Dictionary of Heuristic" with more than 60 specific recommendations for what he calls *heuristic* reasoning.

Polya counsels us to be wary, "if you go into detail, you may lose yourself in the details." His "heuristic"[18] is similar to, though not quite the same as, the heuristics that we talked about in Chapters 11 and 12. For example, his first heuristic is *Analogy*: make an analogy between a difficult problem that you're trying to solve and a simpler one that you know more about. I've mentioned (Chapter 2) the experiment in which Paul Fatt and Bernard Katz were trying to figure out how synaptic transmission at the neuromuscular synapse worked. In essence, they made an analogy between heart muscles and skeletal muscles in the body and hypothesized that the way that nerves signaled to heart muscles—by releasing a chemical signal—was the way that nerves signaled to skeletal muscles. They also followed another Polya rule to *Decompose* a complicated problem into separate parts and deal with them separately. I don't know if Fatt and Katz had read Polya (his book is copyrighted 1945), but, in breaking down the stream of small electrical signals that they observed in skeletal muscles into two groups—"bursts" versus the low-frequency, regular ones—and ignoring the bursts, they were doing exactly what he recommended.

14.E.3 Fast and Frugal Hypotheses

In Chapter 12, I introduced Gerd Gigerenzer's view of heuristics as adaptive cognitive tools that help us cope quickly and efficiently with complexity. Heuristics are "fast-and-frugal" procedures—quickly applied and sparing of time and effort—and, while they often act as stand-alone solutions to difficult problems, they can also be used to build up the more elaborate devices that are hypotheses.

In reviewing the *bias-variance dilemma* (see Box 12.1), I noted that bias, perhaps in the form of a straight line through a mass of data points, preserves the general trend of the data and offers predictive power at the cost of losing the information contained in the variance of the data points. This is a key element of Gigerenzer's vision of "less is more"[19] (i.e., that less detail sometimes yields more information). The bias-variance dilemma also shows how the philosophy of less is more generates a hypothesis. The straight line that fits the data relating height to age embodied a hypothesis. There is no line in the raw data; you express your interpretation of how age and height are related when you draw one. The line does not mean that height is *explained* by age but suggests that there is a common mechanism that relates them linearly. The process of line fitting could even lead to multiple hypotheses if, in examining the graph, we decided to try fitting the points more closely with another function. We'd give up some of the straight-line bias in favor of paying greater attention to the variance. Experienced experimenters recognize complicated trends in data—exponentials, polynomials—that lead to different hypotheses about underlying mechanisms.

Gigerenzer wants to understand how we form heuristics; how, at a fundamental level, we come up with the "straight-line hypothesis." To appreciate his approach, let's start with the observation by Herbert Simon that "intuition is nothing more and nothing less than recognition"[20] (e.g., the expert intuition of a chess master of what would be a good move in a chess game). The chess master's intuition results from having studied and practiced chess for tens of thousands of hours and does not necessarily represent native genius. His vast experience allows him to recognize positions on the board and their outcomes at a glance.

The overarching motif of Gigerenzer's school of thought is that we have a storehouse of mental constructs—he calls them *probabilistic mental models*—that serve us well and that we successfully rely on for many tasks.[21] As products of evolution and experience, the constructs are adaptable cognitive tools that we often use unconsciously. There are many rules for "one-reason decision-making" that, operating on the basis of limited information, often provide workable solutions to problems by *satisficing* rather than *optimizing*. These rules can be incorporated into algorithmic process models; however, even in their simplest form, heuristics can lead to practical hypotheses.

Heuristics, Gigerenzer and colleagues point out, allow you to make choices based on familiar patterns. For instance, you use the *take-the-best* heuristic to make choices based on a property—a *cue value*—that you know about. Suppose you need to guess which of two cities is larger. The take-the-best method is to rank order cues that might be relevant: Does it have an airport? A major-sports team? Then compare the cities according to each cue in order, stop as soon as you find a cue that discriminates between the cities, and pick the city with the higher cue value. This uncomplicated strategy is surprisingly successful where you have at least some relevant information. (You're at random chance otherwise. Heuristics are not magic.) Scientists can develop their own intuitive insights based on heuristics like this.

Take-the-best, for instance, will often generate parsimonious hypotheses. For instance, when my colleagues and I first noticed the disappearing I-cell signals, we thought that some change in the P cell itself must be responsible. There was already a great deal of information about seemingly similar effects that took place in P cells, and, knowing what we knew then, we quickly took the best of the known explanations. As it so happened, that hypothesis was false, but the fact that it was simple, concrete, and testable told us where not to look for answers and gave us a good place to start from.

While heuristics assist in *generating* hypotheses, they cannot take the place of reasoning about or *testing* hypotheses. We should keep in mind the Russian proverb that President Ronald Reagan liked to quote: "trust, but verify"; that is, be willing to believe, but don't be naïve. We should consciously welcome the attempts of our unconscious minds to understand the world while remaining skeptical of its trustworthiness. Your brilliant inspiration notwithstanding, you must still subject your ideas to critical scrutiny as you continue to consult your intuition for new ones.

14.E.4 Develop Your Insight

In *The Art of Insight in Science and Engineering: Mastering Complexity*,[22] Sanjoy Mahajan substantially expands on the theme of discovering solutions from practice and approximation. For Mahajan, the great challenge is that, as scientists, we need to develop insight if we are not to "drown in complexity." He quotes William James's remark that "the art of being wise is the art of knowing what to overlook." Mahajan teaches how to organize, as well as discard, complexity via a series of nine thinking tools for simplifying complex problems and getting plausible, approximate answers to problems where the exact answer might be hard or impossible to come by.

If you had to estimate the volume of a dollar bill, how would you go about it? Use the *divide-and-conquer* strategy to break the problem down into manageable bits: you can easily make good guesses as to the bill's width and length (~ 6 cm × 15 cm, say), but what about its thickness? A dollar bill is a piece of paper and you know, roughly, how thick a ream (500 sheets) of paper is (~5 cm), so divide 5 cm by 500 and you get 0.01 cm for its thickness. Put your guesses together and—Voila!—the volume of a dollar bill is approximately 1 cm^3. By simplifying a problem and estimating an approximate solution, you will often be able to find a hypothesis. Neuroscientists do this when they try to understand a neuron's electrical activity by collapsing its complex structure into a spherical cell body with a single tapering branch representing all the dendrites sticking out of it. Its easy to calculate the volume of the ball-and-stick model, and it is good enough for many purposes.

Or consider that often the first step in generating a hypothesis is noticing something unusual that needs explaining. This means that you need a sense of how things normally are. There isn't always a lot of background information around for comparison, so it's good to have ways of guesstimating what to expect. Suppose you're studying neurons in the mouse brain and you come across a neuron firing steadily along at a rate of 20 action potentials per second (20 Hz). Is it unusual or not? You might start by recalling that firing at 100 Hz is considered quite fast—only a few cells do that; on the other hand, you might sense that a cell's firing once every 10 seconds, 0.1 Hz, would seem slow. Should you take the average between them for comparison? The usual arithmetic mean of 0.1 and 100 Hz is about 50 Hz, but you know the arithmetic mean tends to follow the extremes. In such a case, Mahajan would recommend picking the geometric mean: the square root of the product of the extremes. The geometric mean here is $(0.1 \times 100)^{1/2} = (10)^{1/2} \approx 3$ Hz. This suggests that, at 20 Hz, your new neuron is firing considerably faster than what you'd expect for an average neuron and that therefore you may have found a phenomenon that needs to be explained.

As a scientist, you should know how to do approximations like this and to "talk to your gut" to develop imprecise but insightful ways of thinking about complex problems. Mahajan has many illustrations of how to do this, although his homey examples—how to estimate the number gas stations in the United States or the total land area of the United Kingdom—soon give way to cases in that require detailed background knowledge and technical expertise. For instance, his explanation of how to estimate the power output of a trained human athlete is neat[23]; you'll want to have all of your college physics right at hand to follow it closely, however. The main point is that there are many times when the sheer complexity of the problem you face will seem daunting, and Mahajan's program can give you the optimism you need to tackle it.

14.F Coda

This chapter suggests that you can improve your scientific thinking skills, especially thinking about the hypothesis, by stepping outside the zone of your familiar habits:

- Practice finding the hidden hypothesis in published papers.
- Diagram the logical relationships between a hypothesis and the experiments that test it.
- Avoid the curse of knowledge, whether in communicating with others or to yourself.
- Take an outside view of what you're doing.
- Ask yourself what your hypothesis can explain that you didn't set out to explain.
- Avoid the sunk cost fallacy; when your hypothesis is falsified, let it go.
- Invent several explicit hypotheses to account for any one phenomenon.
- Think like a trader to keep from being too invested in any one hypothesis.
- Train your insight and intuition with heuristics and approximations.

Notes

1. Richard Feynman, *The Meaning of It All: Thoughts of a Citizen-Scientist* (Reading: Perseus Books; 1998), p. 18.
2. Thomas Kuhn, *The Structure of Scientific Revolutions, 2nd Edition* (Chicago: The University of Chicago Press; 1970). The powerful influence of social and other irrational factors on scientists' decisions to accept certain findings as reliable facts is a dominant theme throughout Kuhn's book, but it gets the most attention at the end in chapters 10–12.
3. J. E. Motelow, W. Li, Q. Zhan, A. M. Mishra, R. N. Sachdev, G. Liu, et al., "Decreased Subcortical Cholinergic Arousal in Focal Seizures," *Neuron* 85:561–572, 2015.
4. Methods for visualizing reasoning processes include Venn diagrams (visit https://en.wikipedia.org/wiki/Venn_diagram), probability trees (see Reid Hastie and Robyn Dawes, *Rational Choice in an Uncertain World: The Psychology of Judgment and Decision Making* [Thousand Oaks: SAGE Publications; 2010]), truth tables (see Galen A. Foresman, Peter S. Fosl, and Jamie Carlin Watson, *The Critical Thinking Toolkit* [Malden: John Wiley & Sons; 2017], knowledge webs, and syntax trees (see Steven Pinker, *The Sense of Style: The Thinking Person's Guide to Writing in the 21st Century* [New York: Penguin Books; 2014]).
5. G. Marsicano, C. T. Wotjak, S. C. Azad, T. Bisogno, G. Rammes, M. G. Casciok, et al., "The Endogenous Cannabinoid System Controls Extinction of Aversive Memories," *Nature* 418:530–533, 2002.
6. Daniel Kahneman, *Thinking, Fast and Slow* (New York: Farrar, Strauss and Giroux; 2011), pp. 59–70.
7. Steven Pinker, *The Sense of Style: the Thinking Person's Guide to Writing in the 21st Century* (New York: Penguin Books; 2014), pp. 57–76.
8. Kahneman, *Thinking*, pp. 245–254.
9. Kahneman, *Thinking*, pp. 264.
10. James Gleick, *Genius: Richard Feynman and Modern Physics* (London: Little, Brown and Company; 1992), p. 369.
11. I. Katona, B. Sperlágh, A. Sík, A. Käfalvi, E. S. Vizi, K. Mackie, and T. F. Freund., "Presynaptically Located CB1 Cannabinoid Receptors Regulate GABA Release from Axon Terminals of Specific Hippocampal Interneurons," *Journal of Neuroscience* 19:4544–4558, 1999.
12. K. J. Argue, J. W. VanRyzin, D. J. Falvo, A. R. Whitaker, S. J. Yu, and M. M. McCarthy, "Activation of Both CB1 and CB2 Endocannabinoid Receptors Is Critical for Masculinization of the Developing Medial Amygdala and Juvenile Social Play Behavior," *eNeuro* 4(1), Jan 27, pii: eNEURO.0344-162017.
13. I. Soltesz, B. E. Alger, M. Kano, S. H. Lee, D. M. Lovinger, T. Ohno-Shosaku, and M. Watanabe, "Weeding Out Bad Waves: Toward Selective Cannabinoid Circuit Control in Epilepsy," *Nature Reviews in Neuroscience* 16:264–277, 2015.
14. Stuart Firestein, *Failure: Why Science Is So Successful* (New York: Oxford University Press; 2016).

15. Richard H. Thaler, *Misbehaving: The Making of Behavioral Economics* (New York W. W. Norton; 2015), pp. 188–189.
16. "Sunk cost" is an economic term; see, Thaler, *Misbehaving*, pp. 64–73; Kahneman, *Thinking*, pp. 343–346 for discussion of how sunk costs can affect our ordinary decision-making.
17. George Polya, *How to Solve It: A New Aspect of Mathematical Method*, 2nd ed. (Princeton, NJ: Princeton University Press; 1956).
18. Polya credits an ancient Greek mathematician, Pappus, who, writing about 1,700 years ago, first developed a system of "heuristics" for teaching the "elements of analysis and synthesis" that were useful in solving problems; Polya, *How to Solve It*, pp. 141–148.
19. G. Gigerenzer and H. Brighton, "Homo Heuristicus: Why Biased Minds Make Better Inferences," in G. Gigerenzer, R. Hertwig, and Thorsten Pachur (Eds.), *Heuristics: The Foundations of Adaptive Behavior* (New York: Oxford University Press; 2011), pp. 2–27.
20. Kahneman, *Thinking*, quoting Herbert Simon on intuition, p. 237.
21. G. Gigerenzer and D. G. Goldstein, "Reasoning the Fast and Frugal Way: Models of Bounded Rationality," in Gigerenzer et al. (Eds.) *Heuristics*, p. 36.
22. Sanjoy Mahajan, *The Art of Insight in Science and Engineering: Mastering Complexity* (Cambridge, MA: MIT Press; 2014).
23. Mahajan, *Art of Insight*, p. 21.

15
The Future of the Hypothesis
The Big Data Mindset Versus the Robot Scientist

15.A Introduction

"The End of Theory: The Data Deluge Makes the Scientific Method Obsolete" claimed *Wired Magazine* Editor Chris Anderson,[1] and he went on to explain the Big Data Mindset that is bringing about the end. We are now in the Petabyte (10^{15} bytes) Age, with the growing mass of data of all kinds creating bigger and bigger headaches for data collection, storage, and management[2] as petabytes are buried under exabytes (10^{18}) and zetta-bytes (10^{21}) of data. For Anderson, the technological hurdles, though truly daunting, are not the most arresting features of the Age. He doubts that we'll be able break the mass of information into chunks small enough to be meaningful and argues that already so much exists that we cannot conceptualize and process it. According to Anderson, we need to "give up the tether of data as something that can be visualized in its totality." Hypotheses and theories will be passé, no longer required or even possible. We will have to develop a new "mindset" and "view data mathematically first and establish a context for it later."

One vision is that the future of science will be shaped by the rise of a Big Data Mindset. A diametrically opposed vision is that Big Data plus artificial intelligence (AI) will enable machines to generate and test hypotheses about highly complex data without human help. In the second scenario, logic and the quest for understanding remain paramount scientific objectives, but "Robot Scientists"[3] gradually assume greater autonomy. In this chapter, I want try for a snapshot of the Big Data phenomenon as it promises—or threatens—to revolutionize scientific thinking and practice. An image of anything moving as fast as the Big Data Revolution is, is like a picture taken from a speeding train, blurred and outdated by the time you look at it. The best we can hope for is a sense of its likely impact on hypothesis-based scientific reasoning.

15.B The Big Data Mindset

What exactly is a Big Data Mindset? It is not a necessary consequence of working with Big Data. As we saw in Chapter 4, Big Data per se is compatible with the hypothesis testing procedures of conventional science. In both Big and Small Science settings, researchers use Big Data try to interpret the world through causal relationships and explanations. We're constantly warned that "correlation is not causation" in traditional contexts. In contrast, advocates of the Big Data Mindset believe that, while we may succeed in finding useful predictive correlations among variables, given millions or billions of variables, with trillions of conceivable interactions among them, it is futile to hope to get beyond correlation to causation.

And that's just fine in Chris Anderson's world where "Petabytes allow us to say, 'correlation is enough.'" In this world, we give up on our conviction that science should help us understand nature, as well as on the quaint belief that we need to understand it. (A saying found on hippie T-shirts from the 1960s, "Reality Is a Crutch," comes to mind.) For many scientists, it would be hard to accept such a seismic shift, but Anderson argues that scientific theories are likely to be wrong anyway, so why worry about not having them? Cut right to the chase: "Correlation supersedes causation, and science can advance even without coherent models, unified theories, or really any mechanistic explanation at all."

Amazon, Inc., recommends books for you to read based on its vast storehouse of data on books that you and people like you have bought and liked in the past and algorithms to translate those data into specific recommendations. Driven by its Big Data Mindset, Amazon's recommendations are frequently helpful, despite being unguided by sophisticated literary advice. "If it produces usable results, what else is there?," could pretty well serve as the motto of the Big Data Mindset.

The Big Data Mindset has ripple effects that affect science in other ways than the decision-making process. The concept of "scientific data" itself will change because the Mindset does not demand the organized, sets of precise measurements that scientists have always relied on.[4] Vast quantities of messy data will come from all over: cameras, laboratory instruments, telescopes, web searches, on-line communications, etc. Sensors for every kind of energy will be embedded everywhere, creating and vacuuming up data where none existed before. Almost anything imaginable can be "datafied" and sent to computers for sorting out. For the Big Data Mindset, data don't have to be exact, neat, or carefully curated in order to be useful as long there is a lot of it. Quantity trumps quality.

To get a feel for the advantages and disadvantages of the Big Data Mindset, we'll start with the notorious case of *Google Flu Trends* (GFT) and its "epic failure."[5]

15.B.1 The Rise and Fall of Google Flu Trends

Google is not a medical services provider, it does not conduct research in epidemiology or virology, and yet it created a computer algorithm, GFT,[6] that, for a time, predicted the spread of the major 2009 flu epidemic in the United States weeks ahead of the Center for Disease Control (CDC), the US government agency that is charged with making such predictions. Google did not try to identify actual or potential sick people. Rather, it sifted through the masses of online search data that it routinely collects every day (40,000 searches per second; 3.5 billion per day in 2018) for patterns correlated with outbreaks of illness during the previous five flu seasons. Google scientists assumed nothing about the correlations they might find: instead, they looked for relationships among the search terms that people used and the spread of flu (as reflected in the corresponding CDC records of flu-related doctor visits). The Google scientists took the 50 million most common search terms that they found and tested them, one at a time, to identify the terms that were most highly correlated with previous flu epidemics. They combined the 45 top search terms to create an algorithm to predict future outbreaks (algorithms that included more predictors were not as good). In the end, they tested 450 million different mathematical models to find the one that most accurately predicted (or "postdicted"; see Chapter 4) the spread of flu during the earlier years, and they called it "Google Flu Trends." They did not worry about why the correlations between GFT and flu existed: predictions for the near future, based on data in the recent past, *nowcasting*—not understanding—was their goal.

How did GFT do? At first, very well. It predicted the seasonal 2009 outbreak and tracked the 2010 and 2011 outbreaks as well as the CDC data did. The fact that it was not better than the CDC, however, was disconcerting, but things really began to fall apart in late summer 2011. For the next two years, from August 2011 onward, GFT overpredicted the number of flu cases in 100 of 108 weeks. And it overshot by more than double the real number during the 2012–2013 season. Google engineers repeatedly tweaked their algorithm to no avail, and eventually the company pulled the plug on GFT. Today, a visitor to the website https://www.google.org/flutrends/about/, reads that "Google Flu Trends . . . are no longer publishing current estimates of Flu. . . . It is still early days for nowcasting and similar tools for understanding the spread of diseases like flu."

15.B.2 GFT: What Went Wrong?

According to data analyst David Lazer and colleagues[7] there were many problems with GFT, and they named "[B]ig [D]ata hubris" and "algorithm dynamics" as the most insidious ones. Big Data hubris (hubris is "excessive pride or self-confidence") gives rise to the fiction that Big Data analyses of massive amounts of data alone, unconstrained by concerns for how to conduct investigations, is sufficient for making scientific progress. Blinded by hubris, you may downplay the *validity* of a test (i.e., whether it measures what it is supposed to measure) or not worry about whether your interpretation of the test results is consistent with well-established theoretical principles.

The problem caused by *algorithm dynamics* is that the initially high correlation between users' search behavior and actual illness may fade. For instance, searches for "flu" might bring up other terms (e.g., "sniffles") that prompt people to do follow-up searches on "sniffles" which they had not initially done. If "sniffles" had been on the original 45-term master list that Google used in its algorithm (we don't know because Google never made it public), this would have created positive feedback. The new searches on "sniffles" could artificially inflate the number of unique flu-related searches that users initiated and cause overestimation of the prevalence of flu.

Spurious correlations were also a concern because, with 50 million potential candidate search terms, the probability of finding correlations that were "structurally unrelated" to flu and hence misleading was quite high, according to Lazer et al. A potential source of spurious correlation that the Google scientists excluded from their algorithm was the term "high school basketball." High school basketball games, like the flu, tend to occur during the winter months, and a correlation between high school basketball and flu would be spurious. An algorithm that included "high school basketball" would have been a "part-flu, part-winter detector." Some spurious correlations are obvious, but you don't know how many others remain hidden within the model. Indeed, GFT missed an unusual out-of-season flu outbreak in 2009, conceivably because another "winter detector" remained buried in the algorithm.

Big Data hubris, algorithm dynamics, and spurious correlations are not the only sources of concern stemming from Big Data Mindset strategies like GFT: *overfitting* is a major one.

15.B.3 Overfitting and GFT: The Bias-Variance Dilemma

You recall the problem of overfitting from Chapter 4: we imagined developing a model to account for the pattern in which a handful of coins tossed into the air landed on the ground. Because of *overfitting*, a model that successfully *post*dicted one pattern would do a terrible job of *pre*dicting the landing pattern of a future group of tossed coins. In general, overfitted models fail because the exact data values used to build them are influenced by random variability that won't be duplicated in new datasets, and Big Data strategies are susceptible to this problem. How do we interpret our data in a simple and robust way without running the risk of overfitting?

15.B.4 The Bias-Variance Tradeoff

The *bias-variance tradeoff* (Chapter 11) is the statistical Achilles heel of mathematical models that try to go from correlations found within enormous datasets to predictions of future behavior. The bias-variance tradeoff first came to prominence as computer scientists began to develop machine learning algorithms,[8] essentially computer programs that used feedback to detect regularities in data and to describe these regularities by rules. The algorithms were first tried out on samples of data, *training sets*, that followed a rule, and they would try to fit the data with a mathematical function—basically, to guess the rule. Guided by feedback on how well its guess matched the true rule, the algorithm would make adjustments and try to make better fits.

We saw (Chapter 12) that bias reflects a scientist's decision about how to characterize data by a certain mathematical function. To get a good fit, she has to weight some points more heavily than others, in effect ignoring variability and introducing bias in the form of her expert intuition of the underlying reality, in order to make a better predictive model.

In the machine learning context *bias* has a slightly different meaning; it is the quantitative difference between an algorithm-generated rule and the natural rule that generated the data in the training sets. If your algorithm is too sensitive to small fluctuations in the training sets, it will *overfit* the data in later test sets. (These two kinds of sets, incidentally, are often created by *splitting* one large, original dataset into two halves and using one half for training and the other for testing.)

Problems arise even if you don't actually force your model to account for every hiccup in the data. Let's take the GFT case as an example. Assuming that there is a true relationship between an individual's web search terms and his state of illness this relationship will be subtly different for each individual, and therefore your web-sifting correlational hunt will average across the searches of all users. This averaging introduces a kind of unpredictable bias error even as it reduces the variance error due to individual differences. The bottom line is that there is a tradeoff between variance and bias, and, since it can't eliminate error, the machine learning solution is to minimize the total error from both sources. While reducing error is always advantageous, this process is primarily useful for applied science—for example, helping Amazon sell books by predicting ones you'll like—rather than understanding nature in the deep sense.

15.B.5 Could "Good Bias" Help Fix the GFT Program?

I've alluded to the idea that when scientists throw out information by fitting messy experimental data with a crisp mathematical function they are exercising "good bias." That is, de-emphasizing or ignoring some aspects of your data does not invariably reflect a moral or ethical lapse. You might wonder if you could improve the overall performance of GFT and similar programs by building in good bias that decreases the influence of the correlations between individual web search terms and observed flu cases. You'd loosen the ties between current data and past flu cases in order to improve your ability to predict future cases.

Loosening these ties, however, would contradict the way the Big Data Mindset approached the creation of GFT in the first place: after all, the Google scientists created GFT by deliberately seeking out search terms that were highly correlated with the numbers of flu cases. And in fact, tweaking the GFT algorithm with apparently reasonable fixes—essentially what we are considering "bias"—did not help at all. The predictive validity of GFT continued downhill despite active intervention. In a sense, this is unsurprising once we remember the impetus behind the Big Data movement: there is much too much complicated data for us to be able to grapple with rationally.

15.B.6 GFT and the Need to Understand

David Lazer et al., argue that a major drawback of the GFT approach is its focus on improving predictions, whereas "What is more valuable is to understand the prevalence of flu." Big Data is not useless: to the contrary, it can markedly enhance the value of online surveys and health reporting, but its goal should be on providing "a deeper, clearer understanding of the world" rather than trying to supplant understanding with correlations. Of course, attempting to provide explanations and understanding is what hypothesis-based science is all about. In other words, the value of the GFT approach might be increased by merging it with traditional scientific methods (although we need to be clear on what GFT's scientific goal was; see Box 15.1). But is the Big Data Mindset compatible with conventional reasoning?

The GFT algorithm was not trying to *explain* sickness by looking at web searches. At best, you might argue that GFT was based on a prediction of the implicit hypothesis that sick people (or people worried about becoming sick) were looking for information that would help them get or feel better. This interpretation is dubious. As Lazer et al. point out, the failure of the GFT algorithm was due in part to the fact that it was not a valid measuring device and therefore couldn't say anything about either the prediction or the hypothesis. If your pH meter is broken, it can't tell you anything one way or another about your acid rain hypothesis.

Likewise, the Google algorithm was not a valid detector of flu. The Google team did not construct GFT by looking for meaningfully "flu-related" terms,

Box 15.1 Was GFT Testing an Implicit Hypothesis or a Prediction?

The authors of Google Flu Trends (GFT) don't take a stand on whether their work has anything to do with a hypothesis; the word doesn't appear in their paper. "Prediction" occurs only once, in "prediction intervals," hence this prediction is also unrelated to a hypothesis. However, a *Nature* editorial on GFT says that it "tested the hypothesis that people will more frequently search the Internet using flu-related terms when they get sick."[9] if you accept the definitions of these terms in Chapter 1, then you may disagree. A hypothesis is an explanation. A prediction is a statement about a future state of the world that may or may not follow from a hypothesis. Even when it does, it does not explain anything.

The *Nature* editorial says that the GFT is testing the "hypothesis" that people will do flu-related searches when they get sick, but this is a prediction, not a hypothesis. It says only what the investigators expect sick people to do. Here's an analogy: "People with coughs will buy cough medicine." This is a prediction about sick people's behavior. The statement that "People buy cough medicine because they have coughs" is a hypothesis that explains a particular kind of purchasing behavior. It predicts that people with coughs will buy cough medicine. Probably hypotheses like this seem so obvious that we don't bother to state them—they are implicit—nevertheless, it's better not to confuse an unstated hypothesis with a prediction.

but merely terms that were correlated with the incidence of flu. You can see that these objectives are not equivalent if you re-consider the correlation between searches for "high school basketball" and number of flu cases. This was plausibly accounted for as a spurious correlation—both high school basketball and flu occur in the winter months—but the fact that the correlation exists at all makes the point: terms that were useful in constructing GFT are not in any obvious way "flu-related."

The blind searches for correlations among variables within massive datasets that are driven by the Big Data Mindset are not themselves tests of hypotheses. Claims that the Big Data Mindset is about "millions of hypotheses" suggest that they, like GFT, are conceptually advancing scientific understanding of the world, when instead they only are trying to detect regularities in it. This is just a futuristic update of familiar old inductivist methods.

A hypothesis-friendly application of GFT-type Big Data searches would go from correlations to testable hypotheses. Essentially, the data analysis could serve as an "inference engine" that could aid scientists in generating hypotheses. Methods developed by Judea Pearl and his colleagues may help create such devices.

15.C An Inference Engine

Science is concerned with causes, not correlations, and yet our data are mainly about correlations among variables. Judea Pearl is a mathematical statistician and A. M. Turing Award Winner[10] who finds this situation manifestly unsatisfactory. He wants to deal directly with causes and causal inferences, so he invented a theory of causation, the *Structural Causal Model* (I'll call it a "causal model"), a calculus for expressing causal relationships and methods for solving problems involving causal relationships.[11] What is a causal relationship? If two variables, say children's ages and their heights, are correlated, it just means that they change together. Their relationship is symmetrical, and it doesn't matter how you plot them on a graph, with age on the x-axis and height on the y-axis or vice versa.

A causal relationship is not symmetrical, if A causes B, then it is not true that B causes A. If being young causes males to engage in risky behavior, you show this by plotting age on the x-axis and risky behavior on the y-axis. You can't flip the axes because it is not the case that engaging in risky behavior makes you young (although you wouldn't know that from observing certain older males).

The causal model incorporates features of Bayesian networks, Bayesian inference, and more. Pearl wants to program "brainless" robots to behave as if they understood causality and act appropriately. As it stands, robots and computers in general do not take hints, infer unstated assumptions, or know "what you mean." How many times have you been frustrated to get no hits in an online search until you realized that one tiny spelling error was to blame—what you meant was so obvious! How could the stupid computer not have known? Enabling machines to operate as though they could think causally would be a giant step toward harnessing the power of Big Data.

What we need is an approach that is not based on black-or-white Boolean rules, but on Bayesian probabilistic rules (see Chapter 6) that can be updated with experience and that respond with shades of gray. Pearl's solution is to create systems of equations, called *Bayesian networks*, of conditional expressions for variables—both observed and "latent"—that interact. This approach expresses relationships quantitatively and relates them to diagrams—directed acyclic graphs (directed graphs)—that illustrate possible causal paths from one state to another. These paths are meant to be actual, asymmetrical, causal relationships, where *A causes B*, not just that *A* and *B* are correlated.

A basic example of a problem that Pearl solves with a Bayesian network is this: you know that the grass will be wet if it either rains or the sprinkler system goes on. You also know that, if it rains, the sprinkler system most likely won't be turned on, though it might (i.e., there is a one-way causal interaction between rain and the sprinkler system, but it isn't perfect). There are prior probabilities that rain, sprinkler operation, or both will take place. This is a *conditional probability*: the chance that one event (rain) happened *given that* another happened (sprinkler on). Both rain and sprinkler can cause wet grass so, if the grass is wet, you can assume that the probability is 1.0 that one or both events occurred. Each state of the world is represented by a node in the graph (Figure 15.1[12]), and each arrow shows a causal relationship. Suppose you want to determine the chance that the grass is wet *because* it rained. The legend to Figure 15.1 sketches out how you would get to the answer from the conditional probability distributions. Now imagine that you want to understand an event and have a dozen, or hundreds,

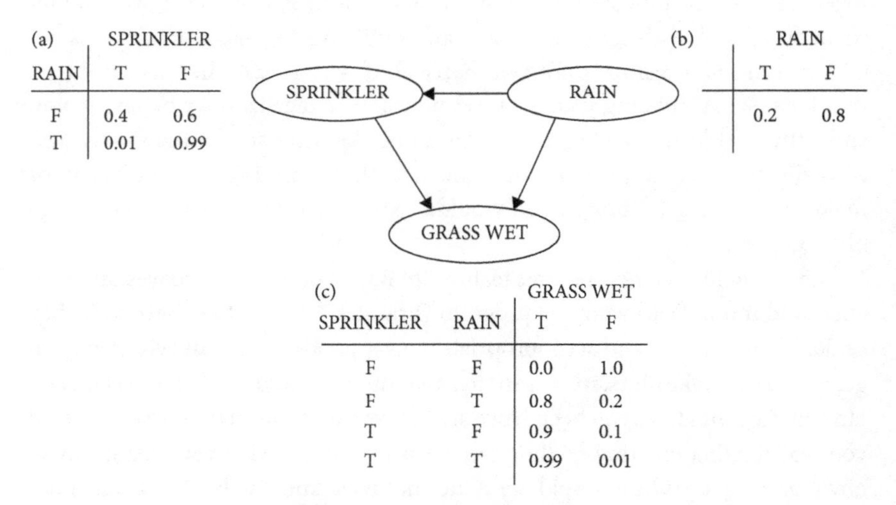

(a)

RAIN	SPRINKLER T	SPRINKLER F
F	0.4	0.6
T	0.01	0.99

(b)

RAIN T	RAIN F
0.2	0.8

(c)

SPRINKLER	RAIN	GRASS WET T	GRASS WET F
F	F	0.0	1.0
F	T	0.8	0.2
T	F	0.9	0.1
T	T	0.99	0.01

Figure 15.1 A basic Bayesian network illustrating how causal interactions among conditions influence outcomes. The tables show the conditional probabilities of the events: "T" means it happens, "F" means it doesn't. Table A says that the probability that if it rains (Rain = T) the chance that the sprinkler is on (Sprinkler = T) is 0.01. Table B says that the chance that it will rain is 0.2. Table C says that if the sprinkler doesn't go on (Sprinkler = F) and it doesn't rain (Rain = F), the grass won't get wet (Grass Wet = F). With these tables you can calculate the probability that the grass will be wet if it rains (sprinkler on or off) and the probability that the grass will be wet given any condition (sprinkler, rain, both). Dividing the probability of "wet given rain" by the probability of "wet in any condition," (i.e., the probability of "wet given rain" plus the probability of "wet given no rain") your answer to the question, "If the grass is wet, what is the probability that it rained?" is approximately 0.36.

of causal variables that interact, like the rain and the sprinkler did. Even with a directed graph, the true causal relationships won't be intuitively obvious, whereas solving the system of equations outlined by the graph will give you a group of probable causes for you to evaluate.

The causal model does more than set up networks of conventional Bayesian equations: it describes a new mathematical framework, a *do calculus* for stating and solving *counterfactual* equations (we encountered counterfactuals in Chapter 11.A.2.c). You'd use counterfactual reasoning in trying to infer "what would happen if" you do some manipulation. The problem is that you can't express "what if" statements in ordinary computer notation, yet that's what you need to make "stupid computers" understand what you mean. The do calculus lets you put such sophisticated intellectual concepts into computer-friendly syntax. For instance, the concept "what if" implies a joint probability distribution that tells you the likelihood that Y event will happen given that a treatment value X occurs. As far as the computer is concerned, this is an asymmetrical, causal relationship because the value of Y does not affect the value of X.

What's more, with Pearl's methods, you can go beyond observable variables to investigate hidden, *latent* variables that you can't measure directly. Say you wondered about switching Snowball's dog food to a cheaper brand than the expensive one he snaps up so readily. You reckon that if Snowball is hungry enough, he'll eat anything, so food deprivation alone, the need to replace the calories he uses up, would make him eat. On the other hand, you recall your grandmother saying "hunger makes the best sauce," thus, in Snowball's case, if his food tastes better he'll eat more of it. But does the more expensive food really taste better to him? Your grandmother's principle suggests the hypothesis that both his state of food deprivation—how long its been since he last ate and how many calories he's burned since then—and the tastiness of the food interact to determine how much he eats. Your model of his eating behavior would include a causal relationship between food deprivation and the latent variable "tastiness"; if he's hungry even bland food will taste better. Constructing a causal model makes it easier to see links between variables that might be crucial for explaining the phenomena that interest you.

The main point for us is that a causal model is upstream of scientific hypothesis formation and can provide clues about hypotheses to test. The hypothesis about Snowball's eating predicts, for example, that he'll eat more cheap food if he's hungry than what he'd need to make up for the calories he's burned. The model highlights the relationships among the variables, which leads to testable hypotheses. Testing the predictions can falsify the hypothesis, although, as usual, if the causal model fails then you won't know which of its assumptions were wrong—maybe you didn't assess Snowball's taste variable appropriately or maybe other factors also influence his eating.

Pearl's approach is a probability-testing procedure that depends heavily on subjective prior probabilities, so it is open to the criticisms of Bayesian approaches that we discussed in Chapter 6. Nevertheless, the causal model is a rational approach to problem-solving with Big Data that is far removed from the correlation-only methods of the Big Data Mindset. Pearl's ideas have contributed to the development of AI, which is devoted to using computers to mimic human reasoning ability. This is the second way in which Big Data will profoundly affect scientific thinking and the hypothesis.

15.D Artificial Intelligence and the Hypothesis

In the beginning, computers were just machines; nobody wondered if they could think. A computer crunching gigantic masses of data and plotting the results is no more than a useful tool. When the IBM computer "Deep Blue" beat the reigning world chess champion, Gary Kasparov, in May 1997,[13] it could be shrugged off because chess is a well-defined, rule-based game, albeit with an enormous number of possible moves and positions. The computer program used brute force augmented by programming heuristics to evaluate 200 million positions per second. Definitely impressive, though since machine performance often exceeds human performance in other ways, the defeat was not especially threatening to the human ego.

When the IBM computer "Watson" beat the champion of the TV game show *Jeopardy*, Ken Jennings, in February 2011,[14] in a test of ability to answer factual questions posed in natural language terms, computer encroachment on human turf became a little more ominous. Although it, too, was relatively structured, the *Jeopardy* competition was markedly less rigid than a chess game and, except for a few notable gaffes, Watson won going away.

And computers are demonstrating humanoid capabilities in far less structured environments. Recently, an IBM computer, *Project Debater*, did well against the 2016 Israeli national debate champion, Noa Ovadia,[15] in a match where neither party knew the topic in advance. Both had to marshal evidence from their background knowledge—no internet hook-ups—and frame their arguments in logically compelling ways while anticipating and counteracting the opponent's arguments. Project Debater didn't win but firmly held its own, as Watson, and many humans, never could have. Oddly, perhaps, success in debate doesn't automatically translate into earning power; IBM is still trying to find a business application for Project Debator that will make money for the company. In any event, the machines are obviously succeeding in handling the demands of increasingly abstract environments.

In some ways, scientific thinking occupies a mid-point between structured and unstructured activity. It has rules and objectives, and it draws on relatively well-defined background information; it also depends on an ability to go beyond the information given and discover new knowledge. Scientific thinking is potentially very fertile ground for the combination of Big Data and AI.

15.D.1 Machine Learning and Neural Networks

"Machine learning" (aka AI) is shorthand for a collection of programming strategies that allow computers to improve their performance in solving problems by trying a solution, comparing that solution with a standard, getting feedback on how they did, and then automatically modifying their programs, trying again, etc. One scheme is a *neural network*[16] that simulates a network of brain neurons. Each conceptual neuron receives and processes incoming bits of information in the form of either plus or minus values, analogous to neuronal excitatory and inhibitory signals. After summing the information each neuron sends its result to other neurons that process the information that they receive and pass it along until the network calculates a final output.

A simple network might have one set, one *layer*, of interconnected neurons. Complicated networks, *deep neural network* (henceforth, I'll use "neural networks" or "networks"), have many layers. The neurons in each layer process a certain kind of information, and the layers communicate, forward and backward, until ultimately their summed activity generates the output of the network. The program compares its output with the concept the neural network is trying to "learn." The difference between the actual and criterion output constitutes a feedback signal that is routed back to the individual neurons and alters both the way they will process new information and the strengths of their interconnections. This is machine learning.

Let's say you want to teach your neural network to recognize a picture of a cat. You feed in many, many pictures of different cats doing all of the adorable things that cats do. This is your *training set* and the bigger, the better. You label each picture as "cat" or "not a cat" and have an algorithm process the data. A neuron in the network gets information from a particular pixel in every picture for all of the pictures in the training set. Every now and then you feed in a new picture that the network has never seen and ask it, "Is this a cat?" The networks output says "yes" or "no" and, depending on whether it was right or wrong, modifies the strengths of its own internal neuronal connections. You keep going like this until you're satisfied that your network can reliably identify a cat. Neural networks can become quite good at pattern recognition, speech perception, etc., and they have been used to guide self-driving cars or write credible newspaper articles.[17] This is all harmless and nonthreatening (unless, maybe, you drive vehicles or write about sports for a living).

"Deep learning" is a merger of deep neural networks and Big Data, and it is ordinary machine learning on steroids.[18] Deep learning is "unsupervised," meaning that if you want your neural network to learn "cat," you don't label the pictures that you use to train it on. Instead, you feed in say, 10 million pictures with a cat in them and instruct the algorithm to find out what's common about them. Eventually it comes up with "catness" all by itself. Unsupervised machine learning like this takes place in a way that, almost by definition, even its human programmers don't understand. Still, it's about cats and we know what cats are. Things grow increasingly problematic as we move to the next level.

15.D.1.a What Deep Learning Can Learn About People that People Don't Know About

Deep learning networks can learn things that people don't know about, according to Stanford social psychology researchers Yilun Wang and Michael Kosinski, who wondered what our faces really say about us[19] and turned to Big Data and a deep learning network to find out. First, they loaded it up with data on 35,326 American men and women—50% heterosexual and 50% gay—collected from online dating sites, complete with pictures and other information including self-identified sexual orientation, dating aspirations, etc. Then they asked their network if there were any correlations between faces and sexual preferences as defined by the online data; the objective was to see if the computer could learn to determine someone's sexual preference by just looking at his or her face. Now, if you ask average people, gay or straight, to guess whether they are looking at a picture of a gay or straight person, they are correct a little more than half of the time, usually from 55% to 61%. After seeing just one photo of a person, the computer program utilizing a deep learning network was right 81% of the time for men and 71% of the time for women. If the program was shown five photos of each person, its success rates went to 91% and 83% for men and women, respectively.

This astonishing result shows that machines can greatly outperform people on a task that, you would have thought, demanded the special perception that only a sensitive, experienced human could have. This feat, while amazing, is not the creepiest part. What is far more unsettling is that the researchers themselves had no idea what specific information their network was picking up in people's faces—obviously there is something in faces that people themselves are oblivious to.

15.D.1.b What, If Anything, Can We Do? The European Union Takes a Step

Wang and Kosinski's study stirred up a lot of anxiety among groups concerned with personal privacy; nevertheless, the profiles that were analyzed had been made public by their owners. The message, you might at first assume, is be vigilant and cautious when you make your personal data public. But is this even possible anymore? Big Data are being scooped up from every nook and cranny of our lives, and cameras are found in more and more places in our society. Computer programs like Wang and Kosinski's would work on millions of faces whose owners have never intentionally advertised themselves publicly. The fact that nobody could say how the program works only made it seem more like a tool of Big Brother.

Privacy concerns, as well as the general question of what the machines are up to, moved the European Union (EU) to enact a law,[20] effective May 2018, that the public has a right to an "explanation" about how a machine makes a decision that affects people. While it may seem straightforward, if you think about the complexity of deep learning networks, the massiveness of Big Data, and the ability of computers to find correlations that the human mind may not be able to grasp, you can see that the requirement for explanations may be a tall order. So tall, in fact, that it has inspired the creation of a new branch of AI, one specifically designed to investigate and let the public know what the original networks are doing.[21] Called "explainable AI" or XAI, the strategy is to put a new algorithm, let's call it xDLN, inside the same computer that is trying to learn, for example, "cat" via the concept-acquiring network, "cDLN." The xDLN will watch the cDLN to see what feature of each image makes each neuron light up as the cDLN gradually develops its concept of catness. By training the xDLN to recognize and label the cDLN's operations, the hope is that the xDLN will be able to tell us what the cDLN is doing so we can understand and explain it. (If you fear that we might need a "yDLN" to understand "why" the xDLN is doing what it's doing, and so on and on, you're not alone.)

Even if XAI functions in the way that it's supposed to, we still won't know whether it can give us an "explanation." In fact, whatever an "explanation" is in this context, it seems be a long way from the naïve notion of "explanation" that we've assumed up to now. If an only dozen experts and a hyper-advanced computer program "get it," has it been adequately explained to the public? The EU law doesn't say, evidently preferring to let the legal system sort things out.

Skeptics think the very principle of explanation must be carefully refined.[22] They distinguish between "subject-centered explanations" that focus on telling us, the technically illiterate masses, approximately what the computers are doing in language we can understand and "machine-centered explanations" that (may) tell the experts in excruciating detail what is happening. Subject-centering sounds palatable, though perhaps broad and fairly superficial, and it will leave the experts in charge. The skeptics doubt that having a "right" to an explanation is the best solution to the looming threats potentially posed by the machines' growing level of influence over our lives. Rather, they say, we'll be better off if we insist on establishing and enforcing workable, but iron-clad privacy rules that set limits on what the machines can know about us. Instead of seeking an explanation of how computers infer people's sexual orientation, people should be enacting laws that will prevent computers from gathering the relevant information in the first place.

The decisions that society will have to make in response to the rise of Big Data and AI mirror the decisions that science itself will have to make. How much do we trust AI? How much autonomy do we turn over to it? What roles will humans have when we can no longer comprehend AI's information processing capabilities or it outputs? An alternative for science lets us put off the thorniest of these issues at least for the time being. This is the way of the Robot Scientist, where technology extends the capacity of humans to do their jobs without completely overturning the current order.

15.D.2 The Robot Scientist

The Big Data and AI revolutions are surely coming to biosciences, cognitive sciences, and more and will change everything when they do. Indeed, the day of the "Robot Scientist" in biology has arrived.[23,24] Guided by AI techniques, Robot Scientist "Adam" has successfully carried out genetic experiments on yeast cells. Not impressed? Consider this: Adam can come up with hypotheses to explain observations, devise experiments to test these hypotheses, physically run the experiments, collect and interpret the data, and then repeat the cycle. And Adam does his work "on time and under budget," meaning he also calculates and conducts the most cost-effective ways of testing his hypotheses. Essentially, after hitting "Enter," humans have no role in Adam's investigations apart from supplying him with raw materials and clearing away experimental waste every five days.

15.D.2.a The Robot Scientist and What He Does

Despite being called a "robot," Adam is a fully equipped mini-lab rather than a humanoid machine; he controls a suite of laboratory instruments and devices, including a −20°C freezer, three cell incubators, two plate readers, three liquid handlers, three robotic arms, two robot tracks, a centrifuge, a washer, an environmental control system, cameras, computers, and more. He can design and carry out approximately 1,000 experiments and make more than 200,000 observations per day. Continuously.

In his basic protocol, Adam selects specified yeast strains from the thousands of mutant strains in his freezer and puts them in tiny plastic wells on plates.

The wells contain "rich" (having all necessary nutrients) growth medium. He measures the cells' growth optically, first on rich medium, then takes a defined quantity of the cells, transfers them to wells lacking a particular nutrient, and measures their growth again.

As an example, Adam's was given the task of identifying the yeast genes that encode several "orphan" enzymes (i.e., nobody knew which genes were necessary to make or "encode" them). The general plan of his experiments was conventional, though a bit technical.[25] We don't need all the details to get a sense of how he works, so here's a sketch: Adam's developers knew of an enzyme that makes the amino acid lysine, but didn't know which gene encoded for that enzyme and they set Adam the task of finding it. Adam started by choosing a mutant strain of yeast that can't make lysine and can't grow without it. He consulted his database to look for biochemical reaction pathways that might lead to the synthesis of lysine and systematically supplied the cells with the ingredients that each pathway required, one ingredient per well of mutant cells. If the cells lacked a particular ingredient to produce lysine, they grew more rapidly when Adam gave it to them, and he detected their increased growth rate. From knowing which reactions require that ingredient, Adam found out which enzymes carried out those reactions. By proceeding in a step-by-step way and by doing many, many experiments, Adam identified all of the enzymes in each biochemical pathway. Then, since one particular gene codes for each enzyme, he deduced which gene or genes had been necessary for each pathway.

Adam then "hypothesized" that each gene he had deduced was responsible for lysine production, and he tested each hypothesis. In the end, he generated and tested 20 hypotheses and ended up concluding that lysine was produced by the activity of three genes that nobody had known were required for lysine production. Human scientists subsequently confirmed Adam's conclusions experimentally.

Adam had copious background information, a logical model of yeast metabolism, a bioinformatics database, and programming software so he could carry out "abductive" (see later discussion) and deductive reasoning and design experiments. He could also record, store, and analyze his results; perform statistical tests; and relate the results to his hypotheses. Rejecting (i.e., falsifying) hypotheses was a crucial step in the process. There was even a dedicated program to translate Adam's computer-coded ideas—he is a robot!—back into natural language to provide a "human-friendly summary of the formalization" of what he did.

15.D.2.b The Robot Scientist and the Hypothesis

The computer programming that guides Adam demands translation of abstract concepts, such as "cause," which I mentioned in the last section, from natural language ("text") into equations that robots can interpret. Research hypotheses must be "represented"[26] in robot-friendly ways. The natural language *textual* definition of hypothesis that Adam works with is more expansive than ours and includes multiple layers of "granularity" that range from informational statements about what is to be tested" to what we've been calling "predictions" (e.g., statements about yeast growth rates).

But what, for Adam, constitutes a hypothesis, and how does he reason? The reality is that Adam *selects* hypotheses as much as he *originates* them. "An annotated genome is essentially a large set of hypotheses," according to two of his developers.[27] Adam needs to have an exhaustive set of potential hypotheses to start with. This does not mean that each hypotheses must be fully formed, but it does require that all of the components that would make up hypotheses that he can use be available for him to manipulate. While this may seem opaque, his general process is fairly straightforward. Let's go over an example.

If Adams knows about metabolites A, B, and C and enzymes D, E, and F, he can work out all possible paths from any enzyme to any metabolite, even paths that have never been proposed before. Once he has a plausible set of metabolites and enzymes, he can deduce the gene(s) responsible by going through the steps outlined in the previous section. He cannot imagine the existence of an unknown metabolite with novel properties outside of his knowledge base, meaning that he could not have been the first to discover, for example adenosine triphosphate (ATP), the energy source that many of his enzymatic reactions depend on. In fact, a major impetus for developing Adam in the first place was that he could execute the huge numbers of mind-numbing logical operations involved in inferring interactions among biochemical pathways, thus freeing up human minds for the truly creative work.

15.D.2.c How Does Adam "Think?"

Fundamentally, Adam uses "abduction," a category of logic invented by the philosopher Charles Peirce that is meant to encompass thinking that is neither inductive nor deductive; basically, Peirce wanted a name for whatever it is that happens when we have a new idea. The philosophical status of abduction "is not settled."[28] As I suggested in Chapter 1, cognitive concepts like abduction generally refer to the external circumstances, the behavioral situation, in which you'd tend to use the term and not to as yet unknown mental operations.

Adam's logic works like this: think of his database of yeast metabolism as a *directed graph*, like the one in Figure 15.2. It has circles for metabolites A, B, and C, and lines for enzymes D, E, and F that convert one metabolite into another. Let's assume that he wants to discover how C is produced in the cell, meaning he needs to find a path through the graph that goes from metabolites and enzymes to C. From his experiments he has deduced that (1) enzyme D produces A, but

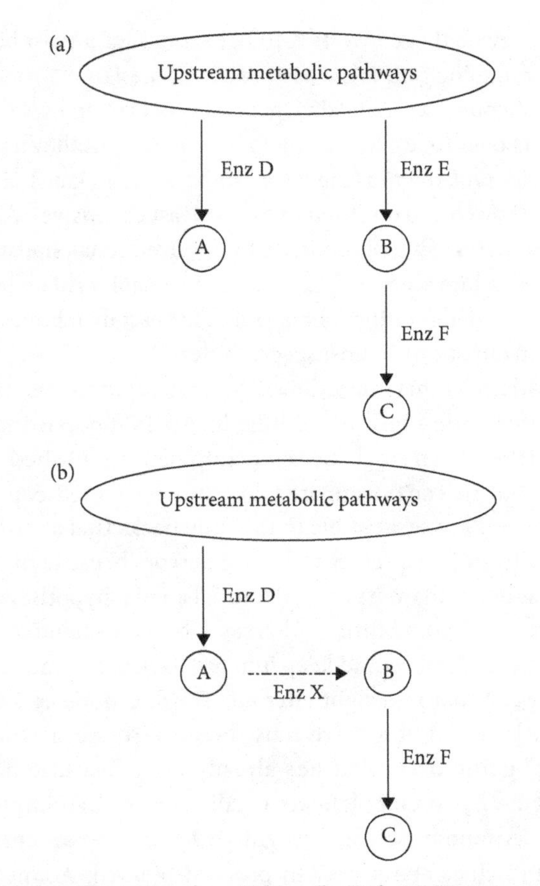

Figure 15.2 Diagram illustrating how Adam, the Robot Scientists, solves problems in yeast genetics.
(a) A tiny part of the graph of all metabolic paths in a cell showing the production of three metabolites A, B, and C (small molecule required for growth) and three enzymes, D, E, and F. Enzymes D and E produce A and B (only) by acting on unknown upstream metabolites. Enzyme F catalyzes the conversion of B into C.
(b) Enzyme E was deleted from this cell, but stimulation of A can produce C indirectly Adam could conclude that a previously unknown enzyme, Enzyme X, converts A into B when Enzyme E is inactive.

not C; (2) enzyme E produces B, and (3) enzyme F converts B, but not A, into C. However, he also finds that, if enzyme E is eliminated, stimulating enzyme D leads to the production of C, which poses a potential problem because there is no known path from enzyme D to C. Adam might hypothesize that a "missing enzyme," enzyme X, converted A to B (and then enzyme F converted B to C, as usual). If his database includes information on genes from any organisms that encode for an enzyme that converts A to B, he looks for a yeast homolog of that gene; the yeast homolog gene is the basis of a hypothesis.

Of the many hypotheses that Adam generates to explain how C could be produced, he selects ones to test according to how consistent they are with the data he has, their *prior probabilities* (the Bayesian parameter again) of being correct, and how much they cost to do (he goes for the least expensive). Adam then tests and rejects hypotheses using rigorous, though conventional, statistical standards. His scientific procedure seems impeccable, and it enables Adam to discover new facts (e.g., a "novel" gene coding for enzyme X in yeast metabolism—not really a new gene, but an old one in an unsuspected role).

Of course, Adam has his limitations. First, his database must have all of the background information he'll need. If he hadn't found existing evidence for enzyme X, he'd have been stuck; he can manipulate established material, even in novel ways, but he can't originate truly unprecedented concepts. Second, Adam's conclusions are susceptible to the drawbacks that statistical testing as a whole suffers from (Chapters 5 and 7; e.g., errors because of noise). Worse, however, is that once Adam has tested and falsified a hypothesis, he deletes it from his universe of possibilities, whereas a human scientist, knowing that falsification is never final, would keep it in the back of her mind because fresh data might suggest that it is right after all. Third, if none of Adam's hypotheses pans out, he concludes there must be an error somewhere and starts "backtracking," going over what he's already done, because his a priori assumption that he has a complete set of all relevant hypotheses leaves him no alternative. A human scientist might go looking in an entirely different direction to find clues about how to proceed. Fourth, Adam is specifically designed to do studies on yeast, and most of biology works with tremendously more complicated organisms, hypothetical explanations, and experimental techniques, and it is far from certain how rapidly other disciplines will be able to adopt Robot Science methods.

Nevertheless, Adam is an impressive achievement and, despite his relatively limited range of present applications, the Robot Scientist project does offer useful lessons for science in general.[29] Science can benefit by more uniform methods of recording and storing data, the kinds of things that Adam does automatically. The project underscores the importance of the hypothesis by showing that science can progress rapidly when scientists express their hypotheses "explicitly, unambiguously, and completely" and that "the hypotheses that have been rejected contain information about the domain of study." The project also emphasizes the importance of "negative data" and the reasoning that leads to it, as these facilitate future work. By taking over the rote aspects of scientific thinking, Robot Science will encourage more humans to think outside the box.

15.E Coda

We've briefly reviewed two very different visions of how Big Data and advanced AI may affect science. In one vision, dominated by the Big Data Mindset, the oceans of data completely overwhelm our mental capabilities and we give up entirely on "theory" and, in fact, on any attempt to understand nature. Highly complicated correlations and predictions made by faster computers accessing colossal datasets become the highest scientific standards. Science progresses by improving correlation and prediction while foregoing explanations and understanding.

In the other vision, translation of abstract scientific, philosophical, and linguistic concepts into computer language becomes ever more sophisticated. The Robot Scientist, Adam, can already perform reasonably complex experiments requiring the manipulation of coordinated suites of laboratory equipment. He can design and execute complicated series of experiments and communicate his results intelligibly to human scientists. While his capabilities are somewhat constrained in the sense that he cannot generate truly original ideas and needs to have a full database of information at his disposal, he is not limited to using the data in previously predetermined ways. Although the scope of Adam's capabilities is presently limited to experiments on yeast, these are early days and we can expect that Adam's siblings or clones will extend their reach into other branches of science in the future.

Notes

1. Anderson C., *Wired Magazine*, www.wired.com/science/discoveries/magazine/16-07/pb_intro.
2. C. D. Borgman, *Big Data, Little Data, No Data: Scholarship in the Networked World* (Cambridge, MA: MIT Press; 2016).
3. R. D. King, K. E. Whelan, F. M. Jones, P. G. Reiser, C. H. Bryant, S. H. Muggleton, et al., "Functional Genomic Hypothesis Generation and Experimentation by a Robot Scientist," *Nature* 427:247–252, 2004.
4. Viktor Mayer-Schönberger and Kenneth Cukier, *Big Data* (Boston: Houghton-Mifflin; 2013). This book, especially chapters 1–4, is an extensive, readable account of the Big Data Mindset.
5. "Epic Failure of Google Flu Trends": https://www.wired.com/2015/10/can-learn-epic-failure-google-flu-trends/.
6. Jeremy Ginsberg, Matthew H. Mohebbi, Rajan S. Patel, Lynnette Brammer, Mark S. Smolinski, and Larry Brilliant, "Detecting Influenza Epidemics Using Search Engine Query Data," *Nature* 457:1012–1015, 2009.

7. D. Lazer, R. Kennedy, G. King, and A. Vespignani, "Big Data. The Parable of Google Flu: Traps in Big Data Analysis," *Science* 343:1203–1205, 2014. doi: 10.1126/science.1248506. See also Steve Lohr, "Google Flu Trends: The Limits of Big Data," *New York Times*, March 28, 2014 Business, Innovation, Technology, Society.
8. https://en.wikipedia.org/wiki/Bias-variance_tradeoff.
9. D. Butler, "Web Data Predict Flu," *Nature* 456:2887–2888, 2008.
10. The A. M. Turing Award is an annual prize given by the Association for Computing Machinery (ACM) to an individual selected for contributions "of lasting and major technical importance to the computer field." The Turing Award is generally recognized as the highest distinction in computer science and the "Nobel Prize of computing"; see https://en.wikipedia.org/wiki/Turing_Award and https://amturing.acm.org/byyear.cfm.
11. Judea Pearl, "An Introduction to Causal Inference," *The International Journal of Biostatistics* 6(2), article 7, 1–59,2010. doi: https://doi.org/10.2202/1557-4679.1203. ((https://www.degruyter.com/view/j/ijb.2010.6.2/ijb.2010.6.2.1203/ijb.2010.6.2.1203.xml). See also

    Kenneth A. Bollen and Judea Pearl, "Eight Myths About Causality and Structural Equation Models," in Stephen L Morgan (Ed.), *Handbook of Causal Analysis for Social Research* (Dordrecht, The Netherlands: Springer; 2013), pp. 301–328.
12. Diagram of simple Bayesian network and calculations, Wikipedia, courtesy of AnAj—Own work (original text: self-made), Public Domain, https://commons.wikimedia.org/w/index.php?curid=19734596.
13. On May 11, 1997, an IBM computer named Deep Blue defeated the world chess champion, Garry Kasparov; see http://www-03.ibm.com/ibm/history/ibm100/us/en/icons/deepblue/.
14. Over the course of three nights in February 2011, an IBM computer named Watson beat human *Jeopardy* Champions Ken Jennings and Brad Rutter: http://www.nytimes.com/2011/02/17/science/17jeopardy-watson.html?pagewanted=all.
15. D. Lee, "IBM's Computer Argues, Pretty Convincingly, with Humans," https://www.bbc.com/news/technology-44531132.
16. Neural networks; https://en.wikipedia.org/wiki/Neural_network. See M. Taylor and M. Koenig, *The Math of Neural Networks Kindle Edition* (Amazon Digital Services LLC/Blue Windmill Media; 2017) for a very basic introduction.
17. Robot Journalists of the Associated Press: https://www.theverge.com/2016/7/4/12092768/ap-robot-journalists-automated-insights-minor-league-baseball.
18. Deep learning: https://en.wikipedia.org/wiki/Deep_learning. See also Deep Learning Symposium Stanford University, https://www.youtube.com/watch?v=czLI3oLDe8M.
19. Yilun Wang and Michael Kosinski, "Deep Neural Networks Are More Accurate Than Humans at Detecting Sexual Orientation from Facial Images," *Journal of Personality and Social Psychology* 114: 246–257, 2018
20. European Data Protection Regulation; https://www.eugdpr.org/. Especially Section 4: Right to Object and Automated Individual Decision Making; Article 21: Right to Object; Article 22: Automated Individual Decision-Making, Including Profiling.

21. See Cliff Kuangnov, "Can A. I. Be Taught to Explain Itself?" an excellent article on "explainable AI" at https://www.nytimes.com/2017/11/21/magazine/can-ai-be-taught-to-explain-itself.html?hp&action=click&pgtype=Homepage&clickSource=story-heading&module=first-column-region®ion=top-news&WT.nav=top-news.
22. Lilian Edwards and Michael Veale, "Slave to the Algorithm? Why a 'Right to an Explanation' Is Probably Not the Remedy You Are Looking For," *Duke Law and Technology Review* 16:18–84, 2018.
23. R. D. King, J. Rowland, S. G. Oliver, et al., "The Automation of Science," *Science* 324:85–89, 2009.
24. L. N. Soldatova and A. Rzhetsky, "Representation of Research Hypotheses," *Journal of Biomedical Semantics* Suppl 2:S9, 2011.
25. Adam makes extensive use of "auxotrophic" mutant strains of yeast, meaning that each strain lacks the capacity to make a certain nutrient that is necessary for growth; see https://en.wikipedia.org/wiki/Auxotrophy.
26. Soldatova and Rzhetsky, "Representation of Research Hypotheses," p. 1.
27. King, Rowand, et al., ibid.
28. I. Douven, "Abduction," in Edward N. Zalta (Ed.), *The Stanford Encyclopedia of Philosophy*, Summer 2017 ed., https://plato.stanford.edu/archives/sum2017/entries/abduction/.
29. King, Rowland, et al., ibid.

Epilogue

Centaur Science—The Future of the Hypothesis

I began this book when I started thinking seriously about the scientific hypothesis in the context of graduate school science education; I end by thinking about how science education affects public perceptions of global climate change and how the hypothesis will fit into the science of the future.

For nearly 400 years, the Scientific Method, anchored by the hypothesis, has guided science. I've said that one of the most important concepts in the philosophy of science is the hypothesis, right up there with empiricism and fallibilism, and I quoted the physicist, Herman Bondi, "There is nothing more to science than its method, and there is nothing more to its method than Karl Popper has said." In retrospect, Bondi's remark seems a bit overstated; we do, after all, have to allow room for asking questions, making observations, Discovery Science, and Big Data. And yet, as I've tried to show, there are links among these other approaches to studying nature and to the hypothesis. They are all part of the method of science, and the purpose of that method is to explain and understand.

Despite the almost incredible successes of human-powered, hypothesis-based science, everything may be changing. With the development of faster supercomputers (the current record is 200×10^{15} operations *per second*,[1] and quantum computing offers the prospect of significantly higher speeds); the availability of limitless, cheap information storage in the "cloud"; and the growth of increasingly more "intelligent" computer programs, the advantages that the human mind has for conducting science are dwindling.

We seem to be approaching a crossroads, with the Big Data Mindset and the Robot Scientists heading off in different directions. The Big Data Mindset promises limitless ability to explore correlations beyond our imagination but forces us deny our nature and forgo the drive for understanding that has gotten us to where we are. The Robot Scientist promises enormous expansion in our ability to conduct laboratory experiments but asks us to accept that the scientific

Having done my best to convey the importance of scientific reasoning founded on the hypothesis, I'm well aware of how imperfect the final product is. If you notice errors or have comments, please send them to me via my website; https://scientifichypothesis.org, Twitter (@BradAlgerLab, or Facebook (https://www.facebook.com/ScientificHypothesis-392008431374835/). Thank you. Bradley Alger.

enterprise can fit within its so far relatively narrow scope. Why can't we have the best of both?

Garry Karparov's ego suffered a blow when he lost to the IBM Computer Deep Blue in the chess championship, but Kasparov did not become one of the greatest chess champions in history by lacking fortitude or imagination. He began working with computers, encouraging the formation of a man–machine hybrid team called "Centaur," after the half-man, half horse beast of Greek mythology. He created "Centaur Chess" and organized a tournament open to all—human grand masters, supercomputers running chess artificial intelligence (AI) (more powerful than Deep Blue), and Centaurs.

And who won? If you guessed a Centaur you'd be right, but not the one you'd probably imagined. The Centaur that won, beating out all of the solo humans and solo supercomputers, was not the human grand master teamed with a powerful AI chess program, but a pair of amateur chess players teamed with three weak AI programs. In Nicky Case's account,[2] the dominant combination was not genius or speed, alone or together: it was the symbiotic marriage of the best of the human mental traits with the strengths of AI. The guiding principle is that, in order to maximize intelligence, each intelligent entity must specialize in what it's best at.

As Case explains, "AIs are best at choosing answers; humans are best at choosing questions." Their respective strengths are coming to the fore in the resurrected field of *intelligence augmentation* (IA), which is the flip side of AI. Rather than emphasizing the ways in which computers supplant human calculational power—a low bar, since we're pretty poor at calculating anyway—computers augment human intuitive intelligence. IA is already happening in art and engineering, where humans "choose questions" including setting goals and constraints. Given the goal, the computers propose an array of possible solution ("choosing answers") that the humans then select from or modify, and the iteration continues until the human decides "we're done."

The future of basic science research can be like that: instead of Robot Scientists, we'll have Centaur Scientists. And the human side of the Centaur Scientist will specialize in inventing hypotheses. Because that's what we do best.

Notes

1. World's fastest Supercomputer (as of June 2018) https://en.wikipedia.org/wiki/Summit_(supercomputer)
2. Nicky Case, "How to Become a Centaur." https://jods.mitpress.mit.edu/pub/issue3-case?version=c847d892-97dc-40a7-a412-315d255b9b2d 1/

Index

Note: Tables and boxes are indicated by *t*, and *b* following the page number

Note: For the benefit of digital users, indexed terms that span two pages (e.g., 52–53) may, on occasion, appear on only one of those pages.